William Henry Flower

Essays on Museums and Other Subjects Connected with Natural

History

William Henry Flower

Essays on Museums and Other Subjects Connected with Natural History

ISBN/EAN: 9783337026790

Printed in Europe, USA, Canada, Australia, Japan

Cover: Foto ©berggeist007 / pixelio.de

More available books at **www.hansebooks.com**

ESSAYS ON MUSEUMS

ESSAYS ON MUSEUMS

AND OTHER SUBJECTS CONNECTED
WITH NATURAL HISTORY

BY

SIR WILLIAM HENRY FLOWER, K.C.B.

D.C.L., D.Sc., LL.D., Ph.D., F.R.S., F.R.C.S., P.Z.S.

CORRESPONDENT OF THE INSTITUTE OF FRANCE, ETC.

London

MACMILLAN AND CO., Limited

NEW YORK: THE MACMILLAN COMPANY

1898

PREFACE

HAVING lately, during an enforced period of restraint from active occupation, had leisure to look back upon the events of a somewhat busy life, it occurred to me that besides the administration, in succession, of two large public museums, and some original scientific investigation, the results of which can be of interest only to the very few, there have been occasions which have involved a contact with wider and more varied circles.

Though the records of these occasions have doubtless been mainly of passing or local concern, it is possible that something may be found among them worth placing in a more accessible and permanent form than in the scattered journals or pamphlets in which they can only be found at present.

Such of these essays and addresses as seemed suitable for republication in the present collection mainly group themselves under three distinct branches of the many into which the intellectual and educational activity of our age is manifested.

1. Those which are here placed first are, more closely than any others, connected with the practical work of my life,—the advancement of scientific knowledge through the development of museums. Here I cannot refrain from expressing my deep sense of the loss this cause has recently sustained by the premature death of Dr. Brown Goode, Director of the United States National Museum. Fortunately for science he was able to formulate his latest views "On the Principles of Museum Administration" in an elaborate paper contributed

to the Newcastle-on-Tyne meeting of the "Museums Association" and published in the report for 1895.

2. The essays on general Biological questions treat very considerably of a subject of even wider interest,—the permanence or transmutation of species, especially in its bearing upon the doctrine of evolution. It might be thought that nothing more need now be said upon this subject, but it is certain that there is still much misapprehension in relation to it; and it may be interesting to some to know the arguments which have been brought forward at various stages of the great controversy by one whose scientific life began before the publication of the *Origin of Species*, and to whom it has fallen to present the principal problems connected with it in various aspects and to various classes of intelligence and education. If some of the essays may not be upon the level of our present knowledge as to facts, the principal lines of thought which run through them are not materially changed by the additions made to that knowledge, and so it has been thought best to leave them as originally written, except for an occasional verbal alteration. The dates at which they were first published should, in all cases, be carried in mind. Sometimes a repetition of ideas or even of verbal expressions found elsewhere may be met with, but these have been allowed to stand, so that each essay may be complete in itself.

3. Upon the third subject, the main point of which is the advocacy of a more systematic study of Anthropology in this country, there has been, as it seems to me, less advance than in either of the other two; and in putting forth its claims for greater recognition I felt for a long time as one crying in the wilderness. I am glad, therefore, to have the opportunity of bringing much which has been said in some of these addresses once more before the notice of those who may have opportunities of forwarding the work, and also acknowledging various recent signs of progress, among which

may be mentioned the establishment of a Professorship of this subject by the University of Oxford.

4. The short sketches of some of the personal characteristics of the contemporary naturalists of our country who have been most prominent in advancing the branches of science to which the essays of the first three sections are devoted form a not unnatural sequel to them.

In conclusion, my best thanks are offered to the various institutions and persons who have kindly given permission for the republication of the essays, the source of which is in every case acknowledged; but I may especially mention the Council of the Royal Society in respect to the biographical notices from their Proceedings; the Council of the British Association for the Advancement of Science, for various addresses published in the reports of their meetings; the Council of the Royal Institution and of the Royal Colonial Institute, for lectures given before them and published in their Proceedings; the proprietors of the *North American Review* and of *Chambers's Journal*, for articles contributed originally to those periodicals; and Messrs. A. and C. Black and Messrs. Macmillan and Co. Ltd., for illustrations of the articles on " Whales," and on " Fashion in Deformity."

W. H. F.

26 STANHOPE GARDENS, LONDON,
  Easter 1898.

CONTENTS

MUSEUMS

ESSAY   PAGE

1. MUSEUM ORGANISATION.—Presidential Address to the British ✓ Association for the Advancement of Science. Newcastle-on-Tyne Meeting, 11th September 1889   .   .   .   1

Choice of subject—Museums of the Ancients and of the Middle Ages—The terms Natural History and Naturalist—Definition of a Natural History Museum—Subdivisions of the sciences represented in a Natural History Museum—Objects of museums—Research and instruction—Methods of arranging specimens for study and for public exhibition — Labels — Guide-books and catalogues — Biological problems of the day — Evolution of organic beings—Natural Selection—Survival of the Fittest—Difficulties arising from the imperfection of our knowledge.

2. MODERN MUSEUMS. Presidential Address to the Museums Association. London Meeting, 3rd July 1893   .   .   30

The Museums Association—Cost of museums—Qualifications and remuneration of curators—Museum literature—The new Museum idea—Separation of specimens required for the advancement of science and for general instruction—The new Natural History Museums of London, Paris, Vienna, and Berlin—American museums—Plan for a model National Natural History Museum.

3. LOCAL MUSEUMS. From a Letter in support of the establishment of a County Museum for Buckinghamshire (24th November 1891), and an Address at the opening of the Perth Museum (29th November 1895)   .   .   .   54

Desirability of having museums in every county to preserve tangible records of local history, customs, and natural history —Necessity of provision for their permanent maintenance and for paid curators—Such museums should be formed in connection with technical schools and free libraries, and supported from the same sources.

ESSAY   PAGE

4. SCHOOL MUSEUMS. Suggestions for the formation and
arrangement of a Museum of Natural History in connection
with a Public School. *Nature,* 26th December 1889   .   58

  Subjects best adapted for School Museums—Zoology, Botany,
  Mineralogy and Geology—What kind of specimens are desirable—
  Every collection or series should be kept perfectly distinct—A
  permanent curator necessary.

5. BOYS' MUSEUMS. *Chambers' Edinburgh Journal,* April
1897   63

  Natural History an absorbing passion with many boys—Often
  a passing phase, sometimes permanent—Natural History as a
  vocation—Love of collecting—Value as a means of education—
  Hints to beginners in forming museums—Personal experiences of
  the author.

6. THE BOOTH MUSEUM. Address at the opening of the Booth
Museum, Brighton. 3rd November 1890   .   .   70

  Formation and object of the Museum—The art of Taxidermy
  —Observation of bird life an unfailing source of pleasure, increasing with increase of knowledge—Such museums promote this
  knowledge and add to human happiness.

7. THE MUSEUM OF THE ROYAL COLLEGE OF SURGEONS OF
ENGLAND. Presidential Address to the Anatomical
Section of the International Medical Congress. London
Meeting, 4th August 1881   74

  John Hunter, the founder of the Museum—His collection
  purchased by Government and placed in the custody of the Royal
  College of Surgeons under certain conditions—Branches of Biology
  represented in the Museum—Foundation and present condition of
  each section—Value of the Museum to the surgical profession and
  the country—Its future maintenance assured by the munificent
  bequest of Sir Erasmus Wilson.

GENERAL BIOLOGY

8. INTRODUCTORY LECTURE TO THE COURSE OF COMPARATIVE
ANATOMY. Royal College of Surgeons of England, 14th
February 1870   97

  The course intended to illustrate the Museum—John Hunter's
  work as a comparative anatomist—The Museum teaches morphology—The study of morphology has led in succession to the ideas

ESSAY   PAGE

of (1) teleology or direct adaptation to purpose: (2) type or common plan; (3) descent from a common ancestor with modification, according to the great principle of orderly evolution—Reception and gradual recognition of the last hypothesis—Some objections to it answered—Explanation it affords of anatomical structures, otherwise accounted for with difficulty—Illustrations—Importance of discriminating essential or fundamental from adaptive characters—Classification—Wider knowledge of morphology the basis of an intelligent study of human anatomy.

9. RECENT ADVANCES IN NATURAL SCIENCE IN RELATION TO THE CHRISTIAN FAITH. Paper read at the Church Congress. Reading Meeting, 2nd October 1883 . . 123

The doctrine of evolution as applied to the life of organic beings—Causes which led to the old belief in the fixity of species, and to the change of opinion upon the subject, now almost universal—Effects of this change of opinion upon our ideas in relation to Creation and the Origin of Man.

10. A PRACTICAL LESSON FROM BIOLOGICAL STUDIES. Reply to Working Men's Address to the President of the British Association. Newcastle-on-Tyne, 14th September 1889 . 135

The survival of the fittest in the struggle for existence, one of the main causes of progress.

11. PALÆONTOLOGICAL EVIDENCE OF GRADUAL MODIFICATION OF ANIMAL FORMS. Lecture at the Royal Institution of Great Britain, 25th April 1873 138

The history of the order *Ungulata*, or hoofed animals, as traced through the Tertiary geological period, shows, as time advanced, a gradual differentiation from more generalised to specialised forms—The existing species are only survivors of large numbers which lived in former times, and are now isolated by the extinction of the intermediate forms.

12. A CENTURY'S PROGRESS IN ZOOLOGICAL KNOWLEDGE. Presidential Address in the Department of Zoology and Botany. British Association, Dublin Meeting, 15th August 1878 153

The last edition of the *Systema Naturæ*, revised by Linnæus, was published in 1766—The first portion of the work, devoted to the class Mammalia examined, and the progress made in the knowledge of each order since that time reviewed—Observations on classification and zoological nomenclature.

ESSAY   PAGE

13. THE ZOOLOGICAL SOCIETY OF LONDON. Address at the
    General Meeting of the Society, held in the Zoological
    Gardens, in celebration of the Fiftieth Anniversary of Her
    Majesty's reign, 16th June 1887   171

      Foundation and growth of the Society—Acclimatisation
    Museum—Scientific meetings—Publications—Lectures—Mena-
    gerie—Recent improvements.

14. WHALES AND WHALE FISHERIES. Lecture at the Royal
    Colonial Institute, 8th January 1895   .   .   .   185

      Fishes and whales—General characters of the latter—Sperm
    whale—Whalebone whales—Mode of capture—Whale fishery
    of the Basques in the Middle Ages—Greenland whale fishery—
    Southern whale fisheries.

15. WHALES, PAST AND PRESENT, AND THEIR PROBABLE ORIGIN.
    Lecture at the Royal Institution of Great Britain, 25th May
    1883   209

      Rudimentary organs giving indications of former conditions
    and origin—Hair—Organs of sight and smell—Teeth—Nature
    and use of baleen or whalebone — Fore-limbs — Hind Limbs —
    Geological history—All evidence in favour of derivation from
    land animals, probably through forms originally inhabiting fresh
    waters.

ANTHROPOLOGY

16. PRESIDENTIAL ADDRESS TO THE DEPARTMENT OF ANTHRO-
    POLOGY. British Association, York Meeting, 1st September
    1881   235

      Importance of the study of race—Natural classification of
    man—Causes of racial difference—Race a vague term—Craniometry
    —Necessity for larger collections and more workers.

17. PRESIDENTIAL ADDRESS TO THE SECTION OF ANTHROPOLOGY.
    British Association, Oxford Meeting, 9th August 1894   .   249

      History of the subject in England—Its scope and extent—
    Societies and collections—Systematic teaching—Anthropometry
    —Value in identification of individuals, especially as applied
    to criminals—Finger-marks—Ethnographic survey—Anthropo-
    logical notes and queries—Our duty to posterity in collecting
    anthropological data.

ESSAY PAGE

18. THE CLASSIFICATION OF THE VARIETIES OF THE HUMAN SPECIES. Presidential Address at the Anniversary Meeting of the Anthropological Institute of Great Britain and Ireland, 27th January 1885 274

The three principal types of man—1. Ethiopian, Negroid, Melanian, or black — 2. Mongolian, Xanthous or yellow — 3. Caucasian or white—Characteristics of each—Subdivisions—Position of the American and Australian races.

19. THE PYGMY RACES OF MEN. Lecture at the Royal Institution of Great Britain, 13th April 1888 . . . 290

Ancient belief in pygmy races—Size a characteristic of race—The smallest existing races—Andaman Islanders—Other Negrito people in the Eastern Archipelago and Mainland—Small races of Africa—Bushmen of South Africa—Negrillos of the forest region of Central Africa—Akkas.

20. FASHION IN DEFORMITY. As illustrated in the customs of Barbarous and Civilised Races, 1880 . . . 315

Origin of fashion — Fashion applied to domestic animals — Fashion in hair—Finger-nails—Tattooing—Noses—Ears—Lips—Teeth—Head—Historic references—Fashion in hands and feet —Action of the foot in walking—Pointed toes—High heels—Tight lacing.

BIOGRAPHICAL SKETCHES

21. BIOGRAPHICAL NOTICE OF PROFESSOR ROLLESTON. *Proceedings of the Royal Society*, 1882 . 357

22. BIOGRAPHICAL NOTICE OF SIR RICHARD OWEN. *Proceedings of the Royal Society*, 1894 . . . 363

23. REMINISCENCES OF PROFESSOR HUXLEY. *The North American Review*, September 1895 . . 381

24. EULOGIUM ON CHARLES DARWIN. Centenary Meeting of the Linnean Society, 24th May 1888 . . 390

MUSEUMS

I

MUSEUM ORGANISATION [1]

IT is twenty-six years since this Association met in Newcastle-on-Tyne. It had then the advantage of being presided over by one of the most distinguished and popular of your fellow-townsmen.

Considering the age usually attained by those upon whom the honour of the presidency falls, and the length of time which elapses before the Association repeats its visit, it must have rarely happened that any one who has held the office is spared, not only to be present at another meeting in the town in which he has presided, but also to take such an active part in securing its success, and to extend such a hospitable welcome to his successor, as Lord Armstrong has done upon the present occasion.

The address which was delivered at that meeting must have been full of interest to the great majority of those present. It treated of many subjects more or less familiar and important to the dwellers in this part of the world, and it treated them with the hand of a master, a combination which always secures the attention of an audience.

When it came to my knowledge that in the selection of the President for this meeting the choice had fallen upon me, I was filled with apprehension. There was nothing in my previous occupations or studies from which I felt that I could evolve anything in special sympathy with what is universally recognised as the prevailing genius of this district. I was,

[1] Presidential Address to the British Association for the Advancement of Science. Newcastle-on-Tyne Meeting, 11th September 1889.

however, somewhat reassured when reminded that in the
regular rotation by which the equal representation in the
presidential office of the different branches of science included
in the Association is secured, the turn had come round for
some one connected with biological subjects to occupy the
chair which, during the past seven years, has been filled
with such distinction by engineers, chemists, physicists,
mathematicians, and geologists. I was also reminded that
the Association, though of necessity holding its meeting in
some definite locality, was by no means local in its character,
but that its sphere was co-extensive, not with the United
Kingdom only, but with the whole of the British Dominions,
and that our proceedings are followed with interest wherever
our language is understood—I may say, throughout the
civilised world. Furthermore, although its great manufacturing
industries, the eminence of its citizens for their skill and intelligence in the practical application of mechanical sciences, and
the interesting and important geological features of its vicinity,
have conferred such fame on Newcastle as almost to have
over-shadowed its other claims to distinction in connection with
science, this neighbourhood is also associated with Bewick,
with Johnson, with Alder, Embleton, Hutton, Atthey, Norman,
the two Hancocks, the two Bradys, and other names honoured
in the annals of biology; it has long maintained a school of
medicine of great repute; and there has lately been established
here a natural history museum, which in some of its features
is a model for institutions of the kind, and which, I trust, will
be a means of encouraging in this town some of the objects
the Association was designed to promote.

There can be no doubt that among the various methods by
which the aims of the British Association (as expressed in its
full title, the *advancement of science*) may be brought about,
the collection and preservation of objects available for
examination, study, and reference—in fact, the formation of
what are now called "museums"—is one of very great
practical importance; so much so, indeed, that it seems to me
one to the consideration of which it is desirable to devote
some time upon such an occasion as this. It is a subject
still little understood, though, fortunately, beginning to attract

attention.   It has already been brought before the notice of
the Association, both in presidential and sectional addresses.
A committee of our members is at the present time engaged
in collecting evidence upon it, and has issued some valuable
reports.   During the present year an association of curators
and others interested in museums has been founded for the
purpose of interchange of ideas upon the organisation and
management of these institutions.   It is a subject, moreover,
if I may be allowed to mention a personal reason for bringing
it forward this evening, which has, more than any other,
occupied my time and my attention almost from the earliest
period of my recollection, and I think you will agree with the
opinion of one of my distinguished predecessors in this chair,
" that the holder of this office will generally do better by
giving utterance to what has already become part of his own
thought than by gathering matter outside of its habitual range
for the special occasion.   For," continued Mr. Spottiswoode,
" the interest (if any) of an address consists not so much in
the multitude of things therein brought forward as in the
individuality of the mode in which they are treated."

The first recorded institution which bore the name of
museum, or temple or haunt of the Muses, was that founded
by Ptolemy Soter at Alexandria about 300 B.C.; but this
was not a museum in our sense of the word, but rather, in
accordance with its etymology, a place appropriated to the
cultivation of learning, or which was frequented by a society
or academy of learned men devoting themselves to philosophical
studies and the improvement of knowledge.

Although certain great monarchs, as Solomon of Jerusalem
and Augustus of Rome, displayed their taste and their
magnificence by assembling together in their palaces curious
objects brought from distant parts of the world,—although it
is said that the liberality of Philip and Alexander supplied
Aristotle with abundant materials for his researches,—of the
existence of any permanent or public collections of natural
objects among the ancients there is no record.   Perhaps the
nearest approach to such collections may be found in the
preservation of remarkable specimens, sometimes associated
with superstitious veneration, sometimes with strange legendary

stories, in the buildings devoted to religious worship. The skins of the gorillas brought by the navigator Hanno from the West Coast of Africa, and hung up in the temple at Carthage, afford a well-known instance.

With the revival of learning in the Middle Ages, the collecting instinct, inborn in so many persons of various nations and periods of history, but so long in complete abeyance, sprang into existence with considerable vigour, and a museum, now meaning a collection of miscellaneous objects, antiquities as well as natural curiosities, often associated with a gallery of sculpture and painting, became a fashionable appendage to the establishment of many wealthy persons of superior culture.

All the earliest collections, comparable to what we call museums, were formed by and maintained at the expense of private individuals; sometimes physicians, whose studies naturally led them to a taste for biological science; often great merchant princes, whose trading connections afforded opportunities for bringing together things that were considered curious from foreign lands; or ruling monarchs in their private capacity. In every case they were maintained mainly for the gratification of the possessor or his personal friends, and rarely, if ever, associated with any systematic teaching or public benefit.

One of the earliest known printed catalogues of such a museum is that of Samuel Quickelberg, a physician of Amsterdam, published in 1565 in Munich. In the same year Conrad Gesner published a catalogue of the collection of Johann Kentmann, a physician of Torgau in Saxony, consisting of about 1600 objects, chiefly minerals, shells, and marine animals. Very soon afterwards we find the Emperor Rudolph II. of Germany busily accumulating treasures which constituted the foundations of the present magnificent museums by which the Austrian capital is distinguished.

In England the earliest important collectors of miscellaneous objects were the two John Tradescants, father and son, the latter of whom published, in 1656, a little work called *Musæum Tradescantianum; or, a Collection of Rarities preserved at South Lambeth, neer London.* The wonderful variety and

incongruous juxtaposition of the objects contained in this
collection make the catalogue very amusing reading.   Under
the first division, devoted to " Some Kindes of Birds, their
Egges, Beaks, Feathers, Clawes and Spurres," we find " Divers
sorts of Egges from Turkie, one given for a Dragon's Egge ";
" Easter Egges of the Patriarch of Jerusalem ; " " Two Feathers
of the Phœnix Tayle ; " " The Claw of the bird Rock, who, as
Authors report, is able to trusse an Elephant." Among
" whole birds " is the famous " Dodar from the Island
Mauritius ; it is not able to flie, being so big." This is the
identical specimen, the head and foot of which has passed
through the Ashmolean into the University Museum of
Oxford ; but we know not what has become of the claw of
the Rock, the Phœnix tail, and the Dragon's egg. Time
does not allow me to mention the wonderful things which
occur under the head of " Garments, Vestures, Habits, and
Ornaments," or the " Mechanick, Artificial Workes in Carvings,
Turnings, Sowings, and Paintings," from Edward the Confessor's
knit gloves, and the famous " Pohatan, King of Virginia's
habit, all embroidered with shells or Roanoke," also still at
Oxford, and lately figured and described by Mr. E. B. Tylor,
to the " Cherry-stone, upon one side S. George and the
Dragon, perfectly cut, and on the other side 88 Emperours'
faces " ; or the other " cherry-stone, holding ten dozen of
tortois-shell combs made by Edward Gibbons." But before
leaving these private collections I cannot forbear mentioning,
as an example of the great aid they often were in advancing
science, the indebtedness of Linnæus in his early studies to
the valuable zoological museums which it was one of the
ruling passions of several kings and queens of Sweden to bring
together.

Upon the association of individuals together into societies
to promote the advancement of knowledge, these bodies in
their corporate capacity frequently made the formation of a
museum part of their function. The earliest instance of this
in our country was the museum of the Royal Society in Crane
Court, of which an illustrated catalogue was published by Dr.
Grew in 1681.

The idea that the maintenance of a museum was a portion

of the public duty of the State or of any municipal institution had, however, nowhere entered into the mind of man at the beginning of the last century. Even the great teaching bodies, the Universities, were slow in acquiring collections; but it must be recollected that the subjects considered most essential to the education they then professed to give were not those which needed illustration from the objects which can be brought together in a museum. The Italian Universities, where anatomy was taught as a science earlier and more thoroughly than anywhere else in Europe, soon found the desirability of keeping collections of preserved specimens, and the art of preparing them attained a high degree of excellence at Padua and Bologna two centuries ago. But these were generally the private property of the professors, as were nearly all the collections used to illustrate the teaching of anatomy and pathology in our country within the memory of many now living.

Notwithstanding the multiplication of public museums during the present century, and the greater resources and advantages many of these possess, which private collectors cannot command, the spirit of accumulation in individuals has happily not passed away, although usually directed into rather different channels than formerly. The general museums or miscellaneous collections of old are now left to governments and institutions which afford greater guarantee of their permanence and public utility, while admirable service is done to science by those private persons with leisure and means, who, devoting themselves to some special subject, amass the materials by which its study can be pursued in detail either by themselves or by those they know to be qualified to do so. Such collections, if they fulfil their most appropriate destiny, ultimately become incorporated, by gift or purchase, in one or other of the public museums, and then serve as permanent factors in the education of the nation, or rather of the world.

It would be passing beyond the limits of time allotted to this address, indeed going beyond the scope of the Association, if I were to speak of many of the subjects which have pre-eminently exercised the faculties of the collector and formed

the materials of which museums are constructed. The various
methods by which the mind of man has been able to reproduce
the forms of natural objects or to give expression to the images
created by his own fancy, from the rudest scratchings of a
savage on a bone, or the simplest arrangement of lines
employed in ornamenting the roughest piece of pottery, up to
the most lovely combinations of form and colour hitherto
attained in sculpture or in painting, or in works in metal or
in clay, depend altogether on museums for their preservation,
for our knowledge of their condition and history in the past,
and for the lessons which they can convey for the future.

Apart from the delight which the contemplation of the
noblest expressions of art must produce in all cultivated minds,
apart also from the curiosity and interest that must be excited
by all the less successfully executed attempts to produce
similar results, as materials for constructing the true history of
the life of man, at different stages of civilisation, in different
circumstances of living, and in divers regions of the earth, such
collections are absolutely invaluable.

But I must pass them by in order to dwell more in detail
upon those which specially concern the advancement of the
subjects which come under the notice of this Association—
museums devoted to the so-called " natural history " sciences,
although much which will be said of them will doubtless be
more or less applicable to museums in general.

The terms *natural history* and *naturalist* have become
deeply rooted in our language, but without any very definite
conception of their meaning or the scope of their application.
Originally applied to the study of all the phenomena of the
universe which are independent of the agency of man, natural
history has gradually narrowed down in most people's minds,
in consequence of the invention of convenient and generally
understood and accepted terms for some of its various subdivisions, as astronomy, chemistry, geology, etc., into that
portion of the subject which treats of the history of creatures
endowed with life, for which, until lately, no special name had
been invented.   Even from this limitation botany was
gradually disassociating itself in many quarters, and a
" naturalist " and a " zoologist " have nearly become, however

irrationally, synonymous terms. The happy introduction and general acceptance of the word "biology," notwithstanding the objections raised to its etymological signification, have reunited the study of organisms distinguished by the possession of the living principle, and practically eliminated the now vague and indefinite term "natural history" from scientific terminology. As, however, it is certain to maintain its hold in popular language, I would venture to suggest the desirability of restoring it to its original and really definite signification, ✓ contrasting it with the history of man and of his works, and of the changes which have been wrought in the universe by his intervention.

It was in this sense that, when the rapid growth of the miscellaneous collections in the British Museum at Bloomsbury ʟ (the expansion of Sir Hans Sloane's accumulation in the old Manor House at Chelsea) was thought to render a division necessary, the line of severance was effected at the junction of what was natural and what was artificial; the former, including the products of what are commonly called "natural" forces, un-affected by man's handiwork, or the impress of his mind. The ⟋ departments which took cognisance of these were termed the "Natural History Departments," and the new building to which they were removed the "Natural History Museum."

It may be worth while to spend a few moments upon the consideration of the value of this division, as it is one which concerns the arrangement and administration of the majority of museums.

Though there is very much to be said for it, the objection has been raised that it cuts man himself in two. The illustrations of man's bodily structure are undoubtedly subjects for the zoologist. The subtle gradations of form, proportion, and colour which distinguish the different races of men, can only be appreciated by one with the education of an anatomist, and whose eye has been trained to estimate the value of such characters in discriminating the variations of animal forms. The subjects for comparison required for this branch of research must therefore be looked for in the zoological collections.

But the comparatively new science of "anthropology" embraces not only man's physical structure; it includes his mental development, his manners, customs, traditions, and languages. The illustrations of his works of art, domestic utensils, and weapons of war are essential parts of its study. In fact it is impossible to say where it ends. It includes all that man is or ever has been, all that he has ever done. No definite line can be drawn between the rudest flint weapon and the most exquisitely finished instrument of destruction which has ever been turned out from the manufactory at Elswick, between the rough representation of a mammoth, carved by one of its contemporary men on a portion of its own tusk, and the most admirable production of a Landseer. An anthropological collection, to be logical, must include all that is in not only the old British Museum but the South Kensington Museum and the National Gallery. The notion of an anthropology which considers savages and prehistoric people as apart from the rest of mankind may, in the limitations of human powers, have certain conveniences, but it is utterly unscientific and loses sight of the great value of the study in tracing the gradual growth of our existing complex systems and customs from the primitive ways of our progenitors.

On the other hand, the division first indicated is as perfectly definite, logical, and scientific as any such division can be. That there are many inconveniences attending wide local disjunctions of the collections containing subjects so distinct yet so nearly allied as physical and psychical anthropology must be fully admitted; but these could only have been over-come by embracing in one grand institution the various national collections illustrating the different branches of science and art, placed in such order and juxtaposition that their mutual relations might be apparent, and the resources of each might be brought to bear upon the elucidation of all the others—an ideal institution, such as the world has not yet seen, but into which the old British Museum might at one time have been developed.

A purely "Natural History Museum" will then embrace a collection of objects illustrating the natural productions of the earth, and in its widest and truest sense should include, as far as they can be illustrated by museum specimens, all the sciences which deal with natural phenomena. It has only been the difficulties, real or imaginary, in illustrating them which have excluded such subjects as astronomy, physics, chemistry, and physiology from occupying departments in our National Natural History Museum, while allowing the introduction of their sister sciences, mineralogy, geology, botany, and zoology.

Though the experimental sciences and those which deal with the laws which govern the universe, rather than with the materials of which it is composed, have not hitherto greatly called forth the collector's instinct, or depended upon museums for their illustration, yet the great advantages of collections of the various instruments by means of which these sciences are pursued, and of examples of the methods by which they are taught, are yearly becoming more manifest. Museums of scientific apparatus now form portions of every well-equipped educational establishment, and under the auspices of the Science and Art Department at South Kensington a national collection illustrating those branches of natural history science which have escaped recognition in the British Museum is assuming a magnitude and importance which brings the question of properly housing and displaying it urgently to the front.

Anomalies such as these are certain to occur in the present almost infantile though rapidly progressive state of science. It may be taken for granted that no scientific institution of any complexity of organisation can be, except at the moment of its birth, abreast of the most modern views of the subject, especially in the dividing lines between, and the proportional representation of, the various branches of knowledge which it includes.

The necessity for subdivisions in the study of science is continually becoming more apparent as the knowledge of the details of each subject multiplies without corresponding increase in the power of the human mind to grasp and deal with them, and the dividing lines not only become sharper, but as knowledge advances they frequently require revision. It might be supposed that such revision would adjust itself to the direction taken by the natural development of the relations of the different branches of science, and the truer conceptions entertained of such relations. But this is not always so.

Artificial barriers are continually being raised to keep these
dividing lines in the direction in which they have once started.
Difficulties of readjustment arise not only from the mechanical
obstacles caused by the size and arrangements of the buildings
and facilities for the allocation of various kinds of collections,
but still more from the numerous personal interests which
grow up and wind their meshes around such institutions.
Professorships and curatorships of this or that division of
science are founded and endowed, and their holders are usually
tenacious either of encroachment upon or of any wide enlargement of the boundaries of the subject they have undertaken to
teach or to illustrate; and in this way, more than any other,
passing phases of scientific knowledge have become crystallised
or fossilised in institutions where they might least have been
expected. I may instance many European universities and
great museums in which zoology and comparative anatomy are
still held to be distinct subjects taught by different professors,
and where, in consequence of the division of the collections
under their charge, the skin of an animal, illustrating its
zoology, and its skeleton and teeth, illustrating its anatomy,
must be looked for in different and perhaps remotely placed
buildings.

For the perpetuation of the unfortunate separation of
palæontology from biology, which is so clearly a survival of an
ancient condition of scientific culture, and for the maintenance
in its integrity of the heterogeneous compound of sciences
which we now call "geology," the faulty organisation of our
museums is in a great measure responsible. The more their
rearrangement can be made to overstep and break down the
abrupt line of demarcation which is still almost universally
drawn between beings which live now and those which have
lived in past times, so deeply rooted in the popular mind and
so hard to eradicate even from that of the scientific student,
the better it will be for the progress of sound biological knowledge.

But it is not of the removal of such great anomalies and
inconsistencies, which, when they have once grown up, require
heroic methods to set them right, but rather of certain minor
defects in the organisation of almost all existing museums

which are well within the capacity of comparatively modest administrative means to remedy, that I have now to speak.

That great improvements have been lately effected in many respects in some of the museums in this country, on the Continent, and especially in America, no one can deny. The subject, as I have already indicated, is exciting the attention of those who have the direction of them, and even awakening interest in the mind of the general public. It is in the hope of in some measure helping on or guiding this movement that I have ventured on the remarks which follow.

The first consideration in establishing a museum, large or small, either in a town, institution, society, or school, is that it should have some definite object or purpose to fulfil; and the next is that means should be forthcoming not only to establish but also to maintain the museum in a suitable manner to fulfil that purpose. Some persons are enthusiastic enough to think that a museum is in itself so good an object that they have only to provide a building and cases and a certain number of specimens, no matter exactly what, to fill them, and then the thing is done; whereas the truth is the work has only then begun. What a museum really depends upon for its success and usefulness is not its building, not its cases, not even its specimens, but its curator. He and his staff are the life and soul of the institution, upon whom its whole value depends; and yet in many—I may say most of our museums—they are the last to be thought of. The care, the preservation, the naming of the specimens are either left to voluntary effort—excellent often for special collections and for a limited time, but never to be depended on as a permanent arrangement,—or a grievously undersalaried and consequently uneducated official is expected to keep in order, to clean, dust, arrange, name, and display in a manner which will contribute to the advancement of scientific knowledge, collections ranging in extent over almost every branch of human learning, from the contents of an ancient British barrow to the last discovered bird of paradise from New Guinea.

Valuable specimens not unfrequently find their way into museums thus managed. Their public-spirited owners fondly imagine that they will be preserved and made of use to the world if once given to such an institution. Their fate is, unfortunately, far otherwise. Dirty, neglected, without label, their identity lost, they are often finally devoured by insects or cleared away to make room on the crowded shelves for the new donation of some fresh patron of the institution. It would be far better that such museums should never be founded. They are traps into which precious—sometimes priceless—objects fall only to be destroyed; and, what is still worse, they bring discredit on all similar institutions, make the very name of museum a byword and a reproach, hindering instead of advancing the recognition of their value as agents in the great educational movement of the age.

A museum is like a living organism—it requires continual and tender care. It must grow, or it will perish; and the cost and labour required to maintain it in a state of vitality is not yet by any means fully realised or provided for, either in our great national establishments or in our smaller local institutions.

Often as it has been said, it cannot be too often repeated that the real objects of forming collections, of whatever kind (apart, of course, from the mere pleasure of acquisition—sometimes the only motive of private collectors), and which, although in very different degrees, and often without being recognised, underlie the organisation of all museums, are two, which are quite distinct, and sometimes even conflicting. The first is to advance or increase the knowledge of some given subject. This is generally the motive of the individual collector, whose experience shows him the vast assistance in forming definite ideas in any line of research in which he may be occupied that may be derived from having the materials for its study at his own command, to hold and to handle, to examine and compare, to take up and lay aside whenever the favourable moment to do so occurs. But unless his subject is a very limited one, or his means the reverse, he soon finds the necessity of consulting collections based on a larger scale than his own. Very few people have any idea of the multiplicity of specimens required for the purpose of working out many of the simplest problems concerning the life-history of animals or plants. The naturalist has frequently to ransack all the

museums, both public and private, of Europe and America in
the endeavour to compose a monograph of a single common
genus, or even species, that shall include all questions of its
variation, changes in different seasons and under different
climates and conditions of existence, and the distribution in
space and time of all its modifications.   He often has to
confess at the end that he has been baffled in his research for
want of the requisite materials for such an undertaking.   Of
course this ought not to be, and the time will come when it
will not be, but that time is very far off yet.

We all know the old saying that the craving for riches
grows as the wealth itself increases.   Something similar is
true of scientific collections brought together for the purpose
of advancing knowledge.   The larger they are the more their
deficiencies seem to become conspicuous; the more desirous
we are to fill up the gaps which provokingly interfere with
our extracting from them the complete story they have to tell.

Such collections are, however, only for the advanced
student, the man who has already become acquainted with
the elements of his science and is in a position, by his
knowledge, by his training, and by his observing and reasoning
capacity, to take advantage of such material to carry on the
subject to a point beyond that at which he takes it up.

But there is another and a far larger class to whom museums
are or should be a powerful means of aid in acquiring knowledge.   Among such those who are commencing more serious
studies may be included; but I especially refer to the much
more numerous class, and one which it may be hoped will year
by year bear a greater relative proportion to the general
population of the country, who, without having the time, the
opportunities, or the abilities to make a profound study of any
branch of science, yet take a general interest in its progress,
and wish to possess some knowledge of the world around
them and of the principal facts ascertained with regard to it,
or at least some portions of it.   For such persons museums
may be, when well organised and arranged, of benefit to a
degree that at present can scarcely be realised.

To diffuse knowledge among persons of this class is the
second of the two purposes of museums of which I have spoken.

I believe that the main cause of what may be fairly termed the failure of the majority of museums—especially museums of natural history—to perform the functions that might be legitimately expected of them is that they nearly always confound together the two distinct objects which they may fulfil, and by attempting to combine both in the same exhibition practically accomplish neither.

In accordance with which of those two objects, which may be briefly called *research* and *instruction,* is the main end of the museum, so should the whole be primarily arranged; and in accordance with the object for which each specimen is required, so should it be treated.

The specimens kept for research, for advancement of knowledge, for careful investigations in structure and development, or for showing the minute distinctions which must be studied in working out the problems connected with variations of species according to age, sex, season, or locality; for fixing the limits of geographical distribution, or determining the range in geological time, must be not only exceedingly numerous (so numerous, indeed, that it is almost impossible to put a limit on what may be required for such purposes), but they must also be kept under such conditions as to admit of ready and close examination and comparison.

If the whole of the specimens really required for enlarging the boundaries of zoological or botanical science were to be displayed in such a manner that each one could be distinctly seen by any visitor sauntering through the public galleries of a museum, the vastness and expense of the institution would be out of all proportion to its utility; the specimens themselves would be quite inaccessible to the examination of all those capable of deriving instruction from them, and, owing to the injurious effects of continued exposure to light upon the greater number of preserved natural objects, would ultimately lose a large part of their permanent value. Collections of this kind must, in fact, be treated as the books in a library, and be used only for consultation and reference by those who are able to read and appreciate their contents. To demand, as has been ignorantly done, that all the specimens belonging to our national museums, for instance, should be displayed in cases in the public galleries, would be equivalent to asking that every book in a library, instead of being shut up and arranged on shelves for consultation when required, should have every single page framed and glazed and hung on the walls, so that the humblest visitor as he passes along the galleries has only to open his eyes and revel in the wealth of literature of all ages and all countries, without so much as applying to a custodian to open a case. Such an arrangement is perfectly conceivable. The idea from some points of view is magnificent, almost sublime. But imagine the space required for such an arrangement of the national library of books, or, indeed, of any of the smallest local libraries; imagine the inconvenience to the real student, the disadvantages which he would be under in reading the pages of any work fixed in an immovable position beneath a glass case; think of the enormous distances he would often have to traverse to compare a reference or verify a quotation, and the idea of sublimity soon gives place to its usual antithesis. The attempt to display every bird, every insect, shell, or plant, which is or ought to be in any of our great museums of reference would produce an exactly similar result.

In the arrangement of collections designed for research, which, of course, will contain all those precious specimens called "types," which must be appealed to through all time to determine the species to which a name was originally given, the principal points to be aimed at are—the preservation of the objects from all influences deleterious to them, especially dust, light, and damp; their absolutely correct identification, and record of every circumstance that need be known of their history; their classification and storage in such a manner that each one can be found without difficulty or loss of time; and, both on account of expense as well as convenience of access, they should be made to occupy as small a space as is compatible with these requirements. They should be kept in rooms provided with suitable tables and good light for their examination, and within reach of the necessary books of reference on the particular subjects which the specimens illustrate. Furthermore, the rooms should be so situated that the officers of the museum, without too great hindrance to their own work, can be at hand for occasional assistance and supervision of the student, and if collections of research and exhibited specimens are contained in one building, it is obvious that the closer the contiguity in which those of any particular group are placed the greater will be the convenience both of students and curators, for in very few establishments will it be possible to form each series on such a scale as to be entirely independent of the other.

On the other hand, in a collection arranged for the instruction of the general visitor, the conditions under which the specimens are kept should be totally different. In the first place, their numbers must be strictly limited, according to the nature of the subject to be illustrated and the space available. None must be placed too high or too low for ready examination. There must be no crowding of specimens one behind the other, every one being perfectly and distinctly seen, and with a clear space around it. Imagine a picture-gallery with half the pictures on the walls partially or entirely concealed by others hung in front of them; the idea seems preposterous, and yet this is the approved arrangement of specimens in most public museums. If an object is worth putting into a gallery at all, it is worth such a position as will enable it to be seen. Every specimen exhibited should be good of its kind, and all available skill and care should be spent upon its preservation and rendering it capable of teaching the lesson it is intended to convey. And here I cannot refrain from saying a word upon the sadly-neglected art of taxidermy, which continues to fill the cases of most of our museums with wretched and repulsive caricatures of mammals and birds, out of all natural proportions, shrunken here and bloated there, and in attitudes absolutely impossible for the creature to have assumed while alive. Happily there may be seen occasionally, especially where amateurs of artistic taste and good knowledge of natural history have devoted themselves to the subject, examples enough to show that an animal can be converted after death, by a proper application of taxidermy, into a real lifelike representation of the original, perfect in form, proportions, and attitude, and almost, if not quite, as valuable for conveying information on these points as the living creature itself. The fact is that taxidermy is an art resembling that of the painter, or rather the sculptor; it requires natural genius as well as great cultivation, and it can never be permanently improved until we have abandoned the present conventional low standard and low payment for " bird-stuffing," which is utterly inadequate to induce any man of capacity to devote himself to it as a profession.

To return from this digression, every specimen exhibited should have its definite purpose, and no absolute duplicate should on any account be permitted. Above all, the purpose for which each specimen is exhibited, and the main lesson to be derived from it, must be distinctly indicated by the labels affixed, both as headings of the various divisions of the series, and to the individual specimens. A well-arranged educational museum has been defined as a collection of instructive labels illustrated by well-selected specimens.

What is, or should be, the order of events in arranging a portion of a public museum? Not, certainly, as too often happens now, bringing a number of specimens together almost by haphazard, and cramming them as closely as possible in a case far too small to hold them, and with little reference to their order or to the possibility of their being distinctly seen. First, as I said before, you must have your curator. He must carefully consider the object of the museum, the class and capacities of the persons for whose instruction it is founded, and the space available to carry out this object. He will then divide the subject to be illustrated into groups, and consider their relative proportions, according to which he will plan out the space. Large labels will next be prepared for the principal headings, as the chapters of a book, and smaller ones for the various subdivisions. Certain propositions to be illustrated, either in the structure, classification, geographical distribution, geological position, habits, or evolution of the subjects dealt with, will be laid down and reduced to definite and concise language. Lastly will come the illustrative specimens, each of which as procured and prepared will fall into its appropriate place. As it is not always easy to obtain these at the time that they are wanted, gaps will often have to be left, but these, if properly utilised by drawings or labels, may be made nearly as useful as if occupied by the actual specimens.

A public exhibition which is intended to be instructive and interesting must never be crowded. There is, indeed, no reason why it ever should be. Every such exhibition, whether on a large or small scale, can only contain a representative series of specimens, selected with a view to the needs of the particular class of persons who are likely to visit the gallery, and the number of specimens exhibited should be adapted to the space available. There is, therefore, rarely any excuse for filling it up in such a manner as to interfere with the full view of every specimen shown. A crowded gallery, except in some very exceptional circumstances, at once condemns the curator, as the remedy is generally in his own hands. In order to avoid it he has nothing to do but sternly to eliminate all the less important specimens. If any of these possess features of historical or scientific interest demanding their permanent preservation, they should be kept in the reserve collections ; if otherwise, they should not be kept at all.

The ideal public museums of the future will, however, require far more exhibition space than has hitherto been allowed ; for though the number of specimens shown may be fewer than is often thought necessary now, each will require more room if the conditions above described are carried out, and especially if it is thought desirable to show it in such a manner as to enable the visitor to realise something of the wonderful complexity of the adaptations which bring each species into harmonious relation with its surrounding conditions. Artistic reproductions of natural environments, illustrations of protective resemblances, or of special modes of life, all require much room for their display. This method of exhibition, wherever faithfully carried out, is, however, proving both instructive and attractive, and will doubtless be greatly extended.

Guide-books and catalogues are useful adjuncts, as being adapted to convey fuller information than labels, and as they can be taken away for study during the intervals of visits to the museum, but they can never supersede the use of labels. Any one who is in the habit of visiting picture-galleries where

the names of the artists and the subject are affixed to the
frame, and others in which the information has in each case
to be sought by reference to a catalogue, must appreciate the
vast superiority in comfort and time-saving of the former
plan.

Acting upon such principles as these, every public gallery
of a museum, whether the splendid saloon of a national
institution, or the humble room containing the local collection
of a village club, can be made a centre of instruction, and will
offer interests and attractions which will be looked for in vain
in the majority of such institutions at the present time.

One of the best illustrations of the different treatment of
collections intended for research or advancement of knowledge,
and for popular instruction or diffusion of knowledge, is now
to be seen in Kew Gardens, where the admirably constructed
and arranged herbarium answers the first purpose, and the
public museums of economic botany the second. A similar
distinction is carried out in the collections of systematic botany
in the natural history branch of the British Museum, with the
additional advantage of close contiguity; indeed, as an example
of a scheme of good museum arrangement (although not perfect
yet in details) I cannot do better than refer to the upper story
of the east wing of that institution. The same principles,
little regarded in former times in this country, and still unknown in some of the largest continental museums, are
gradually pervading every department of the institution,
which, from its national character, its metropolitan position,
and exceptional resources, ought to illustrate in perfection the
ideal of a natural-history museum. In fact, it is only in a
national institution that an exhaustive research collection in
all branches of natural history, in which the specialist of every
group can find his own subject fully illustrated, can or ought
to be attempted.

As the actual comparison of specimen with specimen is the
basis of zoological and botanical research, and as work done
with imperfect materials is necessarily imperfect in itself, it
is far the wisest policy to concentrate in a few great central
institutions, the number and situation of which must be
determined by the population and the resources of the country,

all the collections, especially those containing specimens already alluded to as so dear to the systematic naturalist, known as author's "types," required for original investigations.   It is far more advantageous to the investigator to go to such a collection and take up his temporary abode there, while his research is being carried out, with all the material required at his hand at once, than to travel from place to place and pick up piecemeal the information he requires, without opportunity of direct comparison of specimens.

I do not say that collections for special study, and even original research, should not, under particular circumstances and limitations, be formed at museums other than central national institutions, or that nothing should be retained in provincial museums but what is of a directly educational or elementary nature.   A local collection, illustrating the fauna and flora of the district, should be part of every such museum ; and this may be carried to almost any amount of detail, and therefore in many cases it would be very unadvisable to exhibit the whole of it.   A selection of the most important objects may be shown under the conditions described above, and the remainder carefully preserved in cabinets for the study of specialists.

It is also very desirable in all museums, in order that the exhibited series should be as little disturbed as possible in arrangement, and be always available for the purpose for which it is intended, that there should be, for the use of teachers and students, a supplementary set of common objects, which, if injured, could be easily replaced.   It must not be forgotten that the zealous investigator and the conscientious curator are often the direst antagonists : the one endeavours to get all the knowledge he can out of a specimen, regardless of its ultimate fate, and even if his own eyes alone have the advantage of it ; the other is content if a limited portion only is seen, provided that can be seen by every one both now and hereafter.

Such, then, is the primary principle which ought to underlie the arrangement of all museums—the distinct separation of the two objects for which collections are made ; the publicly exhibited collection being never a store-room or magazine, but only such as the ordinary visitor can understand and profit by, and the collection for students being so arranged as to afford every facility for examination and research. The improvements that can be made in detail in both departments are endless, and to enter further into their consideration would lead me far beyond the limits of this address. Happily, as I said before, the subject is receiving much attention.

I would willingly dwell longer upon it—indeed I feel that I have only been able to touch slightly and superficially upon many questions of practical interest, well worthy of more detailed consideration; but time warns me that I must be bringing this discourse to a close, and I have still said nothing in reference to subjects upon which you may expect some words on this occasion. I mean those great problems concerning the laws which regulate the evolution of organic beings, problems which agitate the minds of all biologists of the present day, and the solution of which is watched with keen interest by a far wider circle—a circle, in fact, coincident with the intelligence and education of the world. Several communications connected with these problems will be brought before the sectional meetings during the next few days, and we shall have the advantage of hearing them discussed by some of those who, by virtue of their special attention to and full knowledge of these subjects, are most competent to speak with authority. It is therefore for me rather delicate ground to tread upon, especially at the close of a discourse mainly devoted to another question. I will, however, briefly point out the nature of the problems and the lines which the endeavour to solve them will probably take, without attempting to anticipate the details which you will doubtless hear most fully and ably stated elsewhere.

I think I may safely premise that few, if any, original workers at any branch of biology appear now to entertain serious doubt about the general truth of the doctrine that all existing forms of life have been derived from other forms by a natural process of descent with modification, and it is generally acknowledged that to the records of the past history of life upon the earth we must look for the actual confirmation of the truth of a doctrine which accords so strongly with all we know of the present history of living beings.

Professor Huxley wrote in 1875 : " The only perfectly safe foundation for the doctrine of evolution lies in the historical, or rather archæological, evidence that particular organisms have arisen by the gradual modification of their predecessors, which is furnished by fossil remains. That evidence is daily increasing in amount and in weight, and it is to be hoped that the comparisons of the actual pedigree of these organisms with the phenomena of their development may furnish some criterion by which the validity of phylogenic conclusions deduced from the facts of embryology alone may be satisfactorily tested."

Palæontology, however, as we all know, reveals her secrets with no open hand. How can we be reminded of this more forcibly than by the discovery announced scarcely three months ago by Professor Marsh of numerous mammalian remains from formations of the Cretaceous period, the absence of which had so long been a source of difficulty to all zoologists? What vistas does this discovery open of future possibilities, and what thorough discredit, if any were needed, does it throw on the value of negative evidence in such matters! Bearing fully in mind the necessary imperfection of the record we have to deal with, I think that no one taking an impartial survey of the recent progress of palæontological discovery can doubt that the evidence in favour of a gradual modification of living forms is still steadily increasing. Any regular progressive series of changes of structure coinciding with changes in time can of course only be expected to be preserved and to come again before our eyes under such a favourable combination of circumstances as must be of most rare occurrence ; but the links, more or less perfect, of many such series are continually being revealed, and the discovery of a single intermediate form is often of immense interest as indicating the path along which the modification from one apparently distinct form to another may have taken place.

Though palæontology may be appealed to in support of the conclusion that modifications have taken place as time advanced, it can scarcely afford any help in solving the more difficult problems which still remain as to the methods by which the changes have been brought about.

Ever since the publication of what has been truly described as the " creation of modern natural history," Darwin's work on the *Origin of Species*, there has been no little controversy as to how far all the modifications of living forms can be accounted for by the principle of natural selection or preservation of variations best adapted for their surrounding conditions, or whether any, and if so what, other factors have taken part in the process of organic evolution.

It certainly cannot be said that in these later times the controversy has ended.   Indeed those who are acquainted with scientific literature must know that notes struck at the last annual meeting of this Association produced a series of reverberations the echoes of which have hardly yet died away.

Within the last few months also two important works have appeared in our country, which have placed in an accessible and popular form many of the data upon which the most prevalent views on the subject are based.

The first is *Darwinism : an Exposition of the Theory of Natural Selection, with some of its Applications*, by Alfred Russel Wallace.   No one could be found so competent to give such an exposition of the theory as one who was, simultaneously with Darwin, its independent originator, but who, by the title he has chosen, no less than by the contents of the book, has, with rare modesty and self-abnegation, transferred to his fellow-labourer all the merit of the discovery of what he evidently looks upon as a principle of overwhelming importance in the economy of nature ; " supreme," indeed, he says, " to an extent which even Darwin himself hesitated to claim for it."

The other work I refer to is the English translation of the remarkable *Essays upon Heredity and Kindred Biological Problems*, by Dr. August Weismann, published at the Oxford Clarendon Press, in which is fully discussed the very important but still open question—a question which was brought into prominence at our meeting at Manchester two years ago—of the transmission or non-transmission to the offspring of characters acquired during the lifetime of the parent.

It is generally recognised that it is one of the main elements of Darwin's, as well as of every other theory of evolution, that there is in every individual organic being an innate tendency to vary from the standard of its predecessors, but that this tendency is usually kept under the sternest control by the opposite tendency to resemble them—a force to which the terms "heredity" and "atavism" are applied. The causes of this initial tendency to vary, as well as those of its limits and prevailing direction, and the circumstances which favour its occasional bursting through the constraining principle of heredity, offer an endless field for speculation. Though several theories of variation have been suggested, I think that no one would venture to say we have passed beyond the threshold of knowledge of the subject at present.

Taking for granted, however, as we all do, that this tendency to individual variation exists, then comes the question, What are the agents by which, when it has asserted itself, it is controlled or directed in such a manner as to produce the permanent or apparently permanent modifications of organic structures which we see around us? Is "survival of the fittest" or preservation by natural selection of those variations best adapted for their surrounding conditions (the essentially Darwinian or still more essentially Wallacian doctrine) the sole or even the chief of these agents? Can isolation, or the revived Lamarckian view of the direct action of the environment, or the effects of use or disuse accumulating through generations, either singly or combined, account for all? or is it necessary to invoke the aid of any of the numerous subsidiary methods of selection which have been suggested as factors in bringing about the great result?

Any one who has closely followed these discussions, especially those bearing most directly upon what is generally regarded as the most important factor of evolution—natural selection, or "survival of the fittest"—cannot fail to have noticed the appeal constantly made to the advantage, the utility, or otherwise of special organs or modifications of organs or structures to their possessors. Those who have convinced themselves of the universal application of the doctrine of natural selection hold that every particular structure or modification of structure must be of utility to the animal or plant in which it occurs, or to some ancestor of that animal or plant, otherwise it could not have come into existence; the only reservation being for cases which are explained by the principle which Darwin called "correlation of growth." Thus the extreme natural selectionists and the old-fashioned school of teleologists are so far in agreement.

On the other hand, it is held by some that numerous structures and modifications of structures are met with in nature which are manifestly useless; it is even confidently stated that there are many which are positively injurious to their possesssor, and therefore could not possibly have resulted from the action of natural selection of favourable variations. Organs or modifications when in an incipient condition are especially quoted as bearing upon this difficulty. But here, it seems to me, we are continually appealing to a criterion by which to test our theories of which we know far too little, and this (though often relied upon as the strongest) is, in reality, the weakest point of the whole discussion.

Of the variations of the form and structure of organic bodies we are beginning to know something. Our museums, when more complete and better organised, will teach us much on this branch of the subject. They will show us the infinite and wonderful and apparently capricious modifications of form, colour, and of texture to which every most minute portion of the organisation of the innumerable creatures which people the earth is subject. They will show us examples of marvellously complicated and delicate arrangements of organs and tissues in many of what we consider as almost the lowest and most imperfectly organised groups of beings with which we are acquainted. But as to the use of all these structures and modifications in the economy of the creatures that possess them, we know, I may almost say, nothing, and our museums will never teach us these things. If time permitted I might give numerous examples in the most familiar of all animals, whose habits and actions are matters of daily observation, with whose life-history we are as well acquainted almost as we are with our own, of structures the purposes of which are still most doubtful. There are many such even in the composition of our own bodies. How, then, can we expect to answer such questions when they relate to animals known to us only by dead specimens, or by the most transient glimpses of the living in a state of nature, or when kept under the most unnatural conditions in confinement? And yet this is actually the state of our knowledge of the vast majority of the myriads of living beings which inhabit the earth. How can we, with our limited powers of observation and limited capacity of imagination, venture to pronounce an opinion as to the fitness or unfitness for its complex surroundings of some peculiar modification of structure found in some strange animal dredged up from the abysses of the ocean, or which passes its life in the dim seclusion of some tropical forest, and into the essential conditions of whose existence we have at present no possible means of putting ourselves in any sort of relation?

How true it is that, as Sir John Lubbock says, "we find in animals complex organs of sense richly supplied with nerves, but the functions of which we are as yet powerless to explain. There may be fifty other senses as different from ours as sound is from sight; and even within the boundaries of our own senses there may be endless sounds which we cannot hear, and colours as different as red from green of which we have no conception. These and a thousand other questions remain for solution. The familiar world which surrounds us may be a totally different place to other animals. To them it may be full of music which we cannot hear, of colour which we cannot see, of sensations which we cannot conceive."

The fact is that nearly all attempts to assign purposes to the varied structures of animals are the merest guesses and assumptions. The writers on natural history of the early part of the present century, who "for every why must have a wherefore," abound in these guesses, which wider knowledge shows to be untenable. Many of the arguments for or against natural selection, based upon the assumed utility or equally assumed uselessness of animal and vegetable structures, have nothing more to recommend them. In fact, to say that any part of the organisation of an animal or plant, or any habit or instinct with which it is endowed, is useless, or even injurious, seems to me an assumption which, in our present state of knowledge, we are not warranted in making. The time may come when we shall have more light, but infinite patience and infinite labour are required before we shall be in a position to speak dogmatically on these mysteries of nature—labour not only in museums, laboratories, and dissecting-rooms, but in the homes and haunts of the animals themselves, watching and noting their ways amid their natural surroundings, by which means alone we can endeavour to penetrate the secrets of their life-history. But until that time comes, though we may not be quite tempted to echo the despairing cry of the poet, "Behold, we know not anything," a frank confession of ignorance is the most straightforward, indeed the only honest position we can assume when questioned on these subjects.

However much we may be convinced of the supreme value of scientific methods of observation and of reasoning, both as mental training of the individual and in the elucidation of truth and advancement of knowledge generally, it is impossible to be blind to the fact that we who are engaged with the investigation of those subjects which are commonly accepted as belonging to the domain of physical science are unfortunately not always, by virtue of being so occupied, possessed of that most precious gift, "a right judgment in all things."

No one intimately acquainted with the laborious and wavering steps of scientific progress (I can answer at least for one branch of it) can look upon that progress with a perfect feeling of satisfaction.

Can it be said of any of us that our observations are always accurate, the materials on which they are based always sufficient, our reasoning always sound, our conclusions always legitimate? Is there any subject, however limited, of which our knowledge can be said to have reached finality?

Or if it happens to any of us as to

> A man who looks at glass
>   On it may stay his eye,
> Or if he pleases through it pass
>   And then the heavens espy,

are not those heavens which are beyond the immediate
objects of our observation coloured by our prejudices, pre-
possessions, emotions, or imagination, as often as they are
defined by any profound insight into the depth of nature's
laws ?   In most of these questions an open mind and a suspended judgment appear to me the true scientific position,
whichever way our inclinations may lead us.

For myself, I must own that when I endeavour to look
beyond the glass, and frame some idea of the plan upon
which all the diversity in the organic world has been brought
about, I see the strongest grounds for the belief, difficult as it
sometimes is in the face of the strange, incomprehensible,
apparent defects in structure, and the far stranger, weird
ruthless savagery of habit, often brought to light by the
study of the ways of living creatures, that natural selection,
or survival of the fittest, has, among other agencies, played a
most important part in the production of the present condition
of the organic world, and that it is a universally acting and
beneficent force continually tending towards the perfection of
the individual, of the race, and of the whole living world.

I can even go farther and allow my dream still thus to
run :

> Oh yet we trust that somehow good
>   Will be the final goal of ill . . .
>
> That nothing walks with aimless feet ;
>   That not one life shall be destroy'd,
>   Or cast as rubbish to the void,
> When God hath made the pile complete.

MODERN MUSEUMS [1]

THE Museums Association is one of the youngest of the numerous social organisations which it is thought expedient at the present day to constitute, in order to give facilities for the interchange of ideas on subjects interesting to a special group of men. It is, indeed, only in the fourth year of its existence, and this is the first time that a meeting has been held in London, the centre in which are gathered the great national collections, and in which reside so considerable a number of persons engaged in their custody. The association claims York as its birthplace, and Liverpool, Cambridge, and Manchester have in succession afforded it hospitality and enjoyed the advantage of its presence.

We all meet with one object in view. We are all impressed with the value—with the necessity, I should say—of the Museum (using the word in its widest sense, as a collection of works of art and of nature) in the intellectual advance of mankind.

How could art make any progress, how could it even exist, if its productions were destroyed as soon as they were created, if there were no museums, private or public, in which they could be preserved and made available to mankind now and hereafter? How could science be studied without ready access to the materials upon which knowledge is built up? In many branches of science the progress is mainly commensurate with the abundance and accessibility of such materials.

[1] Presidential Address to the Museums Association at the Meeting in London, 3rd July 1893.

Though the first duty of museums is, without question, to preserve the evidence upon which the history of mankind and the knowledge of science is based, any one acquainted with the numerous succession of essays, addresses, lectures, and papers, which constitute the museum literature of the last thirty years, must recognise the gradual development of the conception that the museum of the future is to have for its complete ideal, not only the simple preservation of the objects contained in it, but also their arrangement in such a manner as to provide for the instruction of those who visit it. The value of a museum will be tested not only by its contents, but by the treatment of those contents as a means of the advancement of knowledge. Though this is the general consensus of opinion, as expressed in the literature just referred to, there is naturally still much divergence as to the best methods by which this ideal may be carried out, and there are still many practical difficulties to be overcome before the views so ably advocated on paper can be reduced to the test of actual performance. It is with a hope of assisting in the solution of these difficulties that this Association has been founded.

If in the few words with which I am expected to preface the real work of the meeting I shall be found to dwell too exclusively upon the subject of natural history museums, I must apologise to many friends and members of the Association who are present. It must be distinctly understood that under the word museum we include collections of all kinds formed for the advancement of any branch of knowledge, except those specially devoted to books, which already are cared for by the "Library Association"—on the model of which ours was formed. I hope that in our papers at this meeting and in future presidential addresses we shall have all branches of museum work fairly represented.

It is my fate to have been born what is commonly called a "naturalist." I hardly remember the time when I was not a possessor of a museum, but it always took a distinctly biological direction. Hence, although by no means unappreciative of other branches of museum work, I shall confine myself chiefly to that part of the subject upon which I can speak from personal experience. Even in this branch, time will compel me to limit myself to observations upon some of the larger questions connected with our subject, leaving details for discussion in our subsequent meetings.

One great difference between the work of the curator of an art museum and that of one devoted to what are called natural history subjects, is that in the case of the former the specimens he has to preserve and exhibit come into his hands very nearly in the condition in which they will have to remain. A picture, a vase, a piece of old armour, or a statue, beyond a certain amount of tender care in cleaning and repairing, which is more or less mechanical in its nature, is ready for its place upon the museum shelves. But this is far from being the case with the greater number of natural objects. Not only do they require special methods of preservation, but very often their value as museum specimens depends entirely upon the skill, labour, patience, and knowledge expended upon them. In specimens illustrating biological subjects the highest powers of the museum curator are called forth. A properly-mounted animal or a carefully-displayed anatomical preparation is in itself a work of art, based upon a natural substratum. In few branches of museum work has there been greater progress in late years than in this, and few offer still further scope for development.

Partly from this cause, and partly from the fact that art has for a longer period and to a greater degree engaged the attention of civilised man than nature, the methods of preservation, arrangement, and exhibition of works of art are on the whole further advanced than are those of natural objects. But no one can deny that there is still in many galleries devoted to the exhibition of works of art of various kinds great room for improvement. There is generally far too great crowding; too many objects so placed that the tallest man cannot see them properly, even when standing on tiptoe; too many others placed so low that they can only be examined by lying down on the floor; too many completely spoiled by the juxtaposition of incongruous objects, or by unsuitable settings. It is only in a very few public museums (I may instance, as a conspicuous example, the splendid museum of antiquities at Naples) that the immense advantage to be gained by ample space and appropriate surroundings in aiding the formation of a just idea of the beauty and interest of each specimen contained in it can be properly appreciated. Correct classification, good labelling, isolation of each object from its neighbours, the provision of a suitable background, and above all of a position in which it can be readily and distinctly seen, are absolute requisites in art museums as well as in those of natural history.   Nothing detracts so much from the enjoyment and advantage derived from a visit to a museum as the overcrowding of the specimens exhibited.   The development of the new museum idea, to be spoken of later on, will be one way by which this can be remedied in the public galleries ; but if museums are what they ought to be, and what I venture to believe they will be in the future, the question of space on a considerably larger scale than has hitherto been thought of will have to be faced.   This is of course mainly a matter of expense, and after all but a small matter compared with expenditure now considered necessary in other directions.   There are persons who think the country made a tremendous effort in building so much as is yet finished of the new Natural History Museum in the Cromwell Road, and shake their heads at the expenditure asked for either to complete that establishment by the erection of the wings at the sides, or to finish the neighbouring South Kensington Museum in such a manner as worthily to hold its collections, both of art and science.   Others would grudge the further expansion of the magnificent series of treasures of ancient and mediæval art in the British Museum at Bloomsbury, of which the country has such just reason to be proud.   Let such persons consider that the largest museum yet erected, with all its internal fittings, has not cost so much as a single fully-equipped line-of-battle ship, which in a few years may be either at the bottom of the sea, or so obsolete in construction as to be worth no more than the materials of which it is made.   Not that I am deprecating the building of ships necessary for our protection, but rather wishing to show that the cost of such museums as are still required for the proper education of the nation is not such as would produce any sensible impression upon its financial position.

I may make a still more apposite comparison, and point to the vast sums of money spent by this nation upon the whole subject of education now and a few years ago. The total estimate for what is called " Class IV., Education, Science, and Art," for the financial year 1883-84, amounted to £4,748,556. In ten years it has grown to nearly double that amount, the estimate for 1893-94 being £9,172,216, the increase being mainly due to what is termed " Public Education." The amount spent upon the development of museums is comparatively insignificant. The British Museum vote (including the library and the natural history branch) has only increased during the same decade from £146,019 to £157,500.[1] The cost of the various museums maintained by the Science and Art Department shows little appreciable augmentation, except in the case of that at Dublin, where I am glad to see £19,035 is now put down instead of the £13,602 of the former period. Compared with the whole amount expended upon other methods of education, national expenditure upon museums and art galleries is at present very small.

In reference to this subject one cannot help considering how much might have been done if only a moderate portion of that large sum of money obtained a few years ago by the tax on brewers, and handed over to the County Councils to spend in promoting technical education, had been used for erecting museums, which might have taken a permanent place in the education of the country. Every subject taught, in order to make the teaching real and practical, should have its collection, and these various collections might all have been associated in the county museum under the same general management. The staff of teachers would assist in the curatorial work, and thus a well-equipped central college for technical education might have been formed in every county, sending out ramifications into the various districts in which the need of special instruction was most felt, and being also

---

[1] The corresponding figures for the year 1897-98 are : Class IV., Total vote, £10,777,537—British Museum, £162,439.

the parent of smaller branch museums of the same kind wherever they seem required.

But it is not only in the buildings that the expense of the museums of the future will have to be met. Another great advance must be made before they can be placed upon a satisfactory footing, and perform the functions that can be legitimately expected of them. This is in the elevation of the position and acquirements of those who have the care of them. As I have said on a previous occasion, " What a museum really depends upon for its success and usefulness is not its building, not its cases, not even its specimens, but its curator."

Speaking in the presence of a number of gentlemen who are curators of museums, do not let me be misunderstood. I do not mean that you are not zealous in the cause and make great sacrifices for it, and do all you can under the often difficult circumstances in which you are placed ; but what I mean is—and I am sure you will one and all agree with me when I say it—you are not properly appreciated by the public, and the importance and difficulties of your position are by no means sufficiently understood.

In a civilised community the necessities of life, to say nothing of luxuries (which we do not ask for), but the bare necessities of a man of education and refinement, who has to associate with his equals, and bring up his children to the life of educated and refined people, involve a certain annual expenditure, and the means afforded by any occupation for this necessary expenditure give a rough and ready test of the appreciation in which such occupation is held.

Now, a curator of a museum, if he is fit for his duties, must be a man of very considerable education as well as natural ability. If he is not himself an expert in all the branches of human knowledge his museum illustrates, he must be able to understand and appreciate them sufficiently to know where and how he can supplement his own deficiencies, so as to be able to keep every department up to the proper level. His education, in fact, must be not dissimilar to that required for most of the learned professions. Skill, manual dexterity, and good taste are also most valuable. He must, in addition, if he is to be a success in his vocation, possess various moral qualifications not found in every professional man—punctuality, habits of business, conciliatory manners, and, above all, indomitable and conscientious industry in the discharge of the small and somewhat monotonous routine duties which constitute so large a part of a curator's life. Such being the requirements of the profession, let us see what are the inducements offered to men to take it up as a means of livelihood. I really am sorry to have to speak of such a sordid subject, but I know it is one you naturally shrink from talking of yourselves. You would be the last people in the world to take the remedy, now so often resorted to by other classes, into your own hands. A strike of curators is hardly to be contemplated. Remember, also, that I am not speaking of this subject in your interest, or the interests of any individuals. Whether any of you personally should have your emoluments, your social position, your opportunities for good, improved, is not now with me an object of concern : it is in the interest of that great question, the advance of the museum as a means of educating, cultivating, and elevating mankind, that I am speaking—an advance that can only be effectively made when the curatorship of a museum is looked upon as an honourable and desirable profession for men of high intellectual acquirements.

Let me take a few examples of the inducements to enter this profession at the present time. I have before me some recent advertisements. The curator of the Museum of the Philosophical and Literary Society of one of the largest and most flourishing of our manufacturing cities is offered £125 a year for his services. In another town, smaller and less wealthy, it is true, "a resident curator, meteorological observer, and caretaker, is wanted for the museum and library buildings, at a salary of £50 per annum, with rooms, coal, and gas. Applicants are to state age and *scientific* qualifications."

In a recent newspaper discussion upon the establishment of a museum in one of the midland counties, after it had been pointed out that one of the prime necessities of such an institution was a provision for the maintenance of a curator, a leading gentleman of the district, a zealous and sympathetic advocate of the cause, perfectly acquiescing in this view, suggested that £100 a year should by all means be set aside for this purpose!

It is frequently my lot to be consulted by anxious parents of sons who develop a taste for museum work, as to what such a taste will lead to if cultivated.   I need hardly say that, however much I may wish our ranks to be recruited by such enthusiastic aspirants,—boys often of great ability and promise, —I cannot conscientiously offer much encouragement.   The best I can say is, that I hope things will be better in the future than they are at present.   As far as the Metropolitan museums are concerned there has been some improvement, and I think that indications are not wanting that this improvement will continue and extend.

I have referred at the beginning of this address to the great amount of recent literature upon the museum question, consisting largely of depreciation of the old ways of arranging museums, of suggestions for their improvement in the future, and mainly of the development of what may be called *the new museum idea*.   What this idea is was tersely expressed nearly thirty years ago by the late Dr. John Edward Gray in his address to the British Association at Bath (1864) as President of Section D, when near the close of his long career as administrator of a collection which by his exertions he had made the largest of the kind in the world.   Dr. Gray laid down the axiom that the purposes for which a museum was established were two—"first, the diffusion of instruction and rational amusement among the mass of the people, and, secondly, to afford the scientific student every possible means of examining and studying the specimens of which the museum consists."   He then continued: "Now, it appears to me that in the desire to combine these two objects, which are essentially distinct, the first object—namely, the general instruction of the people—has been to a great extent lost sight of and sacrificed to the second without any corresponding advantage to the latter, because the system itself has been thoroughly erroneous."

This was a remarkable admission, coming from a man who had been brought up in, and had acted throughout the greater part of his life upon, the old idea; but it clearly expressed

what was then beginning to be felt by many who turned their
unbiassed attention to the subject, and it is the keynote of
nearly all the museum reform of recent date.   During the
long discussion which followed, the new idea found powerful
advocates in Huxley, Hooker, Sclater, Wallace, and others;
but Owen, whose official position made him the chief scientific
adviser in the construction of the new National Museum of
Natural History, never became reconciled to it, and, unfortunately, threw all the weight of his great authority into
the opposite scale.

   The method of application of this principle depends entirely
upon the general nature of the museum, whether that of a
nation, a town, a school, or a society or institution established
to cultivate some definite branch of knowledge.   It is mainly
of national museums that I am speaking at present, and it is
only in national museums that the fulfilment of both functions
in fairly equal proportions can be expected.   In almost all
other museums the diffusion of knowledge, or popular education,
will be the primary function; and if the true principles of
arrangement of such museums be once grasped, this is a
function which can be carried out upon the largest or the
smallest, or any intermediate scale, according to the means of
the institution and the requirements of the locality.

   The collections for the advancement of science, on the other
hand, are of value mainly in proportion to their size, and no
museum at present existing has come anywhere near what
is required for the exhaustive study of natural history.   If
any one were now to endeavour to write a complete monograph of any family in the animal kingdom, he would search
in vain for materials for doing so, not only in any one
museum, but in all the museums in the world put together.

   Soon after the arrival in our National History Museum
of the great and carefully selected and labelled collection of
Indian birds, presented by Mr. A. O. Hume, containing
upwards of 60,000 specimens, a well-known ornithologist
commenced the volumes devoted to birds in the excellent
series of manuals on the fauna of British India, edited by
Mr. Blanford.   I am told that when he began the work, he
was seen sitting at his table rubbing his hands with delight

at the prospect of success in his labours guaranteed by such
an unprecedented mass of material. But after a few weeks
the scene had changed. He was pacing up and down the
room, wringing the same hands in despair at the hopeless-
ness of solving the tangled problems of the variation according
to age, sex, season, and locality, the geographical distribution, and the limits and relationship of any single species,
owing to the absolutely insufficient number of properly
authenticated specimens at his command. Every zoologist
will recognise this as a scarcely exaggerated description of
what he meets with at every step of his work. Except,
perhaps, for some special and limited groups, which may be
taken up in private collections, a national museum alone can
possibly attempt to bring together the materials required for
such exhaustive work; but it is undoubtedly the duty of all
national museums to endeavour to do this. There should be,
in every great nation, one establishment at least where such
problems may be attacked with some prospect of success, and
the only conditions upon which collections for this purpose
can be maintained are, that they should be so arranged as to
occupy the smallest possible space compatible with their
proper preservation and convenience of access, that they
should be removed from all the deteriorating influences of
light and dust, and at the same time be perfectly available
for the closest examination by all those whose knowledge is
sufficient to enable them to extract any information from
them. This means that they cannot be *exhibited* in the
ordinary sense of the word; although it must not be supposed
that they are on that account in less need of orderly and
methodical arrangement. There is certainly a danger of
collections which are not generally exhibited becoming
neglected, and degenerating into the condition of mere accumu-
lations of rubbish. Anything of the kind is absolutely
incompatible with the true requirements of specimens kept
for research. They specially need to be arranged in an
orderly and methodical manner, and to be thoroughly well
catalogued and labelled, so that each may be found directly it
is wanted, and they must be frequently inspected to see that they
are free from moth or other deleterious influence. The object

of keeping them in this condition is, indeed, that they should be preserved and not destroyed, as ultimately happens to so many exhibited specimens. Much curatorial ingenuity may be exercised in the methods of stowing and arranging such specimens to the best advantage. The conditions of access to them will be precisely those now accorded to books or manuscripts in a library, prints and drawings in an art museum, or the records and public documents in the Rolls Office or Somerset House.

As the actual comparison of specimen with specimen is the basis of zoological and botanical research, and as work done with imperfect materials is necessarily imperfect in itself, it is far the wisest policy to concentrate in a few great central institutions, the number and situation of which must be determined by the population and resources of the country, all the collections which are required for the prosecution of original research. These are especially those containing author's types or the actual specimens upon which species have been established, and which must be appealed to through all time to settle vexed questions of nomenclature. It is far more advantageous to the investigator to go to such a collection, and take up his temporary abode there while his research is being carried out, with all the material at his hand at once, than to travel from place to place and pick up piecemeal the information he requires, without opportunity of direct comparison of the specimens one with another.

On the other hand, in local museums such collections are not only not required, but add greatly to the trouble and expense of the maintenance of the institution, without any compensating advantage. Here it will be the duty of the curator to develop the side of the museum which is educational and attractive to the general visitor and to all who wish to obtain that knowledge which is the ambition of many cultivated persons to acquire without becoming specialists or experts. The study of the methods by which such museums may be made instructive and interesting offers an endless field for experiment and discussion, and the various problems connected with it are treated of not only in the literature I have referred to, but in a more practical manner in many museums in various parts of the world.

Without pursuing this question further at the present time, I should like to repeat, from a previous address on the same subject,[1] certain propositions which are fundamental in the arrangement of collections of the class of which I am now speaking.

The number of the specimens must be strictly limited, according to the nature of the subject to be illustrated and the space available. None must be placed either too high or too low for ready examination. There must be no crowding of specimens one behind the other, every one being perfectly and distinctly seen, and with a clear space around it. If an object is worth putting into a gallery at all, it is worth such a position as will enable it to be seen. Every specimen exhibited should be good of its kind, and all available skill and care should be spent upon its preservation, and rendering it capable of teaching the lesson it is intended to convey. Every specimen should have its definite purpose, and no absolute duplicate should on any account be admitted. Above all, the purpose for which each specimen is exhibited, and the main lesson to be derived from it, must be distinctly indicated by the labels affixed, both as headings of the various divisions of the series and to the individual specimens.

These are the principles of what may be called the New Museum idea as applied to national museums of natural history. It is a remarkable coincidence that since they were first enunciated, and during the time of their discussion, but before they had met with anything like universal acceptance, the four first nations of Europe almost simultaneously erected in their respective capitals—London, Paris, Vienna, and Berlin—entirely new buildings, on a costly, even palatial scale, to receive the natural history collections, which in each case had quite outgrown their previous insufficient accommodation. In the construction of neither of these four edifices can the guardians of the public purse be accused of

---

[1] Presidential Address to the British Association for the Advancement of Science. Report of Newcastle-on-Tyne Meeting, 1889.

want of liberality. Each building is a monument in itself
of the appreciation of the government of the country of the
value and interest of the natural history sciences. So far
this is most satisfactory. Now that each is more or less
completed, at all events for the present, and its contents
in a fair way towards a permanent arrangement, it may
not be without interest on the present occasion to give
some comparative account of their salient features, especially
with a view to ascertain whether and to what extent their
construction and arrangement have complied with the
requirements of the modern idea of such institutions.

It may seem ungrateful to those who have so liberally
responded to the urgent representations of men of science by
providing the means of erecting these splendid buildings,
to suggest that if they had all been delayed for a few years
the result might have been more satisfactory. The effects of
having been erected in what may be called a transitional
period of museum ideas is more or less evident in all, and all
show traces of compromise, or rather adaptation to new ideas
of structures avowedly designed for old ones. In none,
perhaps, is this more strikingly shown than in our own,—
built, unfortunately, before any of the others, and so without
the advantages of the experience that might have been gained
from their successes or their shortcomings. Though a building
of acknowledged architectural beauty, and with some excellent
features, it cannot be taken structurally as a model museum,
when the test of adaptation to the purpose to which it is
devoted is rigidly applied. But to speak of its defects is an
ungracious and uncongenial task for me. If it were not taking
me too far away from my present subject I would rather speak
of the admirable manner in which the staff are endeavouring
to carry out the new idea under somewhat disadvantageous
circumstances.

The new zoological museum in the Jardin des Plantes at
Paris is a glorification of the old idea pure and simple. It
consists of one huge hall, with galleries and some annexes, in
which every specimen is intended to be exhibited, more or
less imperfectly, on alternate periods, to students and to the
general public. The building and cases are very handsome in

style, and there are endless rows of specimens of all kinds
neatly mounted in a uniform manner.   There are no storerooms, no laboratories, no workrooms, connected with the
building.   These are all in other more or less distant parts
of the establishment, separated from it in most cases by the
whole breadth of the garden.   Of course this can only be
looked upon as a temporary condition of affairs.   Fortunately
there is still room on the site of the old museum behind the
new building, and if this is utilised by erecting upon it a
commodious set of workrooms, laboratories, rooms for reserve
collections, and administrative offices directly in connection
with each other and with the main building, which might
then be emptied of a considerable portion of its contents, an
extremely good working museum may be evolved.   But if
this space, as I believe was the original design, is used for the
further extension of the already disproportionately large public
galleries, the opportunity will be lost.

The new museums at Vienna, the one for natural history,
the other for art, placed one on each side of a handsome public
garden in one of the most important quarters of the city,
exactly alike in size and architectural features, are elegant
buildings, and present many excellent features of construction.
The natural history museum, which was alone finished when I
visited Vienna three years ago, is a quadrilateral structure
with a central court, and consists of three stories and a basement.   Each story is divided into a number of moderately-sized
rooms, opening one into another, so that by passing along in
the same direction the visitor can make an inspection in
systematic order of all the collections arranged in each story,
returning to the point from whence he started ; or, if need be,
breaking off at the middle, where a passage of communication
runs across the central court.   An admirable feature in the
design of this museum is, that the public galleries of each
story, lighted by windows from the outside of the building,
have on their inner side other rooms communicating with
them, and lighted from the court within, which are devoted
to the private studies of the curators and to the reserve
collections belonging to the same series as the exhibited
collections in the public galleries, with which they are in

connection. Thus the public collections, the reserve collections, and the officers in charge, are in each section of the museum brought into close relation—a most advantageous arrangement, and one greatly facilitating the new museum idea. The only drawback is that these rooms, occupying the inner side of the quadrangular range of galleries, are necessarily small, and, as the collections grow, will be found insufficient for the purpose. This has, in fact, already proved to be the case in several departments, and a remedy has been found by devoting the whole upper story of the building to the reserve collections of insects, shells, and plants, and the working library of the institution,—an arrangement which gives excellent accommodation for these important departments, at all events for the present. A great difficulty will, however, arise in the future owing to the building being externally architecturally complete and visible on all four sides from the public grounds in which it stands; it therefore admits of no extension, and the public galleries already contain as many specimens as can possibly be placed in them with any advantage. These are in most sections, especially the invertebrata, displayed in an extremely tasteful and instructive manner, but the series is by no means over-large for a national museum. The limitation of space is partly due to the somewhat singular division which has been made between the art and the natural history collections. Instead of taking the dividing line adopted at the British Museum between specimens in a state of nature and those fashioned by man's hand, the pictures, the splendid collection of European mediæval armour, the classical and Egyptian antiquities, are treated as works of art; but the so-called ethnological collection, containing the specimens of Mexican, Peruvian, Japanese, Chinese, Polynesian, African, and prehistoric European art, are placed in the Natural History Museum, taking up a large portion of the space which the curators of the zoological, mineralogical, and geological departments hoped to have had at their disposal for the display of their specimens. Whether room could have been found for them in the Art Museum or not, I cannot say; but certainly their actual position is incongruous, and it is difficult to understand why a Peruvian mummy should find its place in a building

professedly devoted to natural history, while the preserved
remains of the ancient Egyptians are treated as works of art.

Before leaving Vienna I should like to refer to the splendid
specimens of taxidermy by the artist Hodek, the choicest
examples of whose work are contained in a special collection,
occupying a small separate room, consisting of sporting
trophies of the late Crown Prince, Rudolph. Otherwise the
general level of the specimens in the galleries is in no wise
remarkable. The birds have the advantage of being mounted,
not upon turned woolen stands of uniform pattern as in Paris,
but upon pieces of natural tree branches, fixed in square or
oblong oak-stands. The exhibited specimens of vertebrate
zoology include skeletons, but no other anatomical preparations, of which there is a distinct collection in the University
Museum. The exhibited fishes and reptiles are exceedingly
well preserved and mounted in spirit. In the Mollusca,
Articulata, Echinoderms, and Corals, great care has been taken
in setting the specimens off to advantage by selecting
appropriate colours for backgrounds. Specimens in spirit are
interspersed in their proper places. All have printed labels.
The cases in which they are displayed are of oak, and of very
handsome and even ornamental construction.

The arrangement of all these collections displays a most
intelligent appreciation of the needs of the ordinary visitor.
Thus in the room appropriated for the exhibition of insects
there are three distinct series—a general systematic series,
a morphological series, and a very fine special collection of
the insects of the neighbourhood of Vienna. The other rooms
are arranged more or less on similar principles. The main
collection of insects is, as I have mentioned before, entirely
apart in rooms very well adapted to the purpose, in the upper
floor of the building, and kept as usual in drawers in cabinets.

The zoological portion of the new museum for "Natur-
kunde," in Berlin, situated in the Invaliden Strasse, is a
remarkable illustration of the complete revolution of ideas on
museum arrangement, which took place between its commencement and its completion. The building, entirely designed
upon the old system, came empty into the hands of the
present director, who has arranged the contents absolutely

upon the new method. It consists of a fine glass-covered hall, and three stories of galleries, all originally intended for a uniform exhibition of all the various groups of specimens which had accumulated in the crowded rooms of the old museum in the University. When Dr. Möbius succeeded to the directorate, he conceived the bold plan of limiting the public exhibition to the ground floor, and devoting the two upper stories entirely to the reserve or working collections. This was a step which required some courage to take, especially as the two great staircases, which are the principal ornamental architectural features of the building, have by it become practically useless. Except, of course, for certain inconveniences always resulting from adaptation of a building to purposes not originally contemplated, especially local disjunction of different series of the same groups, the result has been eminently satisfactory; and if the arrangement is completed upon the lines laid down by the Director, as explained to me on my last visit, this will be the most practical and conveniently arranged museum of natural history at present existing. As much attention appears to be bestowed upon making the exhibited portion attractive and instructive, as on making the reserve collections complete and accessible to workers. In the former, the characteristics of the native fauna were being specially developed. For instance, the fish collection (of which the individual specimens are beautifully displayed in spirit, fastened on to glass plates in flat-sided jars) consists of a general representative systematic series, and three special faunistic collections, one of the German freshwater fishes, one of the north and east sea fishes, and one of the Mediterranean fishes. One room is devoted to German mammals and birds, and the recently added specimens show indications of an improvement in taxidermy which would have been impossible in the old days of wholesale bird-stuffing. Excellently prepared anatomical specimens, diagrams, explanatory labels, and maps showing geographical distribution, are abundantly introduced among the dried specimens of which such collections are usually composed, and a commencement has been made of illustrations of habits and natural surroundings. On the other hand, in marked contrast to Vienna, everything in the way of architecture and furniture and fittings is severely plain and practical, and a uniform drab colour is the pervading background for all kinds of specimens. All danger from fire seems to have been most carefully guarded against. The floors are of artificial stone, the cases, and even the shelving, are constructed of glass and iron. Wood is almost entirely excluded, both in the structure and fittings. The ground floor, as I have said, is entirely devoted to the public exhibition, the first story to the reserve collection of vertebrates, and the upper story to the invertebrates; and the basement contains commodious rooms for unpacking, mounting, preparing skeletons, etc. The construction of the building allows of considerable extension backwards, whenever more space will be needed, at small cost and with little interference with existing arrangements. I should also mention that the zoological department of the University, with its admirably appointed laboratories and lecture rooms, and excellent working collections for teaching purposes, is in immediate contact with the museum, and the two institutions, though under different direction, are thus brought into harmonious co-operation.

Any one who wishes to compare and contrast the two systems upon which a national zoological museum may be arranged cannot do better than visit Paris and Berlin at the present time. He will see excellent illustrations of the best of both.

Of the museums of the United States of America much may be expected. They are starting up in all directions, untrammelled by the restrictions and traditions which envelop so many of our old institutions at home, and many admirable essays on museum work have reached us from the other side of the Atlantic, from which it appears that the new idea has taken firm root there. In Mr. Brown Goode's lecture on "The Museums of the Future" (Report of the National Museum, 1888-89) it is said: "In the National Museum in Washington the collections are divided into two great classes, —the exhibition series, which constitutes the educational portion of the museum, and is exposed to public view with all possible accessories for public entertainment and instruction; and the study series, which is kept in scientific laboratories, and is scarcely examined except by professional investigators. In every properly constructed museum the collections must, from the very beginning, divide themselves into these two classes, and in planning for its administration provision should be made not only for the exhibition of objects in glass cases, but for the preservation of large collections not available for exhibition, to be used for the studies of a very limited number of specialists."

The museum of comparative zoology at Harvard, founded by the late Louis Agassiz, and now ably administered and extended by his son, Alexander Agassiz, is a conspicuous example of the same method of construction and arrangement. But as I can say nothing of these from personal knowledge, I am obliged to leave out any further reference to them on the present occasion.

From what has just been said it will be gathered that in Europe, at least, an ideal natural history museum, perfect in original design as well as in execution, does not exist at present. We have indeed hardly yet come to an agreement as to the principles upon which such a building should be constructed. But as there are countries which have still their national museums in the future, and as those already built are susceptible of modifications, when the right direction has been determined on, I should be glad to take this opportunity of putting on record what appears to me, after long reflection on the subject, the main considerations which should not be lost sight of in such an undertaking.

In the first place, I have endeavoured to work out in detail, in its application to natural history, that most original and theoretically perfect plan for a museum of exhibited objects in which there are two main lines of interest running in different directions and intersecting each other, which we owe to the ingenuity of General Pitt-Rivers. This was explained in his address as President of the Anthropological Section of the British Association at Bath in 1888, and again in a lecture given about two years ago before the Society of Arts.[1] Upon this plan the museum building would consist of a series of galleries in the form of circles, one within the

[1] *Journal of the Society of Arts*, 18th December 1891.

other, and communicating at frequent intervals. Each circle, would represent an epoch in the world's history, commencing in the centre and finishing at the outermost, which would be that in which we are now living.   The history of each natural group would be traced in radiating lines, and so by passing from the centre to the circumference, its condition of development in each period of the world's history could be studied. If, on the other hand, the subject for investigation should be the general fauna or flora of any particular epoch, it would be found in natural association by confining the attention to the circle representing that period.   By such an arrangement, that most desirable object, the union of palæontology with the zoology and botany of existing forms in one natural scheme, could be perfectly carried out, as both the structural and the geological relations of each would be preserved, and indicated by its position in the museum.   Such a building would undoubtedly offer difficulties in practical construction; but even if these could be got over, our extremely imperfect knowledge of the past history of animal and plant life would make its arrangement, with all the gaps and irregularities that would become evident, so unsatisfactory that I can scarcely hope to see it adopted in the near future.

I have therefore brought before you a humbler plan, but one which, I think, will be found to embody the practical principles necessary in a working museum of almost any description, large or small.

The fundamental idea of this plan is that the whole of the building should be divided by lines intersecting at right angles, like the warp and the woof of a piece of canvas.

The lines in one direction (see Plan, next page) divide the different natural sections of which the collection is composed, and which it is convenient to keep apart; the lines crossing these separate the portions of the collection according to the method of treatment or conservation.   Thus, the exhibited part of the whole collection will come together in a series of rooms, occupying naturally the front of the building. The reserve collections will occupy another, or the middle, section; and beyond these will be the working rooms, studies, and administrative offices, all in relations to each other, as

E

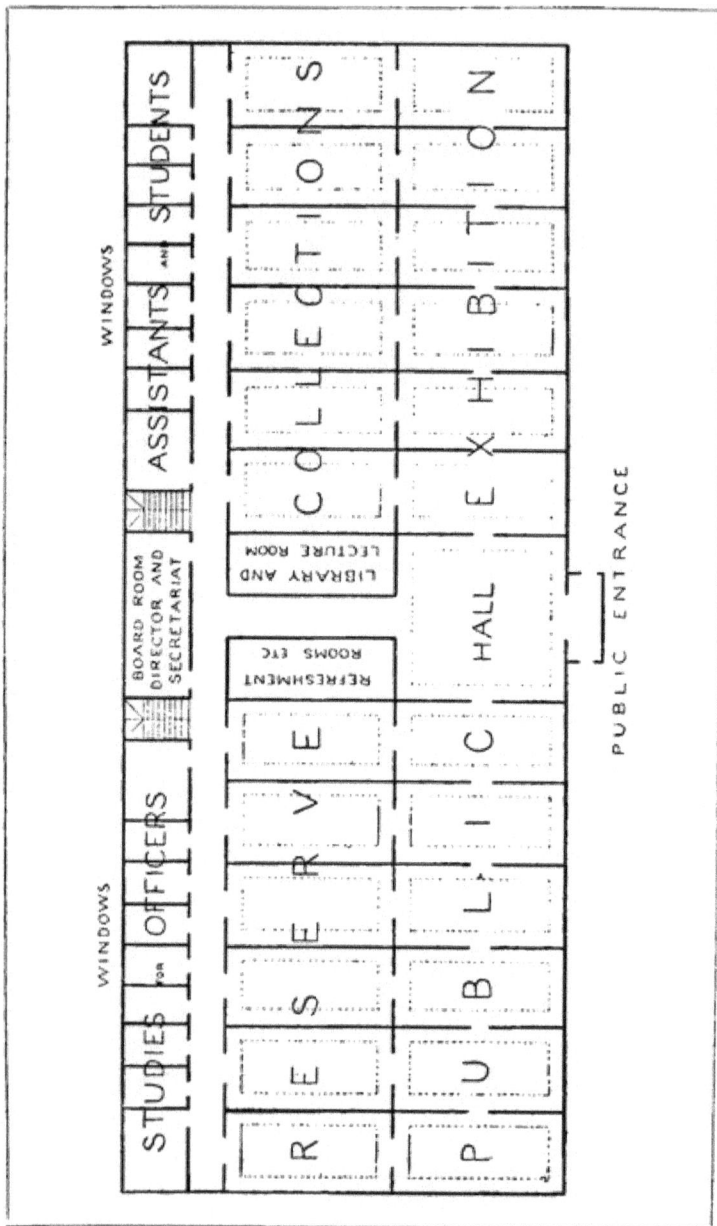

WINDOWS

STUDIES FOR OFFICERS

ASSISTANTS AND STUDENTS

WINDOWS

BOARD ROOM
DIRECTOR AND
SECRETARIAT

RESERVE

COLLECTIONS

LIBRARY AND
LECTURE ROOM

REFRESHMENT
ROOMS ETC

PUBLIC

HALL

EXHIBITION

PUBLIC ENTRANCE

well as to the particular part of the collection to which they belong. A glance at the plan will show at once the great convenience of such a system, both for the public, and still more for those who work in the museum.

This plan, of course, contemplates a one-storied, top-lighted building, as far as the main rooms are concerned, although the workrooms and studies will be in two or more stories. The main rooms should all have a good substantial gallery running round them, by means of which their wall space is doubled. There is no question whatever that an evenly-diffused top light is far the best for exhibition rooms. Windows not only occupy the valuable wall space, but give all kinds of uncomfortable cross lights, interspersed with dark intervals, On the other hand, for doing any kind of delicate work, a good north light from a window, as provided in the plan, is the most suitable. The convenience of having all the studies in relation with each other, and with the central administrative offices, while each one is also in close contiguity with the section of the collection to which it belongs, will, I am sure, be appreciated by all who are acquainted with the capriciously scattered position of such rooms in most large museums, notably in our own. Among other advantages would be the very great one, that when the daily hour of closing the main building arrives, the officers need no longer, as at present, be interrupted in whatever piece of work they may have at hand, and turned out of the building, but, as arrangements could easily be made for a separate exit, they could continue their labours as long and as late as they find it convenient to do so, without any fear for the safety of the general collections.

It will be observed that provision is made for a central hall, which is always a good architectural feature at the entrance of a building, and which in a large museum is certainly useful in providing for the exhibition of objects of general interest not strictly coming under any of the divisions of the subject in the galleries, or possibly for specimens of too great a size to be conveniently exhibited elsewhere. There is also provision in the central part of the building for the refreshment-rooms, as well as for the library and a lecture room, the first being an essential, and the latter a very useful adjunct to any collection intended for popular instruction, even if no strictly systematic teaching should be part of its programme.

I may point out, lastly, as a great advantage of this plan, that it can be, if space is reserved or obtainable, indefinitely extended on both sides on exactly the same system without in any way interfering with the existing arrangements; a new section, containing exhibition and reserve galleries and studies, can be added as required at either end, either for the reception of new departments, or for the expansion of the old ones. With a view to the latter it is most important that the fittings should be as little as possible of the nature of fixtures, but should all be so constructed as to be readily removable and interchangeable. This is a point I would strongly impress upon all who are concerned in fitting up museums, either large or small.

The modifications of this plan, to adapt it to the require-ments of a municipal, school, or even village museum, will consist mainly in altering the relative proportion of the two sections of the collection. The majority of museums in country localities require little, if anything, beyond the exhibition series. In this the primary arrangement to be aimed at is, first, absolutely to separate the archæological, historical, and art portions of the collection from the natural history, if, as will generally be the case, both are to be represented in the museum. If possible they should be in distinct rooms. The second point is to divide each branch into two sections: (1) a strictly limited general or type collection, arranged upon a purely educational plan; (2) a local collection, consisting only of objects found within a certain well-defined radius around the museum, which should be as exhaustive as possible. Nothing else should be attempted, and therefore reserve collections are unnecessary. Even the insects and dried plants can be exhibited on some such plans as those adopted for the Walsingham collection of Lepidoptera in the Zoological Department, or the collection of British plants in the Botanical Department of our Natural History Museum.

I have elsewhere indicated my views as to the objects most suitable for, and the best arrangement of them in, school museums,[1] so I need say nothing further on the subject now. Indeed, I fear I have exhausted your patience, so I will conclude by expressing an earnest hope that this meeting may prove a stimulus to all of us to continue heartily and thoroughly at our work, which I hardly need say is the only way to ensure that general recognition of it which we all so much desire.

[1] *Nature*, vol. xli. p. 177, and *infra* p. 58.

III

LOCAL MUSEUMS [1]

ATTENTION has already been directed, in several letters which have recently appeared in the newspapers, to the desirability of a central institution in which the numerous historical documents and objects of interest connected with the county might not only be preserved from destruction, but also made available for study and reference. Many of these, which may be considered trifles now, will be of great value in after time, as illustrating the history and mode of life of generations passed or passing away. Many customs change with great rapidity, and all evidences of their existence disappear, or only remain in literary allusions, often difficult to understand without actual illustrations. Take for instance the old flint and steel and brimstone match, the universal source of illumination in even my early days. Most living people, accustomed to the daily use of one of the greatest triumphs of applied science, know nothing of the difficulties their grandfathers had to contend with in supplying this most necessary want of common life. I doubt if many of our generation were to see the old apparatus, whether they would know what it was for or how used, and though one must have existed in every cottage in the county fifty years ago, it is possible that it would be very difficult to procure one, even for a museum, at the present time. The candle-snuffers, without which we could not exist in my childhood, are now as extinct as the Dodo. Take again the threshing-flail, an instrument in

[1] From a letter in support of the establishment of a County Museum for Buckinghamshire (24th November 1891), and an address at the opening of the Perth Museum (29th November 1895).

universal use, with little change for thousands of years, now rapidly disappearing in many districts; or the stocks, once such an important social institution in rural life, but of which I know now only one in the neighbourhood of Aylesbury, —that which still remains in the picturesque village of Dinton. Specimens of all these, and many others which will readily suggest themselves, should be preserved in every local museum.

Then again, the natural history, the birds, the butterflies, the wild flowers, and the fossils and minerals of the neighbourhood, so arranged and named that any one can identify every creature or plant he may chance to meet with in his walk, are essential features in a local museum.

But now let me give a necessary warning. There are museums and museums. A good one well arranged, and well kept, clean, neat, and attractive, may be the means of conveying instruction and giving interest and pleasure to the lives of thousands of our fellow-creatures. But such museums do not grow of themselves; money, time, knowledge, and loving and sympathetic care must be expended upon them, both in their foundation and their maintenance, and unless all these can be provided for with tolerable certainty, it is useless to begin. Voluntary assistance is, no doubt, often valuable. There are many splendid examples of what it may do in country museums, but it can never be depended on for any long continuance. Death or removals, flagging zeal, and other causes tell severely in the long run against this resource. A museum must have an endowment adequate to defray its expenses, and especially to ensure the staff of intelligent, educated, and paid curators required to maintain it in a state of efficiency. You might as well build a church and expect it to perform the duties required of it without a minister, or a school without a schoolmaster, or a garden without a gardener, as to build a museum and not provide a competent staff to take care of it.

It is not the objects placed in a museum that constitute its value, so much as the method in which they are displayed, and the use made of them for the purpose of instruction.

The scope of the museum should be strictly defined and

limited; there must be nothing like the general miscellaneous collection of all kinds of "curiosities," thrown indiscriminately together which constituted the old-fashioned country museum. I think we are all agreed as to the local character predominating. One section should contain antiquities and illustrations of local manners and customs; another section local natural history, zoology, botany, and geology. The boundaries of the county will afford a good limit for both. Everything not occurring in a state of nature within that boundary should be rigorously excluded. In addition to this, it may be desirable to have a small general collection designed and arranged specially for elementary instruction in science. This part might be brought into connection with the technical instruction given in the county, and will be a valuable—indeed, a necessary adjunct to it. Every branch of such instruction should have its special collection of objects to illustrate it; the teaching will then be made far more real and practical than it otherwise would be.

Agricultural chemistry and geology, dairy-farming, fruit-growing, and such like subjects, might each have its collection. And these various collections, though kept quite distinct, in different rooms if possible, might be all associated in the County Museum and under the same general management. Thus some of the funds devoted by the county to the purposes of technical education might most profitably assist in the formation and maintenance of the County Museum; indeed, I think myself that if a portion of this money had been in the first place directly allocated to the endowment of a museum in each county more good would have been done for the advancement of education than by most of the schemes at present under discussion. At any rate, the classrooms and laboratories required for the teaching now contemplated should be associated with the museum, if possible under the same roof, and the staff of teachers should assist in the curatorial work, thus forming a sort of central college in every county for technical education, which might send out ramifications into the various districts in which the need of special instruction was most felt, and also might be the parent of smaller branch museums of the same kind wherever they seem required.

This is the kind of institution which I hope will, before long, be established in every county in the land. The marvellous spread of state - supported and rate - supported libraries which has taken place during the last few years appears to be only the prelude to museums supported in the same way. The underlying idea of a library and a museum is precisely the same. They are both instruments of intellectual culture, the one as much as the other. We have this illustrated on a magnificent scale in our great National Library and Museum in London. Before long a well-arranged and well-labelled museum will be acknowledged as a necessity in any well-considered scheme of educational progress. The museum and library will go hand in hand as essential complements to each other in the advancement of science and art, and intellectual development generally. A book without illustrations is of comparatively little value in teaching many of the most important subjects now comprised in general education. A museum should be a book or rather library of books, illustrated not by pictures only, but by actual specimens of the objects spoken of. The great principle of expending public money upon purposes of education, though a comparatively new one, is now conceded upon all sides. The cost of supporting really efficient museums in all our principal local centres would be but a trifling addition to the millions spent upon other methods (some probably of far less value) of educating and elevating the people. But in all cases let us remember, as Professor Brown Goode, the Director of the United States National Museum, in an admirable essay on "The Museums of the Future," says, "One thing should be kept prominently in mind by any organisation which intends to found and maintain a museum, that the work will never be finished, that when the collections cease to grow they begin to decay. A finished museum is a dead museum, and a dead museum is a useless museum."

IV

SCHOOL MUSEUMS [1]

HAVING lately been asked by Dr. Warre, Head Master of Eton, to give him some assistance in the fitting up, arrangement, and management of the museum about to be inaugurated at that College, I put down some notes which he was pleased to think might be of use in pointing out the lines that should be followed with most advantage. As these notes are equally applicable to other school museums, I venture to publish them for the information of those who may be in a position to profit by them, premising that they are mere outlines, which are susceptible of much elaboration in detail, and of some modifications, according to special circumstances.

The subjects best adapted for such a museum are zoology, botany, mineralogy, and geology.

Everything in the museum should have some distinct object, coming under one or other of the above subjects, and under one or other of the series defined below, *and everything else should be rigorously excluded*. The curator's business will be quite as much to keep useless specimens out of the museum as to acquire those that are useful.

The two series or categories under which the admissible specimens should come are the following:—(1) Specimens illustrating the teaching of the natural history subjects adopted in the school, arranged in the order in which the subjects are, or ought to be, taught. (2) Some special sets of specimens of a nature to attract boys to the study of such branches of natural history as readily lie in the path of their

[1] Suggestions for the formation and arrangement of a museum of Natural History in connection with a Public School. *Nature*, vol. xli. p. 177 (1889).

ordinary life, especially their school life, and to teach them some of the common objects they see around them.

The specimens of the first class should be all good of their kind, carefully prepared and displayed, and fully labelled. They should also be so arranged that they can be seen and studied without being removed from their position in the case or in any way disturbed or damaged. It would be best that they should never be taken out of the museum, but if it is necessary to remove them for the purpose of demonstration at lectures or classes, special provision should be made by which a whole tray or case can be moved together, with due precautions against disturbing the individual specimens. As a rule, the teachers should either bring the classes into the museum for demonstrations, or they should rely upon a different set of specimens kept in store in the classrooms, and only brought out when required, and which may be handled and examined without fear of injury. Really good permanent preparations may be looked at, but not touched except by very skilled hands.

In zoology the collection should consist of illustrations of the principal modifications of animal forms, living and extinct, a few selected typical examples of each being given, showing the anatomy and development as well as the external form. The series now in the course of arrangement in the Central Hall of the Natural History branch of the British Museum in the Cromwell Road may, as far as it is complete, be taken as a guide, but for a school museum it will not be necessary to enter so fully into detail as in that series.

In botany there should be a general morphological collection, showing the main modifications of the different organs in the greater groups into which the vegetable kingdom is divided, and illustrating the terms used in describing these modifications. Such a collection may also be seen (although still far from complete) in the same institution.

For a teaching collection of minerals, an admirable model has for several years past been exhibited in the Mineralogical Gallery of the Natural History Museum, being, in fact, the various paragraphs of Mr. Fletcher's *Introduction to the Study of Minerals* cut up, and with each statement illustrated by a choice specimen.

The geological collection would best be limited mainly to a series illustrating the rocks and characteristic fossils of the British Isles, arranged stratigraphically. There would be no difficulty in making such a series on any scale, according to the space available, and if well selected and arranged, it would be extremely instructive and form a complete epitome of the whole subject. It should be placed in a continuous series along one side of the room, beginning with the oldest and ending with the most recent formations. It might be preceded by some general specimens illustrating the various kinds of rock structures, etc.

Mineral and fossil specimens are generally to be procured as wanted from the dealers, and as they require little or no preparation, collections illustrating these subjects can be quickly made, if money is available for the purpose. This is not, however, the case with zoological and botanical specimens, most of which require labour, skill, and knowledge to be expended upon their preparation before they can be preserved in such a manner as to make them available for permanent instruction.

We will next proceed to consider what objects may be included under the second head, many of which need not be constantly exhibited, but may be preserved in drawers for special study. These may be—

(1) A well-named collection of the commoner British insects, especially those of the neighbourhood in which the school is situated, with their larvæ, which should (if means will allow) be mounted on models of the plants upon which they feed. All should have their localities and the date of capture carefully recorded. These are best kept in a cabinet, with glass-topped drawers, with a stop behind, so as to allow them to be pulled out for inspection, but not entirely removed. Such a collection, formed of specimens prepared and presented by Lord Walsingham, can now be seen in the British Room of the Natural History Museum.

(2) A similar collection of British shells, especially the land and freshwater shells of the neighbourhood.

(3) If space and means allow, a collection of British birds, especially the best-known and more interesting species. Rare

and occasional visitors, reckoned in the books as British, which are the most expensive and difficult to procure, are the least important for such a collection. Variations in plumage in young and old, and at different seasons, should be shown in some common species. Every specimen must be good and well mounted, or it is not worth placing in the museum.

(4) The principal British mammals of smaller size, especially the bats, shrews, and mice.

(5) The British reptiles, Amphibia, and commoner fishes, so shown that their distinctive characters may be recognised.

(6) A collection, as complete as may be, of British plants, or at all events of the plants of the neighbourhood. By far the best way of preserving and exhibiting such a collection is in glazed frames, movably hinged upon an upright stand, as may be seen in the Botanical Gallery of the Natural History Museum. A collection arranged in this manner should find a place in every school museum of natural history.

(7) A collection of the fossils found in the quarries of the neighbourhood, should there be any.

Every collection or series should be kept perfectly distinct from and independent of the others, and its nature and object clearly indicated by a conspicuous label.

The exhibited specimens should be arranged in upright wall-cases or in table-cases on the floor of the room. For the latter a high slope is preferable, and in all the exhibition space should not extend too high or too low for comfortable inspection. Between three to six or seven feet from the floor should be the limits for the exhibition of small objects. The three feet nearest the floor may be enclosed with wooden doors forming cupboards or fitted with drawers. Glass in this situation is liable to be broken by the feet or knees.

The museum should have a permanent curator—a man of general scientific attainments, and who is specially acquainted with, and devoted to, museum work, and who might also be one of the teachers, if too much of his time is not so occupied. But, as he is not likely to have special knowledge of more than one branch of natural history, the teachers of the other branches represented in the museum would probably each give advice and assistance with regard to his own department. It is also probable that some of the boys may be sufficiently interested in the work to render valuable aid in collecting and preparing specimens.

If ethnographical, archæological, historical, or art collections be also part of the general museum scheme, they should be kept quite distinct from the natural history collections, preferably in another room.

Above all things, let the following words of Agassiz be remembered : " The value of a museum does not consist so much in the number as in the order and arrangement of the specimens contained in it."

V

BOYS' MUSEUMS [1]

It is a strange and interesting fact in human nature that among thousands of boys who do not take the slightest interest in anything pertaining to what is commonly called "Natural History," there are here and there, at all events among all cultivated nations, some few to whom it is an absorbing passion, affording more delight than anything else in life. Very often this is only a passing phase, affecting boys chiefly between the ages of fourteen and sixteen, and then entirely dying away, but with some it persists through life, materially modifying the whole course of existence. This curious condition of mind, or "idiosyncrasy" as physicians call it, is not confined to particular races or nations; the Japanese have it as well as Europeans and Americans. Nor is it limited to any particular position in the social or educational scale. No one could have had it in a more intense form than the poor Scotch shoemaker, Thomas Edward, child of some of the humblest people of the land, whose biography by Samuel Smiles I presume all readers of this magazine are acquainted with. On the other hand, there have been few keener naturalists and collectors than the late Crown Prince Rudolf of Austria, and in England at the present time few, if any, can yield in this respect to the heir of the wealthy house of Rothschild, the owner of the finest private zoological museum in the world.

One great peculiarity of this condition of mind is that, though it does occasionally run in families, more often it arises as it were spontaneously in one member of a family,

[1] Published under the title of "Natural History as a Vocation," in *Chambers's Journal*, 10th April 1897.

without any inherited predisposition or any external circumstance, as far as is known, leading to it. Indeed, it is often most strongly developed when the circumstances seem most adverse, and where no encouragement whatever has been given to pursue it. I am especially in a position to become acquainted with all the symptoms of what I may call this affection, as from my official position fathers and mothers of boys attacked by it continually consult me as to how the inclination or passion for natural history should be treated in view of the future career of the boy—whether it should be peremptorily suppressed, or whether and in what direction it should be developed, and above all, what are the prospects of its leading to obtaining a livelihood. Having a strong sympathy with boys of such tastes, I have generally availed myself of the opportunity of a little talk with them, and have given them such advice as seemed best in the particular case. As to natural history as the regular occupation of one who has no other means of living, I have little to say that is favourable, as it is about the worst paid and least appreciated of all professions. The only thing that I can say for it is that the prospects are brightening, surely if slowly, both in Europe and America. It is, I firmly believe, a profession of the future. The only way to judge of what is coming is to look back at the past, and note the changes that have been recently and are still taking place. The opportunities for pursuing natural history with some sort of remuneration, small as they are, are undoubtedly greater now than they were twenty or thirty years ago.

With the general spread of education, lectureships, demonstratorships, and curatorships are every year becoming more numerous, and there is no reason to suppose that this excellent state of things will not continue. But still, before advising any one to take up natural history as a profession, I must be convinced of his really intense and abiding interest in the subject, and of his zeal and determination to pursue it at whatever sacrifice of ease or comfort, his readiness " to scorn delights and live laborious days " in fact. Some boys seem to be so devoted to it as to be incapable of applying themselves to anything else. If this is really the case, I generally advise parents to let them go on if possible, giving them the best education in the subject that is available, and letting them take their chance of obtaining an appointment when they are fit for one. The difficulty with such cases is that to pursue natural history with any chance of profit in these days a considerable knowledge of other subjects, especially modern languages, is absolutely indispensable. The continually-increasing amount of scientific literature which must be taken account of in every branch of natural history has much changed the conditions of study required for it, as has also the growing tendency to give appointments solely on the results of competitive examination. In no case would I discourage the taste altogether, but I more often advise some other means of making a living, holding on to natural history as a recreation and relaxation. To a soldier or sailor, for instance, a love of natural history is the greatest possible blessing, and still more to the man of independent fortune. They are often saved by it from all kinds of evil which want of wholesome occupation engenders. Their life becomes a continuous delight, instead of being a burden to themselves and others. Especially to those mainly engaged in absorbing money-making pursuits, the refreshment of an occasional excursion into the realms of nature need not be insisted on. It is perfectly obvious to all who have ever had an opportunity of observing it.

With an early love of natural history is almost always associated a love of collecting, and probably there is no better way of becoming familiar with a subject than by making a collection of objects illustrating it. The value of all knowledge depends a great deal upon the amount of labour and time spent in acquiring it. The easy methods of which we make too much boast in the present day, handbooks, pictures, lectures, well-arranged public museums, etc., have their drawbacks and snares as well as their advantages. They are all helps if properly used, but they will not supersede, and nothing will ever supersede, the downright hard personal work by which all solid, lasting knowledge must be gained. The value of making a collection of any kind of specimens about which you wish to know something is that you are forced to spend time and thought over them, to look at them, carefully to prepare them and compare them, to arrange and name them. In proportion as a collection has had all this done to it will be its value. That a museum depends for its utility, not upon its contents, but upon the mode of arrangement of its contents, is now a trite saying. An ill-arranged museum has been well compared to the letters of the alphabet tossed about indiscriminately, meaning nothing; a well-arranged one to the same letters placed in such orderly sequence as to produce words of counsel and instruction. Far more, however, than the intrinsic value of the collection, in the case of the beginner, is its value as a means of education to the owner. The arrangement of a collection not only teaches the nature and properties of the objects contained in it; it also stimulates a desire to know more of the similar objects not contained in it, but to be found in other and larger collections. Still more important than this, as an educator, it calls out many valuable practical qualities : originality, order, neatness, perseverance, taste, power of discriminating small differences and resemblances, all of which will be found useful in other spheres of life.

It matters less what are the contents of a museum than that there should be some definite object in bringing them together. To be a mere "snapper-up of unconsidered trifles" is not forming a museum. The subject chosen to be illustrated by the specimens collected should not be of indefinite extent, but have some natural limit. This alone will enable the collector to attain to the highest goal in the happiness of his occupation: the filling in of all the vacant spaces when the framework of his collection is completed. The richer a series is, the greater joy there is in adding to it what still remains wanting. Limits to collections of natural history objects are of two kinds. Either a particular natural group of animals or plants may be selected, as shells, butterflies, mosses, seaweeds, etc., or any subdivision of these great groups; or else the products of particular localities, preferably of course that in which the collector lives, may only be regarded. A combination of these two methods of limitation will generally lead to the most manageable and profitable amateur museum. Suppose, for instance, our young friend were to set himself the task of collecting and preserving all the fossils or all the land and freshwater shells, or all the birds' eggs, or all the beetles, to be found within a radius, say, of ten miles round his dwelling-place, what a fund of knowledge he would acquire not only of the appearance of the individual specimens, but of their natural surroundings and habits! And what delightful rambles he would have in the open air, with eyes and ears intently appreciative of all the varied beauties of the lovely world in which we dwell, lost, unfortunately, to so many who pass through it with none of these interests and pleasures!

Although a collection, with a definite object, of specimens obtained, prepared, and arranged by one's self is the ideal of a boy's museum, I do not say that the possession of a few miscellaneous articles, which are sure to be given by kind friends as soon as the taste for possessing them is recognised, may not sometimes be an advantage, especially as a help in stimulating inquiry and knowledge. My first "museum" was, as I recollect, very much of this nature. It was contained in a large, flat, shallow box with a lid, and I made cardboard trays which filled and fitted the bottom of the box, and kept the various specimens separate. Everything was carefully labelled, and there was also a manuscript catalogue in a copy-book. No boy should ever be allowed to keep any sort of museum without a catalogue in which the history of every specimen and the date at which it came into his possession are carefully entered. When the box was outgrown it was superseded by a small cabinet with drawers, then by a cupboard; but before I had left the parental home for college, an entire small room was dignified by the name of my "museum." It was the love of curatorship which thus grew up within me, without the remotest external influence or inherited predisposition towards it, as none of my relatives had any interest in such pursuits, that determined my after career, and led to such success as I have met with in it. My boyish fondness for dissecting animals and preparing their skeletons at that time could find no nearer outlet in any academic career than in the pursuits of a medical student; and the anatomical museum of my college was at first to me a much greater subject of interest than the wards of the hospital—so much so in fact, that while still in my second year of studentship the curatorship falling vacant, I was asked to undertake the office. Here I was in my glory, and although later on the more practical work of the surgical profession had its attractions also,—attractions which at one time nearly carried me off into the stream of London hospital practice,—I finally returned to the old love, and through a succession of fortunate incidents, the museum under my care, instead of the one little box with which I began, is now the largest, most complete, and magnificently housed in the world.

I need hardly say that in all my subsequent career I have always looked back at my early attempts at curatorial work with especial satisfaction. The educational power of all work done when young can never be overestimated. The sooner knowledge is acquired the more valuable it is. You have it so much longer and it becomes so much more a part of yourself. One of the first specimens I possessed was a little stuffed bird with a brown back and white underneath, and a very short tail. I saw it in the window of a pawn-broker's shop in my native town, Stratford-on-Avon. I often passed the shop, and looked at it with wonder and admiration. At last I summoned up courage to ask its price. "Three pence," was the answer. This was a serious consideration; but the financial difficulty being overcome, I carried the bird home in triumph. Having access to a copy of Bewick's *British Birds*, I identified it as the dipper or water ousel, and even learnt its scientific name, *Cinclus aquaticus*. It was wretchedly stuffed. Though more than fifty years have passed since I saw it last, for during an absence at school, it, with many other treasures, fell into places where "moth and rust do corrupt," its appearance is still fixed in my mind's eye, with its hollow back and crooked legs sticking out of impossible parts of its body; but I was not then so critical as I have since become. My only reason for mentioning it is because that bird became part of my permanent stock of ornithological knowledge, and ever since, whether by a mountain stream in the Highlands of Scotland, or a rocky river in the Harz or Thuringian Forests in Germany, when I see a dipper flitting over the rushing water or diving beneath the surface, it seems an old familiar friend of my childhood.

Another specimen of which I have a very vivid recollection was labelled in the handwriting of the kind donor, " Bone from Kirkdale Cave, Yorkshire." It was given to me by an old gentleman who lived in the neighbourhood, and whose large collection of geological specimens was a great delight to me to look over. It was a perfectly valueless fragment from a scientific point of view, not having characters enough to identify the animal to which it belonged, being little more in fact than a chip from the surface of a long bone, of which thousands were found in the cave. But it woke up a train of interest in me, leading to the whole subject of caves and their bygone inhabitants, and the reading of Buckland's *Reliquiæ Diluvianæ* ; while ever since Kirkdale Cave has stood out among all other caves, with a sort of romantic halo conferred upon it in my mind by the present of that fragment of bone with all its early associations. These are only two instances out of any number I could tell of the ways in which a boy's museum may become a source of knowledge and of interest that may last through life.

ADDRESS AT THE OPENING OF THE BOOTH MUSEUM AT BRIGHTON [1]

WE are assembled in a room that contains a collection in many respects unequalled by any other in the world. In the first place, it has been entirely the work of one man in a life of no great length,—he had only reached the age of fifty when he died,—but who devoted an extraordinary amount of energy and perseverance, and also expended a very considerable amount of money, in making it as perfect as possible of its kind. You must not suppose for a moment that Mr. Booth was the only man who ever made a collection of British birds. Birds have always been favourites, and the birds of our own islands, though far less remarkable for form, size, and brilliancy of plumage than those of other lands, have for many reasons been peculiarly attractive. The national pride which causes us to love our countrymen better than foreigners includes birds as well as other bipeds. It has, therefore, been the aim of many public museums, as well as private lovers of natural history, to make as complete a collection of British birds as possible. But it is one thing to have a collection adapted and conveniently arranged for reference and study by the learned ornithologist, or consisting of as many specimens as can be crammed into the smallest space they could occupy, without regard to their condition or their order, and quite another thing to have a collection under such circumstances and so arranged as to convey the fullest possible amount of information and instruction, and to excite the greatest possible interest in the minds of those who, like the majority of us, are not in a

[1] 3rd November 1890.

position to devote any large portion of our scanty leisure to their study. For this latter purpose I have no hesitation in saying it would be difficult to imagine a collection so complete, and so admirably arranged and displayed, as the one in which we are. I purposely avoid comparison with the beautiful series, showing the nesting habits of birds, now being arranged in the National Museum in London under the supervision of Dr. Günther, because the objects of the two are in many respects different, as are also the methods in which they have been carried out. We must recollect that this collection was formed by one who was an intense lover of bird-life, one who spent the greater part of his own life, night and day, summer and winter, in watching their manners and actions in their native haunts, and who knew from his own keen observation exactly what were their favourite surroundings, what kind of soil or of rock, or of tree or of flower, each species would be most likely to be found among or near. Most collections contain only the birds themselves. Here we have not only birds, but the home in which the birds dwelt, most carefully and accurately reproduced, and on such a scale and in such a manner as has never been done anywhere before. As for the birds themselves, not only are they the finest and best specimens of their kind that could be procured, and in many cases showing various stages of plumage, at different ages and different seasons of the year, but far more care, knowledge, and artistic skill has been expended on their mounting than is generally the case either in museums or in private collections. The art of taxidermy, though quite an old one in Europe, extending back certainly three or four hundred years, has made very little progress until very recent times, and even now, though there is so great a development of nearly all branches of art, it has had far too little attention bestowed upon it. Very few people seem to know the difference between a really well-mounted bird or mammal and an inferior one, but there is as much difference between them as between a picture of a lion by Landseer or Rosa Bonheur, and a picture of the same animal depicted by a village artist on the sign of a public-house. And yet so little do people understand this, that they go on filling museums and collections

with wretched examples, and continue to pay the unfortunate
bird-stuffer a miserably inadequate sum for work which should
be the work of a real artist, and which can only be done by a
man who not only has devoted great care and attention to the
subject, but has also the rare gift of inborn genius.   I am
very critical, indeed, as you can see, on the subject of bird-
stuffing, and I am happy to be able to say, and every time
I enter this museum it strikes me more, that the greater
number of specimens in it, though, of course, they are unequal,
are admirable specimens of the art of taxidermy.   Many of
them are very fine indeed, all are above the average, and I
believe there is not a single bad one among them.   The
collection is eminently adapted for public instruction.   If the
advantages to be derived from it were only the momentary
pleasure of looking at it here, it would be of comparatively
little account, but if properly used it may be the means of
spreading knowledge and interest which will affect the whole
course of some of our lives.   We all have cares and troubles
enough in our passage through this world, and we are so often
brought into contact with so much that is mean, disagreeable,
and ugly, that we ought not, for our own sakes, as well as for
the sakes of those among whom we live, to neglect any sources
of joy and gladness that might be offered to us.   The man
who walks through life with his eyes open to beauty, wherever
it can be found, is by so much a happier and a more useful
man than one whose eyes are closed to it.   The observation of
bird-life is one among many of unfailing sources of pleasure.
We cannot walk upon our downs, or along our cliffs, or on our
sands, with our eyes open without seeing birds innumerable,
though I believe that many never do see them.   This
museum, however, should teach us to see them, to know them
one from another, to make them our friends and companions,
and, rightly used, it may be a source of making many lives
happier, and sweeter, and purer than they otherwise would be.
For such a collection as this ever to be dispersed or destroyed
would be a national misfortune.   The Mayor has alluded to
the fact that it was first offered to the British Museum, and
if I had had any idea that, if we did not accept it, it would
be destroyed or dispersed, I should have felt it my duty to

advise the Trustees to take it over.   I received, however, an
intimation that the Corporation of Brighton was not only
willing, but most anxious to take charge of the collection,
and to maintain it.   Although it would have been a great
privilege to me to have been its official guardian, I rejoice
to think that it is going to remain here, and that you
have expressed your determination to maintain it for the
benefit of your fellow-townsmen and for visitors.   The great
central national collections are the fitting repositories of
many specimens of natural objects and works of art.   Such
as are unique, and such as are necessary for the researches
of advanced students who require facilities for their investigations which can only be obtained by the direct comparison of
a large series of specimens one with another, ought to be in
them.   But, on the other hand, the more collections like this
—adapted for general instruction—are to be met with in
other great centres of population, the better it will be for the
welfare of the country generally.   Apart from the fact that
the collection was made at Brighton, and a large number of
the birds in it obtained in the immediate neighbourhood,
Brighton seems to be a particularly suitable place for preserving
a museum like this.   You have a vast number of visitors, who
come for the purpose of seeking repose or health, for whom
such a light, interesting, and easy occupation as is afforded
by learning what this museum can teach, ought to be the
best that can be found.   It only needs to be better known to be
very much more widely appreciated than it has been hitherto.
In conclusion, I feel sure that I am expressing the feelings of
friends around me who are interested in the advancement of
the Natural History sciences, and of many more who are
unable on account of other engagements to be present, in
thanking the Mayor and Corporation for asking us to come to
this interesting meeting, and in congratulating them on
possessing such a valuable addition to the many attractions of
the town.

THE MUSEUM OF THE ROYAL COLLEGE OF SURGEONS OF ENGLAND[1]

WHILE thinking over various subjects for an address with which to open the business of the Section over which I have the honour to preside, it has occurred to me that the time which has been allotted by the arrangements of the Congress for the purpose may be made most useful to my hearers if, instead of entering upon a discussion of any abstract question, I were to ask your attention to a subject upon which I may possibly be able to give a little information of practical use to members of the Congress during their visit to this city.

No class of persons can appreciate so fully the importance and value of museums as those whose occupation it is to study the form and relations of the various parts of the body, whether of plants, animals, or man.

Our science would make little progress if the objects of our inquiries, once used for examination or description, were then thrown aside, and those coming after were denied the opportunity of which we have availed ourselves. A museum is a register, in a permanent form, of facts, suitable for examination, verification, and comparison one with another.

Hence, ever since serious attention has been awakened to the interest of anatomical studies, museums have always been important adjuncts to their successful prosecution, and the preservation of the various structures of the body has occupied the attention of very many anatomists, since the time of the

[1] Presidential Address to the Anatomical Section of the International Medical Congress. London, 4th August 1881.

great Italian teachers of the early part of the seventeenth century, with whom apparently the art commenced.

We have in London, as you are all aware, a museum which stands, in some respects, in a peculiar position, differing perhaps from any in the world in its origin, its scope, its method of maintenance, and its relation to the profession and to the State, in which, for very nearly twenty years, it has been my privilege to pass my days. It has occurred to me that a few words in explanation of the history, arrangement, and contents of that museum might add to the interest and profit of those visits which I trust every one here will find time to pay to it during the meeting of the Congress.

The great mind of John Hunter, far in advance of his age —and, it may be, even of ours—saw at one glance the vast importance of biological science, and the best means to further its pursuit. To this end he founded his museum, and directed by his will that it should always be maintained in its integrity. Wherever civilised men are gathered together, there are now minds who feel what Hunter felt. The wants of such minds have created in every country in Europe, and the enlightened parts of the New World, museums designed to serve in their different degrees the same functions as our Hunterian collection. Such museums are evidently national needs; they have already come, though not by any means to the extent they will in future come, to be looked upon as an essential portion of the educational machinery of the State. Such museums are, in almost every capital of Europe, supported directly at the expense of the State, or are connected with some great educational institution dependant upon Government for aid. In England alone the need has been supplied first by a private individual, and secondly by a private, or semi-private institution, composed of members of a single profession, with only occasional assistance from the State. In this country the State (and therefore every individual composing it) is indebted to John Hunter and the Royal College of Surgeons for relieving it of the burden which must otherwise have fallen upon it, of providing that portion of the national education afforded by a biological museum.

The period occupied by John Hunter in the formation of his collection was all comprised between thirty years—1763, the date of his return from service with the army in Portugal, and 1793, that of his death. The labour which he accomplished during this time was something prodigious, as has often been recounted in various biographies and Hunterian "orations." Notwithstanding all that has been written and said, it is impossible to do justice to his wonderful activity and industry. In nothing, however, were these qualities so conspicuous as in the formation of his museum.

Public museums in his time scarcely existed. The British Museum was little more than a library and gallery of art; the small cabinet of natural history, reinforced by the old collection of the Royal Society, scarcely made any show. Anatomical specimens, even bones and teeth, were looked upon with disfavour. Some that had accidentally found their way into the collection were, even within the present century, treated as intruders, and turned out without much ceremony.

Teachers of anatomy were forming their own private collections, but these were all eclipsed by those of the two Hunters, William and John. That of the latter especially grew to such an extent as to become in some sort a national and public institution. He built a large room to contain it in Castle Street, at the back of his house in Leicester Square, and when finally arranged there, so much interest was taken in it that he found it necessary to open it to public inspection at certain stated times. Still it was maintained entirely at his own cost, and it is stated that by the time of his death he had spent upwards of £70,000 upon it. Whether this estimate be correct or not, his expenditure on it must have been very great, as though he had for many years made one of the largest professional incomes in London, his museum was the sole property he left behind.

John Hunter was a very miscellaneous collector—minerals, coins, pictures, ancient coats of mail, weapons of various dates and nations, and other so-called "articles of *vertu*," engaged his attention. These, however, and his furniture and books, had to be sold to meet the most pressing needs of the family. What would be now called the "biological" part of his collection was kept intact, during the six years which elapsed between his death and its purchase by the English Government in 1799. The preservation of the collection during this period is mainly due to the devotion of William Clift, Hunter's last assistant, whose services were retained for this purpose at a very small salary by the executors, Sir Everard Home and Dr. Matthew Baillie, and whose fidelity was rewarded by his being appointed the first " Conservator " of the collection after it came into the possession of the College of Surgeons.

The story of the negotiations with a Government whose interests and energies were then concentrated upon the great Continental war, and the answer of the Prime Minister Pitt, when applied to on the subject, " What ! buy preparations ! Why, I have not money enough for gunpowder," are well known. These difficulties were, however, overcome, and on the recommendation of a committee of the House of Commons, appointed to inquire into the subject, the sum agreed to by the executors, viz., £15,000, was voted for its purchase on the 13th of June 1799. Then came the question what was to be done with it. There was at that time no department of Government under the care of which such a collection could be placed. The condition of the British Museum has been already alluded to. The now flourishing and all-absorbing " Department of Science and Art " had not been invented. There was one body in London which might be supposed to have some special interest in the maintenance of such a collection,— the venerable and dignified College of Physicians,—but that body, it is commonly reported, demurred to accept it on the ground of want of funds to meet the annual expense of its maintenance. With reference to this report, Dr. Pitman has been kind enough, in response to my inquiries, to examine the archives of the College, and finds that there is no record of any such offer having been made or refused. If any negotiations were entered into, they must, therefore, have been of a purely informal nature.

There was still another corporate body—a comparatively obscure one at that time—the Corporation of Surgeons, which had only separated itself some fifty-four years before from the old City Company of Barbers and Surgeons,[1] and although it had thrown off the connection which restrained its members from assuming the position of cultivators of a liberal profession, it had as yet done little to raise itself in public estimation, and had few resources from which to provide for the expenses of such a collection.   Nevertheless, the Court of the Corporation determined by an unanimous vote on 23rd December 1799 to accept the museum on the terms proposed by the Government, and almost simultaneously obtained a new charter, under which it became "The Royal College of Surgeons," a body accredited by Government to examine all persons wishing to practise surgery in the kingdom, and migrated from its old quarters in the City to the house in Lincoln's Inn Fields, round which the present establishment has grown up.

Thus John Hunter's museum and the College of Surgeons of England, though of entirely independent origin, have had their fortunes inextricably intermixed, since the former became national property, and the latter took the title and position it now holds.

The College is still the principal examining body for those who practise surgery throughout the kingdom.   It takes no part directly in professional education, though it exercises a considerable indirect influence by the manner of conducting its examinations, and by the curriculum it requires from candidates.   Its revenues are mainly derived from the fees paid for the diplomas which it grants, which, for the last ten years, have averaged 383 a year.   In former times these fees

---

[1] By an Act of Parliament, passed in the 18th year of the reign of George II., entitled, "An Act for making the Surgeons of London and the Barbers of London two separate and distinct Corporations," it was enacted that the union and incorporation of the Barbers and Surgeons of London, made by the Act of the 32nd year of King Henry VIII., should from and after the 24th day of June, 1745, be dissolved, and that such of the members of the said united Company who were Freemen of the said Company, and admitted and approved Surgeons, within the Rules of the said Company, and their successors, should from thenceforth be made a separate and distinct Body Corporate and Commonalty Perpetual, which at all times thereafter was to be called by the name of "The Master, Governors, and Commonalty of the Art and Science of Surgeons of London."   The first Charter of the Company dates from the first year of the reign of King Edward IV. (A.D. 1461).

considerably exceeded the expenses of the comparatively slight examination required from candidates, and the surplus, besides defraying the current expenses of the Museum and Library, was devoted to the erection of the present buildings, and the acquisition of the freehold property and invested capital of the College.   It says much for the personal disinterestedness of the eminent members of the surgical profession who have constituted the Court of Examiners, and who until very lately were practically the ruling body of the College, that they fixed their own remuneration at so low a rate as to permit an expenditure during the present century upon the purposes just indicated, of a sum which cannot be estimated at less than £400,000.   Now, owing to the more searching and practical character of the examinations, the expenses of conducting them have augmented to such an extent as to be scarcely more than covered by the payments of the candidates ; and but for the proceeds of the investments made under different circumstances, the College would not have the means of carrying on the scientific work it has undertaken.

The various professorships and lectureships that are attached to the College have grown up chiefly in consequence of one of the conditions under which the Hunterian Collection was entrusted to it by Government—that a course of no less than twenty-four lectures shall be delivered annually by some member of the College upon comparative anatomy and other subjects, illustrated by the preparations.   Other lectureships have been founded by private benefactions, but these are of limited number, or on special subjects, and are intended, not so much for the education of students, but rather as the means of introducing new discoveries or ideas to members of the profession and others interested in scientific pursuits, to all of whom they are freely open without payment.

Besides the museum, the College has added to its means of benefiting its own members and the profession generally, a library containing every important work and periodical upon surgery, medicine, anatomy, and the collateral sciences.

During the first six years after the collection came into the possession of the College, it remained in the gallery in Castle Street, which had been built by Hunter for its reception; but in 1806, the lease of the premises having expired, it was removed temporarily to a house in Lincoln's Inn Fields, adjoining the College of Surgeons, while the building in which it was destined to be lodged was preparing for its reception. This building, towards the erection of which, Parliament contributed the sum of £27,500, was completed and first opened to visitors in 1813.

The museum was greatly enlarged entirely at the expense of the College in 1835, and a still more important addition, that of the great Eastern hall, was completed in 1855. Towards the expense of this, Parliament contributed a further grant of £15,000, the whole of the rest of the expenses of the purchase of the site, the building and the annual maintenance of the museum, having been borne by the College.

In accepting the Hunterian Collection, the College of Surgeons undertook a heavy responsibility, weightier perhaps than was contemplated at the time. Although not required by the letter of the contract to do more than preserve Hunter's specimens, the College undertook the charge in the spirit of the founder, and thus made itself responsible for maintaining such a collection as should meet the requirements of the ever-expanding and vigorous young science to which it ministers. Hunter's collection was held to be the nucleus of a national biological museum, and its preservation and augmentation by the College has certainly prevented the formation of such a collection by the State.

Hunter was no specialist, and even after eliminating the non-biological subjects before alluded to, a very miscellaneous collection remained; illustrations of life in all its aspects, in health and in disease; specimens of botany, zoology, palæontology, anatomy, physiology, and every branch of pathology; preparations made according to all the methods then known; stuffed birds, mammals and reptiles, fossils, dried shells, corals, insects and plants; bones and articulated skeletons; injected, dried, and varnished vascular preparations; dried preparations of hollow viscera; mercurial injections, dried and in spirit; vermilion injections; dissected preparations in spirit of both vegetable and animal structures, natural and morbid; undissected animals in spirit, showing external form, or awaiting leisure for examination ; calculi and various animal concretions ; even a collection of microscopic objects, prepared by one of the earliest English histologists, W. Hewson.

It is very difficult to compare the present Hunterian Museum, as it is still often called, although officially only recognised as the Museum of the Royal College of Surgeons of England, with any other existing collection, as its nature and the character of its contents have been determined by several accidental circumstances rather than by any very settled purpose. Originally a private collection, embracing a large variety of objects, it has been carried on and increased upon much the same plan as that designed by the founder, with modifications only to suit some of the requirements of advancing knowledge. The only portions of Hunter's biological collection which have been actually parted with are the stuffed birds and beasts, which, with the sanction of the trustees appointed by Government to see that the College performs its part of the contract as custodians of the collection, were transferred to the British Museum, and a considerable number of dried vascular preparations, which having become useless in consequence of the deterioration in their condition, resulting from age and decay, have been replaced by others preserved by better methods. Of the various departments of which the museum now consists, very few,—in fact only the collection of illustrations of skin diseases, and the collection of surgical instruments,—are not the direct continuation of series founded by John Hunter.

To find an analogous institution to the Museum of our College of Surgeons, in Paris, for instance, we should have to combine the collections of Comparative Anatomy and Anthropology at the Jardin des Plantes, and even a portion of the separate palæontological collection at that establishment, the collection of human anatomy of the Musée Orfila, and that of pathological anatomy of the Musée Dupuytren. If these were all brought together under one roof, and somewhat compressed and rearranged, we should have something in its nature resembling the museum of which I am now speaking.

G

In this combination on one spot, and under one management, of so many diverse collections, we have a survival of a condition of scientific knowledge more characteristic certainly of the last century than of the one in which we live ; but in this age of specialities it is well perhaps to be reminded by such an institution of the essential unity of biological knowledge, and of the important illustrations which one branch of it may afford to another, especially when the detailed facts are to be combined for the purpose of philosophical generalisation.

In visiting the museum, and in the comparison which may be instituted between it and others of its kind, it is important to recollect this origin and history, as they will account for many shortcomings. It must not be forgotten that to its comparative antiquity (for it is certainly the predecessor and prototype of all the anatomical museums of this country and of America, and to most of those on the Continent) is due many faults of construction and arrangement which should not be found in a building designed with the knowledge and experience of recent years. I have elsewhere pointed out what I consider the chief of these.[1]

Though the large size of the principal rooms allows of a fine *coup d'œil*, such a construction does not permit of that separation and distribution of the different series which is desirable for the purposes of study. Human anatomy, invertebrate zoology, and pathology, for instance, come into such near juxtaposition as to produce some confusion in the minds of strangers, though familiarity with the arrangement soon disperses the difficulties at first met with in finding the situation and limits of the particular department required. The narrowness and unprotected condition of the shelves in the galleries is also a radical defect now unfortunately irremediable. Furthermore, the indulgence of those who have the happiness to live elsewhere than in the absolute centre of a population of four millions of coal-burning people, must be asked for certain dusky results of such a situation, which no amount of care and expense can obviate.

I must now ask leave to be your guide to some of the contents of the museum, as it is at present arranged, and will

[1] *Journal of Anatomy and Physiology*, vol. ix. May 1875.

take the different branches of biology which are illustrated
in it in some kind of order, beginning with the part which
relates to life in a normal condition.   Hunter's collection
and observations were not limited to the animal kingdom.
Wherever any physiological process could be illustrated by
vegetable life, vegetables were pressed into the service, as may
be seen in the physiological gallery, and by the Memoranda
on Vegetation, left by him in MS. and printed by the College
in 1860.   In his collection were many portions of various
recent plants, and a series, amounting to 184 in number,
of fossil woods, fruits, and impressions of stems and leaves.
These specimens are arranged in the large wall-case on the right-
hand side (on going in) of the entrance door of the first or
western hall.   With them are some additions made in former
years, but since the great development of the parts of the
museum more essential to the general purposes of the institution, it has been necessary to restrict the growth of such branches
as are more fully and advantageously illustrated elsewhere.

The zoology of invertebrate animals largely attracted
Hunter's attention.   Many of the treasures collected in the
famous voyages of Captain Cook came into his possession
through his friend, Sir Joseph Banks.   He purchased, whenever opportunity offered, as at the sale of Mr. Ellis's famous
collection of corals and zoophytes.   In 1786, at the sale of
the Duchess of Portland's museum, he bought for fifteen
guineas the fine *Pentacrinus*, now in the museum, of which
very few examples had then been found.   Of insects, especially
Lepidoptera, he had a large series.   Of fossil invertebrates, as
many as 2092 specimens are now recorded in the catalogue
as Hunterian.   The series of fossil cephalopods is remarkably
rich.

Such invertebrate animals as are dissected, or illustrate
any special anatomical fact, are arranged in the so-called
physiological series in the gallery, to be described presently,
but beyond these there remained a vast number of specimens
only showing external form, which by selection and arrangement have been lately formed into a special zoological
collection, intended to introduce the student to a general
knowledge of the principal forms of animal life, and to the

mode in which they are grouped. This series, arranged in the floor cases on the left side of the western museum, includes selected specimens of nearly all the orders, and in many cases of the families, both of the living and extinct forms; illustrated both by their hard and imperishable parts as the "corals" or stony skeletons of the Actinozoa, the shells of Mollusca, and the tegumentary structures of the Articulata, and by the softer and more destructible parts of the bodies preserved in spirit. The various groups are distinctly separated from each other and clearly named. Students who desire to pursue the study of any of the sections more deeply than the small selected series of exhibited specimens will allow, will find the remainder of the specimens mentioned in the catalogues, arranged in drawers below the cases. The series does not extend beyond the invertebrata, as the peculiarities of the remaining classes of the Animal Kingdom are abundantly illustrated in other parts of the museum.

Although locally far removed, occupying one portion of the upper gallery of the middle museum, a small but interesting special collection, illustrating the subject of Helminthology, may be mentioned here. It was thought that the importance in a medical and social point of view of those animals which infest the interior of man and the principal domestic and other animals, justified a more extended exhibition of their modifications than could be assigned to any other group of animals of such inferior organisation, and by the aid of the well-known helminthologist, Dr. Spencer Cobbold, the present collection was arranged and catalogued in 1866; the materials being mostly already in the collection, though scattered in other series or hidden in the storerooms. The collection contains upwards of 200 specimens, and may still be somewhat extended. The intention is to show every parasitic animal which, under any circumstances, can affect the human body, and a selection of the principal types of those that inhabit the lower animals, especially such species as are associated with man. If increased beyond these limits, the collection would become interesting only to the student of detailed systematic zoology, and therefore not a legitimate object for our museum.

I will pass next to the section of the museum which is, perhaps, altogether the most characteristic, and is certainly the most eminently Hunterian. It was specially the creation of the founder, is still arranged almost exactly as he left it, and, notwithstanding the very numerous additions, still contains a larger proportion of Hunterian specimens than any other department. This is the collection which is called *Physiological*, because the specimens in it are classified mainly according to their supposed function. Physiology, as we know it now, is scarcely a subject which can be illustrated in a museum. The processes and actions which take place in the living body are not to be shown in bottles, but the organs, through the medium of which physiological processes are performed, can be, and it is these which are illustrated in this collection. It is more truly a collection of comparative anatomy, or morphology as we should now call it. It shows the variations in form which the different organs undergo either in different species, or in the same species under different conditions, as age and sex or season. Many of these modifications clearly have relation to function, as we see in the difference of form and relative size of the compartments of the stomach of the young ruminant, which is nourished by milk, and the adult which feeds on grass, the periodic variations in the size of the testis in birds, etc. But in a vast number more we can see no special adaptation to purpose, but merely variation, apparently for variety's sake. Look, for instance, at all the differences of the form of the liver throughout the mammalian series, which, as far as we know, have no relation to its action as a secreting gland. Though of little interest to the physiologist, modifications of this kind are of the highest importance to the morphologist. They throw light upon one of the great biological problems, classification, which, when rightly interpreted, means nothing more or less than a statement of the order in which living beings have been evolved one from another. From such variations of form most precious indications of the relationship of one animal to another can be obtained, and the less these variations are related to adaptation to some particular function, the better they can be relied on for this purpose.

But Hunter's ideas were far different. He tried to bring together analogous parts according to their uses—organs of progressive motion adapted for flying—eyes modified for seeing in water—eyes modified for seeing in air, and so forth. Practically, such a system could not be logically carried out. Too many modifications of form were found to occur, to which no special modification of function could be assigned, a compromise had to be made, and in the large number of cases the organs had to be arranged according to the affinities of the animals to which they belonged—brains of fishes, brains of birds, brains of mammals, etc. As the collection continues to advance, the classification according to homology is gradually superseding that according to analogy, with which it began.

This collection at present contains 6982 specimens mounted in bottles, of which 3745, or more than one-half, are Hunterian. It may be convenient to know that these are distinguished by the figures upon them which refer to the catalogue being painted in black. The specimens added since Hunter's time are lettered in red. The greater number of the former must be fully a century old, and being still in as perfect preservation as when first put up, afford a fair guarantee of the absolute permanence, with proper care, of specimens preserved in alcohol. The skill displayed in dissecting, injecting, and mounting the majority of these preparations has scarcely ever been surpassed in modern times, and this collection alone, if it were all that Hunter had left, would be a grand monument to his industry and zeal for anatomical knowledge ; as is its valuable and instructive descriptive catalogue, published in five volumes, and completed in the year 1840, a lasting evidence of the same qualities on the part of Mr. Clift's eminent successor in the conservatorship of the museum, Professor Owen.

Many points in comparative anatomy can be illustrated quite as efficiently, and more economically, by dried preparations, which require neither spirit nor bottles to preserve them in. Though we have not attained in this country the art of making such preparations in the elegant and instructive manner pursued in several of the museums in Italy, notably in that of Pisa, and though nearly all the original Hunterian dried preparations have perished long ago, or become partially

useless, there will still be found some worthy of attention in
the rail cases round the galleries which contain the spirit
preparation. While speaking of the contents of these cases
I would specially call attention to the series showing the
modifications of the small bones of the ear, throughout the
mammalian class, arranged a few years ago by Mr. Alban
Doran, one of the assistants in the museum, which is probably
not surpassed in extent or variety and method of arrangement
anywhere else.

The Histological Collection is contained in a separate small
room adjoining the physiological galleries, and consists of
upwards of 12,000 specimens, illustrating the minute structure
of the tissues of plants and animals, mostly prepared under
the direction of Professor Quekett, the third conservator of the
museum, who devoted the greater part of his life to this work.
Since his death in 1861, it has been rearranged and kept in
order; but the additions have not been numerous, chiefly in
consequence of the practical difficulties in exhibiting such a
collection to visitors to a public museum.

Although the anatomy of man naturally takes its place
among that of other species in the Physiological series, the
preparations illustrating it were chiefly confined to viscera—
the details of regional anatomy, and of the arrangement and
distribution of muscles, vessels, and nerves, not finding a
natural place in the scheme upon which that department of
the museum was organised. It was, however, a few years ago
thought desirable that human anatomy, in consideration of its
great importance to our profession, should be exhibited on a
much more extended scale than it had been hitherto, and that
a ready demonstration should be afforded by means of permanent preparations of the structure of all parts of the human
frame. To those who have already learnt their anatomy, and
who wish to refresh their memory, or verify a fact about
which some passing doubt may be felt, or those who are pre-
cluded by circumstances from visiting the dissecting-room, the
preparations of this series must prove of great value. The
series of dissections already made with this end, commenced by
a former able assistant in the museum, Dr. J. Bell Pettigrew,
and carried on to their present perfection by Mr. W. Pearson,

are arranged on shelves over the floor cases on the western side of the western museum, contiguous to the series of human osteology, to which they form the natural sequel.

No portions of the structure of vertebrate animals can be preserved with greater facility than the bones and teeth. Moreover, the skeleton being the framework around which the rest of the body is built up, gives, more than any other system, an outline of the general organisation of the whole animal, and it has this special importance, that a large number of species—all those in fact which are not at present existing upon the earth—can be known to us by little beyond the form of the bones. Osteology has, therefore, always had many votaries, as a special branch of study, and it is one which finds much favour in the eyes of curators of museums, from the satisfactory manner in which it can be illustrated by specimens. Hunter's osteological collection was considerable, quite in advance of any other in this country. The two small whales (*Balænoptera rostrata* and *Hyperoodon rostratus*) which formed part of it were almost the only skeletons of animals of their order which existed in any museum at the time of his death. This fact alone shows the marvellous change that has taken place within less than a century in the facilities for the study of comparative anatomy. How great the contrast to what may now be seen here in the College of Surgeons, in the British Museum, in Oxford, Cambridge, Edinburgh, Dublin, in a score or more of museums on the European continent, in America, even in Australia and New Zealand! Richly supplied osteologial collections have sprung up in every considerable centre of scientific culture over the world; but as ours was one of the first in point of time, we may also claim for it a high position in point of completeness. Others, such as that at the British Museum, the Jardin des Plantes at Paris, and the famous Leiden collection, may be larger, but this is because the College Museum has been designedly limited rather to selected illustrations of all the most important modifications of structure, than to numerous examples of closely allied species, which can only be looked for in a purely zoological museum. When important forms have become extinct, their characters are shown by their fossilised remains, which, though at present most illogically arranged in a distinct room apart from their existing allies, will soon be incorporated in the general osteological series, where alone they can find a reasonable position in an anatomical museum.

The value of a collection is not to be estimated only by the number of specimens it contains, nor by even their rarity or judicious selection, but also by the condition of the specimens, and the facility by which they may be made available for study and reference. On this head we claim to be somewhat in advance of other museums, on account of the improvements which have been made in late years in preparing and articulating entire skeletons, and in displaying portions of the bony framework in an instructive manner. Formerly all the bones were rigidly fixed together, so that their articular surfaces, if not actually destroyed, were completely concealed; and no bone could possibly be removed and separately examined. The aim of a series of changes in the method of mounting skeletons introduced here, and now adopted more or less completely in many other museums (the details of which were carried out with great skill by our late able articulator, Mr. James Flower) has been to obviate all these difficulties, and to make each bone, as far as possible, independent of all the rest, whilst preserving the general aspect and form of the entire skeleton.

Another improvement in the osteological series introduced within the last twenty years has been the formation of a special collection designed to show the principal modifications of each individual element of the skeleton throughout the vertebrate classes, by placing the homologous bones of a number of different animals in juxtaposition. For convenience of comparison, the specimens of this series are all placed in corresponding positions, mounted on separate stands, and to each is attached a label bearing the name of the bone and the animal to which it belongs. This series is especially instructive to the students of elementary osteology, and forms an introduction to the general series.

As in other departments of the museum, the more nearly man is approached in structure, the more complete do the illustrations of anatomical modification become, and, as might

be expected, the osteology of man is far more thoroughly
shown than that of any other species. The specimens of
human osteology (of which a revised catalogue, enumerating
1306 specimens, was published two years ago) begin by
illustrations of the development of the bones; these are
followed by the normal skeleton, exhibited under various
aspects, then by individual variations, among which may be
mentioned one of the most remarkable objects in the museum,
the skeleton of the celebrated Irish giant, O'Brian, who died
in London in 1783, and about the preservation of whose
remains so many legends are told in the biographies of John
Hunter. Finally, the special osteology of man or illustrations
of the osteological characters of the various races of mankind.
In this important subject Hunter was a long way in advance
of most of his contemporaries, as the origin of his collection
dates almost, if not quite, as far back as that of the founder
of physical anthropology, the celebrated Blumenbach. The
series has been greatly augmented of late years, and completely
rearranged, and the splendid addition made to it last year by
the purchase of the great private collection of the late Dr.
Barnard Davis has brought it up in point of completeness to
truly national importance.

As forming a transition from the department of normal
anatomy and physiology to that of pathology, may next be
mentioned the teratological series, or collection of congenital
malformations of man and the lower animals, which necessarily
forms part of every general biological museum. This difficult,
mysterious, and, as far as the light it throws upon the
workings of the laws of nature, still unsatisfactory subject,
had considerable attraction for Hunter, and many of the
specimens in the series form part of his museum. It has
been steadily, though not very rapidly increasing ever since,
and had the advantage, a few years ago, of being thoroughly
revised, rearranged, and catalogued by Mr. B. T. Lowne. It
is arranged in the upper gallery of the middle museum.

The pathological series is the section of the museum to the
study of which, in the eyes of Hunter and his successors, all
the others form an introduction. It occupies the whole of
the two galleries and part of the ground-floor of the western

hall. As the museum of the College differs from those attached
to the various medical schools, in having no hospital or *post-
mortem* room in connection with it, from which to draw the
supplies for completing this collection, it has been increased by
the acquisition from time to time, when opportunity afforded,
of various private collections, as those of Mr. Heaviside in
1829, Mr. Langstaff in 1835, Mr. Howship and Mr. Taunton
in 1841, Mr. Liston in 1842, and Sir Astley Cooper in 1843,
obtained by purchase; and the collections of Sir William
Blizard in 1811, Sir Stephen Love Hamminck in 1851, and
Dr. Peacock in 1876, presented to the College. Contributions
of recent specimens are also constantly received from numerous
individual donors, the acquisitions from this source having
greatly increased of late years. The total number of specimens
now in the catalogue amounts to 5148, of which 1672 are
Hunterian. As in the physiological galleries, the latter are
distinguished by their numbers being painted in black. The
descriptive catalogue of this series, written by Sir James
Paget, and published in five quarto volumes between the years
1846 and 1849, is one of the best-known and most valuable
of all the publications of the College, and has always been
looked upon as a model upon which other pathological
catalogues should be formed. The additions made to the
collection since that time have been so numerous that the
necessity of a new catalogue has long been felt. Under these
circumstances, it is a matter of great congratulation to all
who are interested in the welfare of this valuable collection,
that the author of the original catalogue has undertaken, with
the co-operation of Dr. Goodhart and Mr. Doran, to make a
new one, in which the old descriptions will be revised, the
new specimens incorporated in their appropriate places, and
such changes introduced into the general arrangement as the
advance of pathological knowledge and greater experience of
the requirements of the museum appear to necessitate. This
great work, especially arduous for one so much engaged in
professional avocations as Sir James Paget, is now far advanced. The prospect of its early completion will doubtless
compensate the members of the Congress who will make an
inspection of this part of the collection for the transitional

and somewhat disarranged condition in which they will find it on their present visit.

As adjuncts to the general pathological series are certain special collections, which have separate catalogues devoted to them. One, which will be examined with interest by those devoting themselves to aural pathology, is the series of preparations illustrative of diseases of the ear, formed by the late Mr. Joseph Toynbee, which came into the possession of the College at his death in 1866. It is a large and probably unique collection of 824 specimens, illustrating all the known morbid conditions of the organ of hearing, such as could only have been brought together by one specially engaged for a considerable number of years in investigating this branch of surgery, and the value of which is greatly enhanced by a complete descriptive catalogue, published during Mr. Toynbee's lifetime. This series is arranged in part of the rail cases of the lower pathological gallery in the western museum. The remainder of the same cases are devoted to the collection of urinary calculi and other concretions, salivary, biliary, and intestinal, both from man and various animals, probably the most complete and best arranged in the world. The careful chemical analysis and description of the whole of these specimens has been the work of Mr. Thomas Taylor.

In a corresponding position in the upper gallery of the same museum is the Dermatological collection, consisting of an extensive series of beautifully executed models, of actual specimens, casts and drawings illustrating the various affections of the skin. This collection was commenced in the year 1870, the whole of the specimens in it, the cases which contain them, and the catalogue describing them having been presented to the College by Mr. Erasmus Wilson, at that time Professor of Dermatology in the College.

Lastly, must be mentioned a collection—for the reception of which a separate room, approached from the end of the eastern museum, was devoted, in 1870—of surgical instruments and appliances, which, though small at present, contains many instruments curious for their antiquity, or interesting for their associations, and doubtless, now that a convenient and appropriate locality has been established for their reception and preservation, will be gradually augmented by additions of a similar nature.   It is mainly to the interest taken in the subject which it illustrates by the late Sir William Fergusson that the establishment of this collection is due.

Such is a general outline of the history and contents of the museum which, for eighty years, the College of Surgeons has maintained for the benefit not only of its own members, but for that of the profession at large, and indeed of all who take any interest in biological science, whether the young student preparing for his examination, or the advanced worker who has here found materials for many an important contribution by which the boundaries of knowledge have been materially enlarged.   To all such it is freely open without any fee or charge.   Even the written or personal introduction of members, still nominally required, is never asked for on the four open days from any intelligent or interested visitor; and on the one day of the week in which it is closed for cleaning, facilities are always given to those who are desirous of making special studies, and to the increasing number of lady students, whether artistic, scholastic, or medical.   Artists continually resort to the museum, to find opportunities of studying the anatomy of man and animals, which no other place in London affords; and of late years it has been the means of a still wider diffusion of knowledge, by the visits which have been organised on summer Saturday afternoons by various associations of artisans, to whom a popular demonstration of some part of its contents is usually given on each occasion by the conservator.

If the knowledge of organic nature is of any value to man, and this is a proposition which I am sure all who attend this Congress will admit, as on such knowledge the whole superstructure of their profession is built, there can be no question but that such an institution as I have here sketched out must be one of pure and simple benefit.   Its maintenance has been a worthy object upon which the College has spent its care and its money, and whatever may be the changes which impending legislation may effect in the organisation of the profession, we may all hope that the great work begun by John Hunter, and carried on by those who, under the guidance

and support of the Council of the College, have followed him
in the care of the collection, may not be impaired or destroyed.
Whether the whole of the charges of maintaining such a
museum in all its parts on a continually extending scale
should be the duty of one institution, like the College of
Surgeons, or even of one profession, may be a question for
future consideration; but, in the meantime, how easily could
its preservation and extension be rendered entirely independent of all the chances and changes of medical education and legislation, or even of Government assistance and
interference ! When we see the immense sums voluntarily
provided every year in this country by donation and bequest ;
when we see, and see with pleasure and gratitude, through the
length and breadth of the land, cathedrals, churches, chapels,
colleges, schools, hospitals, and asylums founded, endowed,
enlarged, and restored,—may we not hope that an old and tried
institution like ours will not be so entirely neglected as it
has hitherto been by members of our profession in search of
some means for the disposal of any surplus wealth they may
possess. Few objects can be so surely productive of good, so
little liable to abuse at any future time, as the preservation,
augmentation, and maintenance of a museum, in which the
facts of the beautiful and wonderful world around us are
displayed for the instruction of mankind.[1]

---

[1] The hope expressed above was soon realised in the magnificent bequest of
Sir Erasmus Wilson, who died in 1884, leaving nearly the whole of his property,
amounting to upwards of £200,000, to the College of Surgeons unfettered by any
conditions. The part that I was permitted to take in bringing about this great
benefit to the Museum cause has always been a source of unmitigated satisfaction to me.

GENERAL BIOLOGY

VIII

INTRODUCTORY LECTURE TO THE COURSE OF COMPARATIVE ANATOMY [1]

MR. PRESIDENT AND GENTLEMEN—I am sure that the feeling that is uppermost in the minds of all here assembled is regret at the cause which has placed me before you on this occasion. Of the able and distinguished men who have filled this chair since it was instituted in the commencement of the century, not one was more able or more distinguished than its last occupant, either as an original investigator in the branch of science he adorns or as a facile expositor of its truths. The loss of Professor Huxley's services to this Institution is indeed to be deplored. The only consolation is that what is loss to us may be gain to others. He but leaves us that he may concentrate his time, his energies, his genius, elsewhere.

Feeling as deeply as I do the responsibilities of the Hunterian professorship, and my incapacity to discharge its duties in a manner worthy of its dignity and importance, I should certainly have declined to undertake it when it was proposed to me, if it had not been for the thought that these lectures are emphatically *museum* lectures, intended to illustrate and explain the treasures of our noble collection, and that from these has been derived almost all I know of the great subject that I am now called upon to teach. I shall, therefore, perhaps more than another might, speak to you directly from those specimens. I am, as it were, *their* mouthpiece. Having lovingly dwelt and worked among them so long, I felt that I could hardly refuse to tell you something of what they are always telling me—to endeavour to put their silent eloquence

[1] Delivered at the Royal College of Surgeons of England, 14th February 1870.

H

in some sort of articulate language. If I can only succeed in awakening an interest in them, and lead any one who may chance to come into this theatre to continue his studies in the museum, where he will meet with far greater return for his time and his attention than here, the delivery of these lectures will not have been altogether in vain.

I have just mentioned the museum. It is impossible to do so without recalling the circumstance that this day is the one on which we have been accustomed to celebrate the anniversary of the birth of the illustrious founder of that collection.

The 14th of February is a red-letter day in our calendar. On this day a series of eloquent tributes have been paid to the merits of that great man by successive Hunterian orators. I will not delay the commencement of the proper subject of this course by endeavouring to speak of Hunter either as a philosopher, a physiologist, or a surgeon, but I cannot refrain from making a passing allusion to his work in relation to the special subject of these lectures—*Comparative Anatomy.* And even this would have been superfluous, after the able analysis of Hunter's work in this branch of science contributed by Professor Owen to the fourth volume of Palmer's edition of his collected writings, if it had not been for the vast elucidation of the nature and amount of that work by the subsequent publication of the two closely-printed octavo volumes of *Essays and Observations on Natural History, Anatomy, Physiology, Psychology, and Geology, by John Hunter, being his Posthumous Papers on those Subjects,* arranged and revised by Professor Owen (1861).

Hunter's reputation has suffered grievously from the extreme difficulty he always met with in giving adequate expression to his ideas. Mr. Clift, who acted for a time as his amanuensis, has told us how he has often "written the same page for him at least half a dozen times over, with corrections and transpositions almost without end;" and those who are familiar with his writings must own that, after all this labour, the result was often far from satisfactory. Hunter was himself painfully aware of the deficiency. This it is which has detracted much from the estimate in which many of his published works are held, and it is this which was, I suspect, the reason why his published writings bear so small a proportion to the vast mass of rough manuscript which he left behind him, especially on the subjects with which we are at present engaged.

The perusal of the work just mentioned will show how, while occupied with a large and anxious practice—in itself labour enough for ordinary men—while cultivating with what I might describe as a passionate energy the sciences of physiology and pathology, while collecting and arranging a museum such as never has been formed before or since by a single individual, he had also carefully recorded a series of dissections of different species of animals which, as his learned editor justly says, if " published *seriatim*, would not only have vied with the labours of Daubenton as recorded in the *Histoire Naturelle* of Buffon, or with the *Comparative Dissections* of Vicq d'Azyr, which are inserted in the early volumes of the *Encyclopédie Méthodique* and in the *Mémoires de l'Académie Royale de France*, but they would have exceeded them both together." [1] In fact they would have established Hunter's fame as by far the greatest contributor to the knowledge of animal structure of his time, and, what is more important, would have aided most materially in the advancement and diffusion of that knowledge.

The work as now published contains notes of dissections, more or less complete, of no less than 129 species of mammalia, 80 species of birds, 20 species of reptiles, 9 of amphibia, and 19 of fishes, besides numerous invertebrated animals of various classes; and these appear to be by no means the entire series of dissections left by Hunter.

The fate of the original manuscripts forms a sad page in the history of our science, and, I must also add, in the history of a man whom our Profession, and this College especially, might otherwise have regarded with feelings of respect and gratitude. Exactly thirty years after Hunter's death, his brother-in-law and executor, Sir Everard Home, who had retained possession of them when the collection was transferred to the College, committed them all to the flames. The

[1] Owen in *Works of John Hunter*, Palmer's edition, vol. iv. p. vi.

use that he had made of them in composing the lectures which he delivered from this chair, though often previously surmised, has become but too obvious since the publication of this work.[1]

Happily, Hunter's devoted assistant and friend, the assiduous and excellent custodian of his collection for nearly

---

[1] An account of this transaction, with some important remarks upon it by Mr. Clift, is appended to Professor Owen's edition of Hunter's *Essays and Observations*, referred to above. The late Sir Benjamin C. Brodie in his autobiography, prefixed to Mr. Charles Hawkins' edition of his works (1865, vol. i. p. 102), speaks of it in the following words : "Some years before his [Sir Everard Home's] death, he got great discredit from having destroyed a considerable portion of John Hunter's manuscripts which had come into his possession as one of Hunter's executors. This act was equally unjustifiable and foolish. It was unjustifiable because the manuscripts should have been considered as belonging to the museum, which Parliament had purchased ; and it was foolish, because it has led to the notion that he had made use of John Hunter's observations for his own purposes much more than was really the case. I had frequent opportunities of seeing these papers during nine or ten years, in which I was accustomed, more or less, conjointly with Mr. Clift, to assist him in his dissections. They consisted of rough notes on the anatomy of animals, which must have been useful to Hunter himself, or which would, I doubt not, have afforded help to Mr. Owen in completing the catalogue of the museum ; but they were not such as could be used with much advantage by another person. In pursuing his own investigations, Home sometimes referred to these ; but I must say that, while I was connected with him, I never knew an instance in which he did not scrupulously acknowledge whatever he took from them, or do justice to his illustrious predecessor. Unhappily, he was led afterwards to deviate from this right course ; and in his later publications I recognise some things which he has given as the result of his own observation, though they were really taken from Hunter's notes and drawings. One of these is a paper on the progressive motion of animals, and another a series of engravings, representing the convolutions of the intestinal canal, and neither of them of much scientific value."—See also a note on the same subject in the obituary notice of Sir B. Brodie in the *Lancet*, 25th October 1862.

It is very possible that the extent to which Home made use of these manuscripts had been much exaggerated ; but Brodie certainly did not fully appreciate either their value or the amount of Home's plagiarism. It is not in his latest writings, but in the first volume of his lectures on Comparative Anatomy, published in 1814, that these chiefly occur, as will be seen by the editorial notes appended by Professor Owen to Hunter's Essays. The only justification which can be set up for Home is, that the observations quoted without acknowledgment were really made by him when acting as Hunter's pupil or assistant, and had become incorporated in the papers of his master ; but nothing of the kind appears ever to have been alleged, and this, of course, would be no excuse for first retaining possession of, and subsequently destroying nearly the whole of the manuscripts.

fifty years, William Clift, had occupied himself in the interval
between Hunter's death and the removal of the manuscripts to
Sir Everard Home's house, in making copious extracts from
these precious documents.   The publication of these extracts,
nearly seventy years after their author had passed away,
however meritorious a labour on the part of their editor, can
unfortunately do little to regain for Hunter that place among
the leaders of science which his own untimely death, and the
negligence, or worse, of his executor had deprived him.
Highly original and valuable as are the observations contained
in these volumes, they have by this time been nearly all
anticipated by others.   A crowd of workers in the same field,
pressing in from all directions, have covered the ground which
ought to have been occupied by the figure of Hunter, and from
which no tardy recognition of his merits can ever dispossess
them.   We shall still continue to look to Cuvier and to Meckel
as the main sources of our modern knowledge of comparative
anatomy, and not to Hunter.   Let us, though, never forget
that our illustrious countryman had, before their time, collected
materials for a work which needed but the finishing touches to
have made it one of the greatest, most durable, and valuable
contributions ever made by any one man to the advancement
of the science of comparative anatomy.

The present course of lectures, as just said, are intimately
associated with the museum ; their annual delivery was one
of the conditions on which the care of the Hunterian collection
was entrusted by the Government to our College, and it was
expressly stipulated that they should be " illustrated by the
preparations."   I am bound, therefore, in my choice of subjects
to consider what the museum teaches, and what it does not
teach.   The museum teaches one subject, and, primarily, only
one subject—namely, the variations in the form and relations
of the different parts or organs of which the bodies of various
animals are composed.   In brief, it teaches *Animal Morphology*,
which is nearly synonymous with *Comparative Anatomy*.

Now, *morphology*, I need hardly say, is not *physiology*,
though it may be one of the foundations on which that
complex science is based.   But, contrary as it may seem to

some of the most cherished notions of the founder of the collection, an anatomical museum can scarcely do more in teaching physiology than a collection of minerals can in teaching chemistry. The physiologist should certainly be well versed in the nature of the materials and organs by which the functions of which he treats are carried on — having, indeed, need of all the help he can obtain from all quarters in the solution of the difficult problems which come before him; but his science is, in the main, experimental. Whether it be from the inherent difficulties that attend such inductions, or from the hasty and illogical way in which they have often been made, hitherto a large proportion of the attempts to solve physiological inquiries by an appeal to morphology alone have ended in failure, often of a mortifying character. We still have to confess our ignorance of the purpose and application of innumerable most obvious and striking modifications of structure, which no amount of reasoning or guessing, without actual observation on the living organism, seems able to remove. I am speaking of physiology in the sense in which the term is ordinarily restricted. There is a more general and higher physiology, to which I shall refer presently, to elucidate which morphology is one of the most essential aids.

When morphology was first cultivated with anything like scientific precision, views since called "teleological," exclusively prevailed. Every animal was looked upon as an isolated machine; every part of that animal was supposed to have been formed expressly for carrying on the economy of that particular species or individual in the most efficient manner, without any reference to other species or individuals. If anything further was looked for in anatomising an animal beyond the mere gratification of curiosity or love of knowledge for its own sake, it was *direct adaptation to purpose*.

Many, indeed, are the curious speculations indulged in by anatomists of this school, impelled, as they appear always to have been, to find an immediate use for every modification of form. In numerous instances these were mere conjectures, which an enlarged knowledge of the habits and economy of the animal treated of, or of kindred species, showed to be fallacious.

As time went on, however, and men began to obtain a
deeper insight into these subjects, resemblances either between
the whole structure or between particular parts of different
animals, which could not be explained on the utilitarian
principle, became strikingly obvious. Moreover, such things
as rudimentary and functionless organs in one animal, representing developed or functional organs in another, became
known in the further prosecution of morphology.   Then the
idea gradually dawned that there was some "secret bond"
which linked creature to creature, and which permitted
deviation only to a certain point from a certain given pattern
of construction.

This was the doctrine of *type* or *common plan*. A "type"
or a "common plan" for each natural group of the animal
kingdom was supposed to have influenced, or to have been
held in view at, the creation of every different species composing that group, the deviations from this type being related
to the special exigencies of the particular species.

This was a great step in advance when regarded as a mere
exposition of the facts of morphology, and when the idea was
not carried out by fanciful imaginations beyond the point
warranted by these facts.   Upon this view many anatomists of
great eminence seemed to rest.   But still it explained nothing,
accounted for nothing.   It gave not even a shadow of a reason
for the resemblances amid diversity found everywhere.   It
only asserted that the Creator had imposed certain apparently
quite arbitrary restrictions to His power ; but, beyond this
almost paradoxical assertion, it gave no clue to elucidate anything like a theory of creation.

In the meanwhile, however, the great results arrived at in
other branches of science, the increasing accumulation of facts
from various sources, all tending to show that the orderly and
harmonious working of the whole universe was due to constantly acting causes or laws, acting now, having acted for
immeasurable time past, and, as far as we can see, about to
act for immeasurable time to come,—the great work of the
astronomers and the geologists, leading to "the general conception of some *great principle of orderly evolution*, according to
which the present as well as past systems of existence have

been produced out of preceding orders of things," [1]—could not pass unnoticed by the biologists.

First with a faint and uncertain sound, but in later times more boldly and confidently, an hypothesis has been propounded which does profess to account for some, at least, of the facts of animal creation, and to afford a guide to the solution of the problem, if not of the original beginnings, at least of the present diversities, as well as resemblances, among the animal and vegetable life on the globe.

This theory has for its basis that the secret bond of union is not " conformity to type," is no ideal to which the operations of Creation were limited, but is one of actual consanguinity, or, as otherwise expressed, " genetic affinity."

The fundamental part of the theory is that, as individuals are known to come into being by a process of generation acting according to fixed and certain laws, the same to-day and yesterday and to-morrow, in like manner have races, varieties, species, and other larger groups of animals and plants come into being—that the species existing at any one period on the earth's surface are, in fact, the direct descendants, modified according to definite laws, of the creatures inhabiting the earth in previous periods.

The theory of orderly evolution already applied to most of the phenomena of the physical world has thus been also applied to organic nature.

This hypothesis, originally promulgated in a comparatively crude form by De Maillet, Erasmus Darwin, and Lamarck, and more recently advocated by the author of the *Vestiges of Creation*,[2] for a long time found little favour with English naturalists of eminence. Indeed, so late as 1856, Baden Powell, in his masterly essay " On the Philosophy of Creation," arguing with great force " that one grand over-ruling principle, the universality of law, order, and continuity, presiding as powerfully over the earlier stages of creation as during its continuance at the present moment, applies

---

[1] Baden Powell, *Unity of Worlds and of Nature*, 2nd edition, 1856, p. 448.

[2] A work in many respects far in advance of the age in which it appeared (1844). It was published anonymously, but is now known to have been written by Robert Chambers of Edinburgh.

equally to organic as to inorganic existence," can find little support from the writings of any who had made the diversities of organic life their special study. It was, in fact, in opposition to the views then held by most naturalists, and chiefly from a profound and philosophical analysis of the laws which govern the physical universe, that, reasoning from the general to the special, he came to the conclusion that "at least, *as a philosophical conjecture*, the idea of transmutation of species under adequate changes of condition, and in incalculably long periods of time, seems supported by fair analogy and probability."

Very shortly after, however, these views received an immense stimulus by the working out of a necessary complement to the main theory, one which was absolutely essential to its general reception—namely, a suggestion of a possible and intelligible *modus operandi*—almost simultaneously by the studies of Wallace amid the exuberant displays of nature in the unexplored forests of the Malayan archipelago, and by the patient accumulation and careful and candid examination of fact upon fact, drawn from every branch of biological science, and all converging on the theory of "*the origin of species by means of natural selection*," by the great naturalist [1] whose name is now so thoroughly identified with the entire transmutation hypothesis, that already in the German booksellers' catalogues "Darwinismus" is made a prominent subdivision in their classification of scientific literature; and in this country "Darwinism" is the term popularly, if not quite correctly, applied to the general doctrine of "organic evolution."

In reference to the effect which the publication of these researches has produced, the learned and judicious President of the Linnæan Society, Mr. Bentham, in his last anniversary address, says that " the investigation of the origin, development, and life history of species or races has been termed *the great problem of the day*"; and the impulse they have given to the study of biology in general, and to our special branch in particular, is not exaggerated by the eminent German anatomist Gegenbaur in stating, in a work just published, that "*the*

---

[1] Charles Darwin was grandson of Erasmus Darwin, mentioned above.

theory of descent will begin a new period in the history of comparative anatomy."

On the other hand, it happened rather unfortunately that popular views of natural theology, probably through the great influence of a famous work published during what I have described as the first or " utility " stage of morphology, have attached themselves more or less to the ideas of that period, and well-meaning but mistaken persons regard with aversion and alarm the modern theories which are, to a certain extent, subversive of those ideas. I have no hesitation in saying *mistaken*, for it is perfectly evident that all arguments as to " the power, wisdom, and goodness of the Creator," derived from an animal structure not miraculously created, but produced by the ordinary laws of generation, as all known animal structures are, must be entirely and equally valid, whether the laws producing that structure have been operating for a shorter or longer period.

As Professor Asa Gray has well put it, " If the argument from structure to design is convincing when drawn from a particular animal, say a Newfoundland dog, and is not weakened by the knowledge that this dog came from similar parents, would it be weakened if, in tracing his genealogy, it were ascertained that he was a remote descendant of the mastiff or some other breed, or that both these and other breeds came (as is suspected) from some wolf? If not, how is the argument for design in the structure of our particular dog affected by the supposition that his wolfish progenitor came from a post-tertiary wolf, perhaps less unlike an existing one than the dog in question is to some other of the numerous existing races of dogs, and that this post-tertiary came from an equally or more different tertiary wolf? And if the argument from structure to design is not invalidated by our present knowledge that our individual dog was developed from a single organic cell, how is it invalidated by the supposition of an analogous natural descent, through a long line of connected forms, from a cell, or from some simple animal existing ages before there were any dogs ? " [1]

[1] *Natural Selection not inconsistent with Natural Theology: a Free Examination of Darwin's Treatise on the Origin of Species.* Boston, 1861.

Those who recognise that the ebb and flow of the tides, the thunderstorms, the rains, and frosts, are beneficent in their effects, although the result not of direct miraculous interference, but of unchanging cosmical laws, have not the right to accuse of want of reverence the men who affirm that the wonderfully complex and beautifully adjusted contrivances of animal structure may also have been brought about through the intervention of agencies of similar nature.

This opposition, moreover, may have done harm in another manner, by evoking the natural tendency that exists in many earnest and intensely truth-loving minds to recoil against any display of unphilosophical dogmatism, and any appeal to passions and prejudices, where reason alone ought to have sway, and may thus have led in some cases to a too warm and partial adoption of theories condemned on such grounds.

The rising school of biologists are destined to live in troublous times. There can be no doubt that many of the questions now opening consequent upon the rapid and widespread acceptance of the evolution theory, will give more disquiet to a large class of persons otherwise not indisposed to welcome the advances of scientific discovery, than was ever given by the promulgation of the astronomical revelations of the sixteenth century, or the more recent establishment of the high antiquity of the earth, and are likely to lead to equally active reprisals.

The astronomers and geologists have in their turn had to confront the storm——they and their sciences have survived and triumphed; and, on the other hand, the faith and morality of the world have not suffered. Now the biologists are standing in the breach. As a part of or necessary sequel to the great doctrine of organic evolution, the question of the origin and position of Man will inevitably obtrude itself. In the coming discussion on this subject, all who take part in it, or alarm themselves about it, would do well calmly to consider this point. It is one similar to that just mentioned with regard to the so-called " natural theology."

Whatever man's place may be either *in* or *out* of nature, whatever hopes, or fears, or feelings about himself or his race he may have, we all of us admit that these are quite uninfluenced by our knowledge of the fact that each individual man comes into the world by the ordinary processes of generation, according to the same laws which apply to the development of all organic beings whatever,—that every part of him which can come under the scrutiny of the anatomist or naturalist has been evolved according to these regular laws from a simple minute ovum, indistinguishable to our senses from that of any of the inferior animals. If this be so—if man is what he is, notwithstanding the corporeal mode of origin of the individual man, so he will assuredly be neither less nor more than man, whatever may be shown regarding the corporeal origin of the whole race, whether this was from the dust of the earth, or by the modification of some pre-existing animal form. This, I conceive, is the ground on which those who would maintain unimpaired the spiritual and moral dignity of the human race may safely stand.[1]

In 1861 one of the most distinguished of American naturalists wrote: "Those, if any there be, who regard the derivative hypothesis as satisfactorily proved must have loose notions as to what proof is. Those who imagine it can be easily refuted and cast aside must, we think, have imperfect or very prejudiced conceptions of the facts concerned and of the questions at issue." The experience of nine years leaves the case precisely, in my judgment, as thus stated by Professor Asa Gray. Anything which assists to throw light upon it, to lead us nearer either to its acceptance or rejection, is of primary importance to the biologist. We must not refuse to take it into earnest consideration.

I think, therefore, that it may be worth while to devote a few words to two objections which are frequently urged, not indeed against this theory, but against the prominent place I would assign to its consideration by all who would enter deeply into the philosophy of biology. Both of these objections are founded upon misconceptions as to the real nature of the present aspect of the theory of development.

[1] For the bearings of the Evolution theory of Creation on this subject, see Baden Powell, *op. cit.*

The first, taking its leading idea from the title of Mr. Darwin's book, is that it is unprofitable work to trouble ourselves about the *origin* of species, or, indeed, of anything else. These events of the very remote past, it is said, are quite beyond our ken; we really can never know anything certain about them, and had far better occupy ourselves with things of the present—things which we can really hope to know much about.

The other, akin to this, has lately been put forth on high authority. Let the "derivative hypothesis" be granted, the studies of the zoologist are not thereby in any way affected. "For all intents and purposes of the descriptive and recording naturalist," it is said, "species are constant; they will last our time. When the existing binomial units of botanical and zoological specific lists cease to show their present distinctive characters, the *homo sapiens* of Linnæus will have merged into another, probably a higher specific form."[1]

As I have said, both these objections are founded on an entire misconception of the main question at issue. If the developmental theory as held by Darwin, Wallace, and their followers is true, the origin of species is as much a thing of the present as of the past. It is an essential part of the theory that the laws which have produced the diversity of organic forms in the world are those by which the world not only has been, but *is* governed. They are as constant, ever-acting, and unchanging as the laws which direct the movements of the clouds or cause the torrents to flow down from the mountain side. There is no proof whatever that the laws of variation and natural selection, if such be the laws which lead to the introduction of new forms and the extinction of old ones, were ever more potent than they are at present. A large class of the arguments by which the theory is supported is derived from observation upon the present phenomena of life. According to the hypothesis, transitional forms and incipient species are to be met with everywhere around us. It is this, in fact, which has given rise to the difficulty zoologists and botanists always meet with in defining the limits of the various so-called species composing so many

[1] Owen on the "Aye-Aye." *Trans. Zool. Soc.* vol. v. p. 92.

groups both of animals and plants. Until it is settled whether there is an insensible blending in the conditions expressed by the terms "variety," "race," "subspecies," "species," or whether the old idea of the immutability of species is to be maintained, zoology can hardly be said to have a philosophical basis.

The second objection just named appears to suppose a simultaneous march of all organic beings from form to form— a transmutation *en masse* of the whole—so that to any one among them they would all appear stationary. Such a view is utterly opposed to the greater number of zoological and palæontological facts, and to all the necessities of the Darwinian hypothesis. One species, either with little inherent capacity for variation, or so circumstanced that such variation as may have occurred in different individuals has never been accumulated by selection, may remain without alteration for ages, while other species differently constituted or circumstanced may have undergone vast transmutations during the same period. In fact, the derivative hypothesis is not a theory of things long ago; it is not a curious speculation into the beginnings of the present condition of things. It may be appealed to every day in the solution of constantly occurring problems in morphology.

I will give an example from the subject of our present course. There is in Australia an animal, the koala (*Phascolarctos cinereus*), rather larger than a cat, but having more the aspect of a small bear (Fig. 1). It is covered with a soft woolly fur, and lives entirely among the boughs of trees, on the leaves of which it feeds. Its tail is a mere rudiment. Its feet are admirably adapted for grasping the branches on which it climbs. The hind feet especially are remarkably modified for this end, being very broad, and having a strong prehensile or opposable inner toe or hallux, like a thumb. The skeleton of that foot is very singular (see Fig. 2). The hallux is stout, and placed nearly at a right angle with the other toes. The next two digits are rather slender, placed close together, and in the living animal are united almost to the claws in a common integument. The two outer toes are free and much stouter than the others, the fourth especially so.

Now, a naturalist of the last century would have looked

upon this foot simply as a piece of machinery beautifully

FIG. 1.—The Koala (*Phascolarctos cinereus*).   From Gould's *Mammals of Australia.*

adapted for the particular work it had to perform in the animal's life.   He would have pointed to the broad palm, the

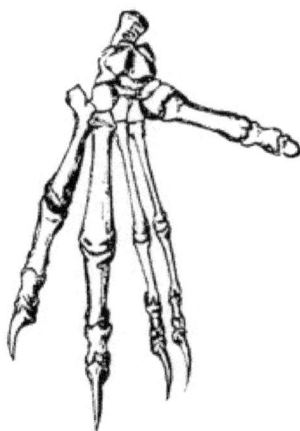

FIG. 2—Skeleton of the hind foot of the Koala (*Phascolarctos cinereus*).

opposable great toe, and would doubtless, if a worthy representative of his school, have developed some ingenious theory

to show that the smaller size of the second and third toes,
and that being wrapped up to their ends in skin, added to
the efficiency of the foot as a climbing organ, though indeed
he might be at a loss to say why the American opossums, of
precisely similar habits, and with similarly opposable great
toes, have all their other four toes free and equal.

The philosophical naturalist of the second of the types that
I have sketched out, in investigating this structure, will

Fig. 3.—Skeleton of the hind foot of the Virginian Opossum (*Didelphys
Virginiana*).

institute a careful comparison between the foot of the koala
and that of other animals belonging to the same group.

Among the marsupial animals of Australia, few are so well
known as the kangaroo. Like the koala, it is a vegetable
feeder, and these two forms have some dental and other
structural characters in common. But the mode of progression, and the limbs which effect this, offer the greatest possible
contrast.

In the kangaroo the hind legs are disproportionately large.
The foot especially is long and narrow, and at first sight
appears composed of a single large toe. Its motion along the
ground consists of a series of leaps or hops effected entirely
by these powerful hind limbs. On looking more closely at

the foot (see Fig. 4), it will be seen that the first digit or
hallux is entirely wanting, but the other four
toes are present with their complete number of
phalanges.   The fourth is immensely developed,
the fifth moderately so, but the second and
third are reduced to the most slender rudiments, and are so united in life by a common
integument, that they look like a single toe
with a double claw.

Now, here are two feet as unlike as possible
in their functions—the one formed for rapidly
hopping along over the arid plains, the other
for slowly and securely climbing among the
boughs of the forests—yet presenting a deep-
seated resemblance in a character not found in
the feet of any other known mammal except
the immediate allies of these two.

We may call this "conformity to type"
without getting much nearer to an explanation
of the phenomenon.   Perhaps it is safest to
rest at this stage.

But is it not powerfully suggestive—I will
not say more, for, of course, by itself it cannot
be considered as a proof—of true relationship,
of inheritance from a common ancestor ?   We
can easily see that in some manner the great
preponderance of the one toe in the kangaroo,
and the reduction of the others, would be
advantageous in its mode of progression, reduc-
ing the foot to a narrow spring-board as it

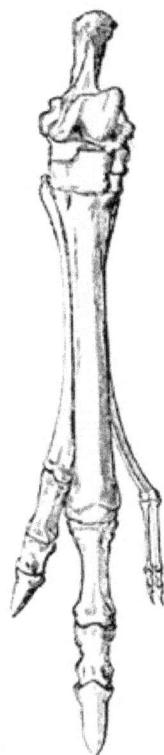

Fig. 4.—Skeleton
of the hind foot of
the kangaroo (*Ma-
cropus major*).

were ; and we can see that this attenuation of the second and
third toes, after the first had quite disappeared, might be a
stage in the direction of their total disappearance.   But while
this was taking place, let us suppose that one branch of the
family took to climbing trees (I must not enter fully into all
that might be supposed on this subject, as I am merely
introducing this as an example of the mode in which morphological problems are illustrated by these general views), and
the variations in a particular direction, tending towards better

climbing, were preserved and accumulated on the Darwinian hypothesis, the hallux (a common part in all mammals, only in abeyance when not wanted) was redeveloped in a form best adapted for its actual purposes, the other digits gradually resuming their normal condition; these two, however, still preserving strong reminiscences of the ancestral state.[1]

It is certainly significant that all the Australian vegetable-feeding climbing opossums have feet constructed on this type, one that is similar to that of the kangaroos of the same country, while the American opossums, further removed geographically, and therefore, in all probability, in actual relationship, with feet functionally the same, show no trace of this deep-seated structural peculiarity—a peculiarity most important for the consideration of the philosophical anatomist, as it evidently depends on some more far-reaching cause than mere adaptation to purpose.

In considering a little more fully the application of such views to the study of morphology, I must say a few words on *classification*. It was felt at the very outset of the study of natural history that without some system of classification the subject was little more than a hopeless chaotic confusion. The first instinct of a zoologist is to arrange in some sort of order the multitudinous objects with which he has to deal. In the beginning of science, with little sound knowledge, such classifications were often mere arbitrary arrangements, founded on some easily accessible peculiarities, and forming nothing

[1] In New Guinea and North Australia there are actually species of kangaroos, constituting the genus *Dendrolagus*, which habitually reside in trees; not climbing among the smaller boughs, as is the manner of the koalas and opossums, but sitting on the larger horizontal branches. Their feet present a marked deviation from those of the common kangaroos of the plains, the deviation being *in the direction* of the feet of the koala. They are shorter and broader; the lateral toes are relatively more developed, and are on the same plane with the fourth or large toe, and although they have no hallux or first toe, the bone of the tarsus (internal cuneiform), which usually supports that digit, is of relatively larger size.

The phalangers, a family of climbing vegetable-feeding marsupials, form another link in the structure of their feet between the kangaroo and the koala, though most nearly resembling the latter. On the other hand, in a very remarkable little ground-dwelling animal, the *Chœropus*, the kangaroo type of foot is modified in the opposite direction, all the digits except the fourth being reduced to excessively rudimentary proportions.

more than a means by which the distinction between certain
objects and certain other objects could be grasped by the mind
of the founder and communicated to others.   Classifications
based on such characters became so general, and were often so
carelessly and ignorantly put together, that they threatened to
bring the whole system into disrepute ; whereas I have little
hesitation in saying that, in reality, *classification* is one of the,
if not the, most important aims and ends of the study of
morphology.   It is the best contribution which we can make
towards the solution of the great biological problem, for a true
classification, viewed by the light of the derivative hypothesis,
is nothing more or less than an expression of the actual
amount of affinity between different objects.   An order, a
family, or a genus is no longer a group of animals linked
together by some arbitrarily selected characters, but a group
supposed to have been descended from a common ancestor, and
to have become, by whatever process, gradually differentiated
from other groups of animals.[1]

As in such groups, when once established, there can be
no crossing with other groups (as in human families, to which
they are sometimes compared), all the resemblances which are
found between members of different groups must either be
characters inherited from the common ancestor, perhaps lost
through many generations and reappearing at a subsequent
period by the process of " reversion," or they must be characters
having merely *analogical* resemblances—*i.e.* characters developed by variation, but which, owing to similarity of conditions
of existence or other causes, have become similar to each other.

To discriminate between these two classes of characters,—
namely, those that are essential and fundamental, or, in the
language of one school, are dependent on conformity to type,
or, according to the derivative hypothesis, are inherited from
remote ancestors, and those that are modifications to suit the

---

[1] In examining into the validity of the derivative hypothesis, much is to be
expected from the study of the geographical distribution of animals and plants,
both in present and in former times ; but such study will be quite in vain unless
morphologists have first determined correctly the affinities of the animals and
plants treated of.   It is obvious that all inferences from geographical or
palæontological research are useless, unless the classification on which they are
based is at least approximately true to nature.

conditions of existence to which the animal is exposed, or, in other words, adaptive characters,—is a constantly recurring problem to the systematic morphologist; and the difficulties that encompass the solution of this problem are the main causes of the little progress hitherto made towards a general agreement as to classification otherwise than in the main groups. Let me illustrate this point by a single and, as it happens, by no means difficult example.

You have before you the skeletons of two animals very similar in their general or superficial characters—one is the common dog, the other the carnivorous marsupial, the thylacine,

FIG. 5.—The Thylacine (*Thylacinus cynocephalus*). From Wolf's *Zoological Sketches.*

sometimes called "Tasmanian wolf." These animals, when wild, have very similar habits. You will see that, in the general characters of the skeleton, the structure of the limbs, and the arrangement of the teeth, a remarkable similarity prevails. You might easily conclude that these animals were related. They were, indeed, formerly placed near together by systematists.

Now, if I place by the dog another well-known animal, a sheep, you will see that in many characters of its bones, its feet, its teeth, etc., it differs far more from the dog than the latter does from the thylacine. Yet zoologists affirm, without the slightest hesitation, that the sheep and the dog are far more nearly related to each other than either is to the thylacine. This assertion is founded on a discrimination of the essential from the adaptive characters. The dog and the sheep belong to one great branch of the mammalian class, although they represent different stocks of that branch.   The thylacine belongs to another primary branch, but in that branch it has taken the place, functionally, of the dog in the other; its organs have become adapted to perform the same work.[1]

It has been asserted, in a recent very able pamphlet by an anonymous writer, "On the Difficulties of the Theory of Natural Selection," [2] "that on the theory of 'natural selection' it is all but impossible, such are the probabilities against it, that identical structures should have arisen independently. Yet many structures undeniably exist which, to all appearance, must have so arisen."   And the "remarkable identity of structure between certain of the teeth of the large predatory marsupial, the thylacine, or Tasmanian wolf, and those of the common dog," is cited as a case in point.   Now, though I am fully willing to admit difficulties in the theory, quite inexplicable with our present knowledge, as many almost as the author quoted, I cannot see this one in the same light.   It appears to me that the probabilities, instead of being against the independent origin of such similar structures (erroneously called *identical*) as those just mentioned, are exceedingly in their favour.   I cannot in the least see why a marsupial animal should not be carnivorous and predaceous as well as a placental animal, and if so, why its teeth should not have come to possess the general attributes essential to animals of such habits—namely, large, pointed, recurved fangs at each of the anterior angles of the jaw; the teeth between them much reduced, so as not to interfere with the piercing and holding action of the canines; the molars more or less scissor-shaped.

---

[1] The two great branches of the mammalia referred to are—(1) the Monodelphous, or placental, comprising the great bulk of the animals of the class; and (2) the Didelphous, or nonplacental.   The latter, which are the animals commonly known as *Marsupialia*, or pouched animals, though varying extremely in the structure of the feet, teeth, etc., also in their food and manner of life, all agree together, and differ from the monodelphia in numerous important characters of the skeleton, brain, heart, and especially of the reproductive organs and processes.   The marsupials are, at the present day, entirely confined to Australia, its neighbouring islands, and the American continent, though, in former times, they had a more extensive geographical range.

[2] Published originally in the *Month*, 1869.

These are plainly adaptive characters, and whatever process has produced them in the one case is just as likely to have produced them in the other—I might almost say, must have produced them in the other. So far, however, are the teeth of the thylacine and the dog from being essentially "identical,"

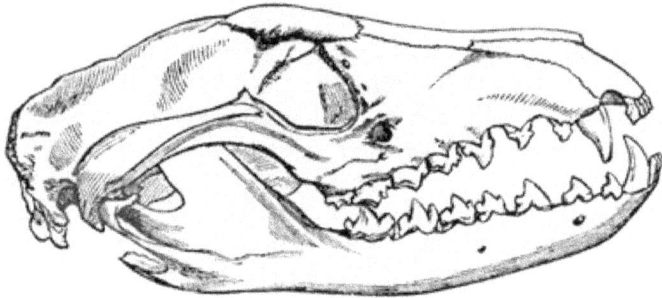

FIG. 6.—Skull of Thylacine.

that in numerous non-adaptive and non-functional characters, as the number of the incisors, premolars, and molars, the mode of succession, and the minute structure (all which will be spoken of at a later period of the course), they widely differ, the dog conforming in all these points with the ordinary placental

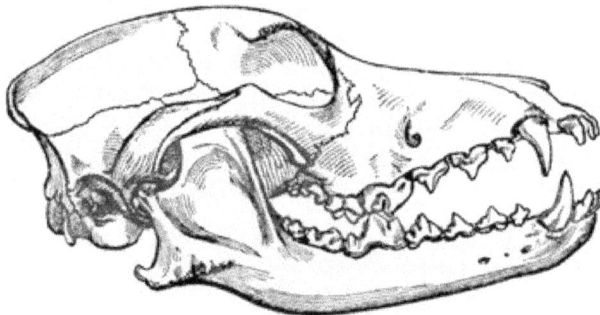

FIG. 7.—Skull of Shepherd's Dog.

mammals, the thylacine with the remaining marsupials. Such characters, underlying, in large groups of animals, the various modifications in relation to use, cannot be too diligently sought out as the true landmarks to guide our steps through the intricacies that beset our path in tracing affinities.

Another valuable guiding principle in morphological studies is this. It is a remark of Gegenbaur's, in his valuable work "On the Structure of the *Carpus* and *Tarsus*," but it must have occurred to any one who has given much thought to these problems. When we wish to discover the distinguishing characters between different organisms, it is necessary to examine them in their most fully developed condition. If, on the other hand, our object is to trace their resemblance, their intimate relationships, we must study them in their early embryonic stages. By these methods we can do much to separate what is secondary or superadded from what is fundamental or essential in the character of an animal. The farther back we can carry our researches, the more prominent do the characters, common to the whole group to which the animal belongs, become. The more completely mature the specimen, the more do the special characters of the species or even of the individual predominate.

It is not my province in these lectures to indulge much in speculations. Indeed, as students of morphology, we are as yet little in a position to do so. I shall not say much more even on the general views to which I have lately referred, for I feel that the acquisition of a sound basis of fact to work with is what is, at present, most needed in comparative anatomy. The difficulties which beset the beginner, and indeed the more advanced student, in this subject are very great. As regards the branch to which I propose more particularly to direct your attention, the anatomy of the class Mammalia, which, we might suppose, was better worked out than any other, the information to be found in books is scattered, fragmentary, unequal, and often untrustworthy; even good elementary treatises are wanting, much more anything like an exhaustive work. Moreover, our museum— superior as I believe it to be to any other of the kind—is, as yet, far from adequate to supply the knowledge frequently sought for in it.

In considering the special subject for these lectures, I have often thought that the greatest permanent benefit would be conferred on our science by collecting together in a systematic form all the available information upon limited groups of the animal kingdom, supplementing, as far as possible, the deficiencies of knowledge by fresh observations, and illustrating the subject by a complete series of preparations. At the present day it is only by working out a definite branch of limited extent monographically, that any solid advances in detailed knowledge can be attained. Courses of lectures on this principle, especially if they were published so as to reach a wider circle of students than this theatre is apt to contain, may do much to *advance* knowledge; but, on the other hand, they might do less than more elementary lectures, giving a general outline of a larger variety of topics, to *diffuse* knowledge, for it is probable that they would be attractive to only a limited number of auditors or of readers.

I think it therefore advisable, in the first course I have the honour of addressing to you, to confine my subject to a survey of the structure of the animals of the highest class of organised beings; and I think that this will prove of the greater importance and interest to the audience assembled here, because it may be presumed that all have already either commenced or completed the acquisition of a knowledge of the structural anatomy of one, the most elevated member, of that class. I propose, therefore, to take human anatomy as a point of departure, and, presuming on your acquaintance with its details, shall refer only to some of its general outlines, and shall point out the deviations from and resemblance to the mammalian structure as we know it in man, in descending throughout the series of animals composing the class.

I trust that some further interest may be given thereby to the daily work of the student of our profession. Human anatomy is too often learned as a mere collection of hard names applied to a complicated network of structures, the form, position, and relation of which have to be got up, probably to be forgotten soon afterwards. It might, however, be made a far more attractive and useful subject if taught by the light of a wider morphology. But I am afraid that very little of our anatomical teaching, either by books or lectures, is of this class at present. If any comparative anatomy is introduced, instead of enlightening and illustrating the subject, it often only adds another load to the already over-burdened memory—for instance, after the usual dry, detailed and technical description of the part treated of, a disquisition is added on what is called its "transcendental" anatomy, extremely incomprehensible to most minds, and consisting chiefly in the imposition of a new set of names to parts of which the student has just succeeded in mastering the old ones.

All this ought to be reversed; the essential nature of the part in question, as deduced from comparative anatomy, should be first announced with a glance at its principal modifications; then its special characters as seen in man will be a subject of intelligent interest. I know from experience that after studying and teaching human anatomy for many years on the ordinary methods, there were many parts the meaning and nature of which I never understood until I began to dissect animals; and increase of knowledge in this direction constantly throws light on apparently unmeaning or incomprehensible parts of human structure.

One word more by way of introduction. I began with the mention of my immediate predecessor in this chair. To him I must once again return. The subject matter of the present course will embrace many points treated of by him, more or less fully, in some of his lectures delivered in this theatre, and many on which I have had the advantage of his conversation and counsel. His teaching has entered so deeply into, and mingled so closely with, observations and reflections that may have suggested themselves in the course of my work in the museum, that in many cases it would be difficult to trace the sources of information that I may have to impart, and I may, perhaps, appear guilty of appropriating what should belong to another. Let me acknowledge then, once for all, how deeply I am indebted to Professor Huxley, not only for the information conveyed in the public manner of lectures and books, but also for the generous way in which, on all occasions, his time, his knowledge, his thoughts, have been freely given to me. And I say this the more willingly, because I know that I am but one of many whose labours have been lightened, whose efforts have been stimulated, whose difficulties have been smoothed, by his encouragement and support, by his candid and judicious criticism, and by the example of his often self-sacrificing devotion to the advancement of scientific truth.

I also desire, in conclusion, Mr. President, to take this opportunity of publicly expressing the obligations which I, as well as numbers of others, feel towards you and your predecessors in that chair, as well as to the former and present members of the Council of the College, individually and collectively, for the immense aid that you have given to the progress of philosophical biology in this country, and I may say, in the whole world, by the maintenance and augmentation of John Hunter's Museum. An epoch of revolution appears to be at hand in our profession, which may lead to a material alteration in the respective positions and opportunities of the various corporate bodies. Whatever changes may take place, the College of Surgeons will always look back with satisfaction to the fact that, for the first seventy years of the century, when these studies were less appreciated than they are now, or will be hereafter, it has, with scanty aid from the national resources, cherished the growth of a truly national institution, the benefits of which are not confined to any one class or profession, but are freely open to the whole community. I can say, moreover, with perfect assurance, speaking both from my own experience and from knowledge of the history of the museum, as recorded in its archives, that any deficiencies which exist in the condition of the collection, any needs which it does not supply, must be due to other causes than want of encouragement to the officers of the establishment, or want of liberality in supplying the requisite funds, on the part of the Council of the College.

RECENT ADVANCES IN NATURAL SCIENCE IN
RELATION TO THE CHRISTIAN FAITH [1]

I HAVE been requested by the Subjects Committee of the Congress to place before you a brief statement of some of the advances which have recently been made in natural science, with a view to open a discussion upon their relations, real or supposed, to religious belief. The particular advances which, as I am given to understand, were especially in the minds of the Committee in proposing this question, are those which have resulted in the more or less general adoption by scientific men of the view of the sequence of events which have taken place and are still taking place in the universe, to which the term " evolution " is now commonly applied.

All that is embraced by this term, the various realms of nature in which its manifestations are traced, the various shades of meaning attached to it by different persons, would constitute far too large and complex a subject to be treated of in the time to which addresses to this meeting are wisely restricted. I will therefore select for special consideration the only point in the application of the theory upon which I can speak with any practical knowledge—one which is, however, in the eyes of many of very vital interest. It is the one, at all events, which at the present moment attracts most attention ; the new ideas upon it being received with enthusiasm by some, and with distrust, if not with abhorrence, by others.

The doctrine of continuity, or of direct relation of event to some preceding event according to a natural and orderly

---

[1] Paper read at the Church Congress (Reading Meeting), 2nd October 1883.

sequence is now generally recognised in the inorganic world; and although the modern expansion of this doctrine as applied to the living inhabitants of the earth appears to many so startling, and has met with so much opposition, it is, in a more restricted application, a very old and widespread article of scientific as well as of popular faith.

Putting aside, as quite immaterial to the present discussion, the still controverted question of the evidences of the production of the lowest and most rudimentary forms of life from inorganic matter, it may be stated as certain that there is no rational and educated person, whatever his religious beliefs or philosophical views, who is not convinced that every individual animal or plant, sufficiently highly organised to deserve such distinctive appellation, now existing upon the world, has been produced from pre-existing parents by the operation of a series of processes of the order to which the term natural is commonly applied.   These processes are also fundamentally the same throughout the whole range of living beings, however much modified in detail to suit the various manifestations under which those beings are presented to us.   We feel absolutely certain, when we see a horse, a bird, a butterfly, or an oak-tree, that each was derived from pre-existing parents, more or less closely resembling itself.   Though we may have no direct evidence of the fact in each individual case, the knowledge derived from the combined observations of an overwhelming number of analogous cases is of such a positive character that we should entirely refuse to credit any one who made the contrary assertion, and should feel satisfied that he had been deluded by some error of observation.   We cannot, indeed, conceive of the sudden beginning of any such creatures, either from nothing, from inorganic matter, or even from other animals or plants totally unlike themselves.

To persons whose opportunities of observation of animal and plant life are limited to a comparatively few kinds, existing under comparatively similar circumstances, and which observations moreover only extend over a comparatively limited period of time, it appears that in each kind of animal or plant such as those just mentioned, individuals of various succeeding generations present a very close resemblance to each other.

That they often vary a little cannot escape careful observation, but the deviations from the common characters of the kind to be noticed by persons whose range of vision is thus limited are not striking, and usually appear not to pass beyond certain bounds. Hence arose the common idea, natural enough under such circumstances, but which gradually developed itself, not only into a scientific hypothesis, but even, it would appear, almost into an article of religious belief, that the different kinds or " species," as they are technically called, of animals and plants, had each its separate origin, its fixed limits of variation, and could not under any circumstances become modified or changed into any other form.

This idea became deeply rooted in the human mind, in consequence of the very long period during which it prevailed, the horizon of observation having remained practically stationary from the time man first began to observe and record the phenomena of nature, until little more than a century ago, when commenced that sudden expansion of knowledge of the facts of the animal and vegetable world, which has been steadily widening ever since. Now it is important to observe that it is strictly *pari passu* with the growth of knowledge of the facts, that the theoretical views of nature have changed, and the older hypothesis of species to which I have referred has gradually given way to a new and different one.

The expansion of the special branches of knowledge affecting our views upon this subject has taken place in many different directions, of which I can here only indicate the most striking.

1. The discovery of enormous numbers of forms of life, the existence of which was entirely unknown a hundred years ago. The increase of knowledge in this respect is something inconceivable to those who have not followed its progress. Not only has the number of well-defined species known multiplied prodigiously, but infinite series of gradations between what were formerly supposed to be distinct species are being constantly brought to light. The difficulty of giving any satisfactory definition of what is meant by the term " species " is increasingly felt day by day by practical zoologists, as evidenced by the introduction of such terms as " sub-species,"

"permanent local variety," etc., into general use, and especially by the wide differences of opinion as to the number or limits of the species included in any given group of animals or plants among naturalists who have made such group their special study.

2. Vast increase in the knowledge of the intimate structure of organic bodies, both as revealed by ordinary dissection, and by microscopic examination, a method of investigation only brought to perfection in very recent years. By the knowledge thus acquired has been demonstrated the unity of plan pervading, under diverse modifications, the different members of each natural group of organisms, at one time attributed to "conformity to type," a so-called explanation which explained nothing, but for which a *vera causa* may be found in descent from a common ancestor. Wonderful gradations in the perfection to which different structures have attained in the progress of their adaptation to their respective purposes have also been shown, and of still greater importance and interest, the numerous cases of apparently useless or rudimentary organs in both animals and plants, which were absolutely unaccounted for under the older hypothesis.

3. The comparatively new study of the geographical distribution of living things, which has only become possible since the prosecution of the systematic and scientific explorations of the earth's surface, which have distinguished the present century. The results of this branch of inquiry alone have been sufficient to convince many naturalists of the unsoundness of the old view of the distinct origin of species, whether created each in the region of the globe to which it is now confined, or, as many still imagine, all in one spot, from which they have spread themselves unchanged in form, colour, or other essential attributes to their present abodes, however diverse in climate and other environments or conditions of existence.

4. Lastly, though most important of all, must be mentioned the entirely new science of palæontology, opening up worlds of organic life before unknown, also showing infinite gradations of structure, but mainly important as increasing our horizon of observation to an extent not previously dreamt of in the

direction of time. Powers of observation formerly limited to
the brief space of a few generations are now extended over
ages, which the concurrent testimony of various branches of
knowledge, of astronomy, cosmogony, and geology, show are
immeasurable compared with any periods of which we hitherto
had cognisance. We are enabled to trace, and every year, as
discovery succeeds discovery, with increasing distinctness,
numerous cases of sequences of modification running through
groups of animals in successive periods of time, such as the
gradual progress in the development and perfection of the
antlers of deer, from their entire absence in the earliest known
representatives of the type, through the simple conical or
bifurcated form, increasing in complexity as time advanced to
the magnificent many-branched appendages which adorn the
heads of some species of recent stags; such also as the
progressive modifications, so often described, beginning in the
short-necked, heavy-limbed, many-toed tapir-like animal of the
Eocene period, and ending in the graceful, long-necked, light-
limbed, single-toed horse of our own age, and numerous others
which time will not allow me even to mention.

It would be impossible here to trace the history of the
effect of this enormous influx of knowledge upon the doctrine
of the separate origin and fixed characters of species; to
narrate the scattered efforts of philosophical minds, dis-
contented with the former views, but not yet clearly seeing
the light ; to describe the slow and struggling growth of the
new views, amid difficulties arising from imperfections of knowledge and the opposition of prejudice, or to apportion to each
of those who by their labours have contributed to the final
result, his exact share in bringing it about. How much, for
instance, is due to the work and the writings of our illustrious
countryman, Darwin ? and how much to those who have
preceded or followed him ? All this forms an episode in the
history of the progress of human knowledge, which has been
abundantly chronicled elsewhere.

The result may, however, be briefly stated to be, that the
opinion now almost, if not quite, universal among skilled and
thoughtful naturalists of all countries, and whatever their
beliefs upon other subjects, is that the various forms of life

which we see around us, and the existence of which we know
from their fossil remains, are the product, not of independent
creations, but of descent with gradual modification from pre-
existing forms. In short, the law of the natural descent of
individuals, of varieties, races, or breeds (which being within
the limits of the previous powers of observation, was already
universally admitted), has been extended to the still greater
modifications constituting what we call species, and consequently to the higher groups called genera, families, and
orders. The barrier fancied to exist between so-called varieties
and so-called species has broken down.

Any one commencing the study of the subject at the
present time without prejudice, and carefully investigating the
evidence upon which to form his conclusions, bearing in mind
that he must look for his proofs, not so much in direct
experiments or absolute demonstrations, which from the nature
of the case are impossible, but in the convergence of the
indications furnished by the interpretations of multitudinous
facts of most diverse kinds, must find it extremely difficult to
place himself in the position of those who held the older view,
so much more reasonable, so much more in accordance with
all that we know of the general phenomena of nature, does
this new one seem. In fact the *onus probandi* now appears
entirely to lie with those who make the assertion that species
have been separately created. Where, it may be asked, is the
shadow of a scientific proof that the first individual of any
species has come into being without pre-existing parents ?
Has any competent observer at any time witnessed such an
occurrence ? The apparent advent of a new species in geological history, a common event enough, has certainly been
cited as such. As well might the presence of a horse in a
field, with no sign of other animals of the same kind near it,
be quoted as evidence of the fallacy of the common view of
the descent of individuals. Ordinary observation tells us of
the numerous causes which may have isolated that horse from
its parents and kindred. Geologists know equally well how
slight the chances of more than a stray individual or fragment
of an individual here and there being first preserved and
afterwards discovered to give any indication of the existence

of the race.   Those who object to the new view complain
sometimes of the frequency with which its advocates take
refuge, as they call it, in the "imperfection of the geological
record."   I think, on the contrary, the difficulty is always to
allow sufficiently for this imperfection.   When we contrast
the present knowledge of palæontology with what it was fifty
or even ten years ago—when we see by what mere accident
as it were, a railway driven through a new country, a quarry
worked for commercial purposes, a city newly fortified, so many
of the most important discoveries of extinct animals have been
made—we must be convinced that all arguments drawn from
the absence of the required links are utterly valueless.   The
study of palæontology is as yet in its merest infancy, the
wonder is that it has already furnished so much, not so little,
corroboration of the doctrine of transmutation of species.

Direct proof is, then, equally absent from both theories.
For the old view it may be said that it has been held for a
very long time by persons whose knowledge of the facts of
nature which bear upon it was extremely limited.   On the
other hand, the new view is continually receiving more
support as that knowledge increases, and furnishes a key to a
vast number of otherwise inexplicable facts in every branch of
natural history, in geological and geographical distribution,
in the habits of animals, in their development and growth,
and especially in their structure.   I will take one instance
from the last named—the anatomy of the whale.   How is it
possible, upon any other supposition than that it is the
descendant of some land animal, with completely developed
limbs and teeth, which has become gradually modified to suit
an aquatic mode of existence, to explain the presence of the
numerous rudimentary, and to their present possessors
absolutely useless, structures found in its body.   Amongst
others, a complete set of teeth, existing only in embryonic life,
entirely disappearing even before birth, and rudimentary hind
legs, with their various bones, joints, and muscles, of which
no trace is seen externally.   It may be asserted that the
whale was originally created so, as it was asserted, and long
maintained, that fossil shells and bones were originally
created as such in the rocks in which they are found.   It

K

took more than two centuries of continuous and most acrimonious discussion to convince the world, especially the theological world, that these were the actual remains of animals which had once lived in a former period of the earth's history. Their evidence is now, however, universally admitted as supplying knowledge of the changed conditions of the surface of the earth, and with equal clearness do these rudimentary organs, hidden in the secret recesses of the whale's body, furnish, to those who inquire, indications that the animal has passed through phases of existence unlike those in which we now see it.

I do not for a moment assert that the new view explains everything that we students of nature are longing to know, or that we do not everywhere meet with obscure problems and perplexing difficulties, facts that we cannot account for, and breaks in the chain of evidence. As to the details and mode of operation of the secondary laws by which variation and modification have been brought about, we are far from being in accord. Happy for us that it is so, or our work would be at an end. I only maintain that the transmutation view removes more difficulties, requires fewer assumptions, and presents so much more consistency with observed facts, than that which it seeks to supersede, and is, therefore, so generally accepted, that there is no more probability of its being abandoned, and the old doctrine of the fixity of species revived, than that we should revert to the old astronomical theories which placed the earth in the centre of the universe, and limited the date of its creation to six ordinary days.

The question of the fixity or the transmutation of species is a purely scientific one, only to be discussed and decided on scientific grounds. To the naturalist, it is clearly one of extreme importance, as it gives him for the first time a key to the interpretation of the phenomena with which he has to deal. It may seem to many that a question like this is entirely beside the business of a Church Congress, as it is one with which only those expert in the ways of scientific investigation, and deeply imbued with knowledge of scientific facts, could be called upon to deal. This would certainly have been my view, if it had not been that some who, from their capacities and education, should have been onlookers in such a controversy, awaiting the issues of the conflict while the lists are being fought out by the trained knights, have rushed into the fray, and by their unskilful interposition have only confused the issues, casting about dust instead of light. In the hope of clearing away some of this dust the present discussion has been decided upon.

It is self-evident that a solid advance of any branch of knowledge must, in some way or other, and to a greater or less degree, influence many others, even those not directly connected with it, and therefore the rapid simultaneous strides of so many branches of knowledge as may be embraced under the term of "Recent Advances in Natural Science," will be very likely to have some bearing upon theological beliefs. Whether in the direction of expanding, improving, purifying, elevating, or in the direction of contracting, hardening, and destroying, depends not upon those engaged in contributing to the advance of science, but upon those whose special duty it is to show the bearing of these advances upon hitherto received theological dogmas. The scientific questions themselves may well be left to experts. If the new doctrines are not true, there are plenty of keen critics among men of science ready to sift the sound from the unsound. Error in scientific subjects has its day, but it is certain not long to survive the ordeal, yearly increasing in severity, to which it is subjected by those devoted to its cultivation. On the other hand, the advances of truth, though they may be retarded, will never be stopped by the opposition of those who are incompetent by the nature of their education to deal with the evidence on which it rests. There is no position so dangerous to religion as that which binds it up essentially with this or that scientific doctrine, with which it must either stand or fall. The history of the reception of the greatest discoveries in Astronomy and Geology, the passionate clinging to the exploded pseudo-scientific views on those subjects supposed to be bound up with religious faith, the fierce denunciations of the advocates of the then new, but now universally accepted ideas, are well-worn subjects, and would not be alluded to but for the repetition, almost literal repetition in some cases, of that reception which has been accorded to the new views of biology.

Ought not the history of those discoveries and the controversies to which they gave rise, to be both a warning and an encouragement?   Those who hoped and those who feared that faith would be destroyed by them, have been equally mistaken ; and is it not probable that the same result will follow the great biological discoveries and controversies of the present day ?

In stating thus briefly what is the issue of these discoveries, as generally understood and accepted by men of science, I have done all that I promised, and must leave, in far more competent hands, the part of the subject especially appropriate for discussion at this meeting.   I may, however, perhaps be allowed to put a few plain and simple considerations before you, which may have some bearing upon the subject, and which have no pretensions to novelty, though being often lost sight of, their repetition may do no harm.

I said at the commencement of this paper that it has long been admitted by all educated persons, whatever their religious faith may be, that that very universal, but still most wonderful process, the commencement and gradual development of a new individual of whatever living form, whether plant, animal, or man, takes place according to definite and regularly acting laws, without miraculous interposition.   Further than this, I believe that every one will admit that the production of the various races or breeds of domestic animals is brought about by similar means.   We do not think it necessary to call in any special intervention of creative power to produce a short-horned race of cattle, or to account for the difference between a bull dog and a greyhound, a Dorking and a Cochin-China fowl.   The gradual modifications by which these races were produced, having taken place under our own eyes as it were, we are satisfied that they are the consequence of what we call natural laws, modified and directed in these particular cases by man's agency.   We have even gone farther, having long admitted, without the slightest fear of producing a collision with religious faith, that variation has taken place among animals in a wild state, producing local races of more or less stable and permanent character, and brought about by the influence of food, climate, and other surrounding circumstances.

The evidences of the Divine government of the world, and of the Christian faith, have been sufficient for us, notwithstanding our knowledge that the individual was created according to law, and that the race or variety was also created according to law. In what way then can they be affected by the knowledge that the somewhat greater modifications, which we call species, were also created according to law? The difficulties, which to some minds seem insuperable, remain exactly as they were; the proofs, which to others are so convincing, are entirely unaffected by this widening of scientific knowledge.

Even to what is to many the supreme difficulty of all, the origin of man, the same considerations are applicable. Believe everything you will about man in his highest intellectual and moral development, about the nature, origin, existence, and destiny of the human soul; you have long been able to reconcile all this with the knowledge of his individual material origin according to law, in no whit different in principle from that of the beasts of the field, passing through all the phases they go through, and existing long before possessing, except potentially, any of the special attributes of humanity. At what exact period and by what means the great transformation takes place no one can tell. If the most Godlike of men have passed through the stages which physiologists recognise in human development without prejudice to the noblest, highest, most divine part of their nature, why should not the race of mankind, as a whole, have had a similar origin, followed by similar progress and development, equally without prejudice to its present condition and future destiny? Can it be of real consequence at the present time, either to our faith or our practice, whether the first man had such an extremely lowly beginning as the dust of the earth, in the literal sense of the words, or whether he was formed through the intervention of various progressive stages of animal life?

The reign of order and law in the government of the world has been so far admitted that all these questions have really become questions of a little more or a little less order and law.   Science may well be left to work out the details as it may.   Science has thrown some light, little enough at present, but ever increasing, and for which we should all be thankful, upon the processes or methods by which the world in which we dwell has been brought into its present condition.   The wonder and mystery of Creation remain as wonderful and mysterious as before.   Of the origin of the whole, science tells us nothing.   It is still as impossible as ever to conceive that such a world, governed by laws, the operations of which have led to such mighty results, and are attended by such future promise, could have originated without the intervention of some power external to itself.   If the succession of small miracles, formerly supposed to regulate the operations of nature, no longer satisfies us, have we not substituted for them one of immeasurable greatness and grandeur ?

A PRACTICAL LESSON FROM BIOLOGICAL
STUDIES [1]

IN the name of the British Association for the Advancement
of Science, I beg to thank you, and those societies which have
joined together in presenting this address, for the kind terms
in which it is expressed and has been delivered to me. It is
most gratifying to us—who believe in the beneficial results
upon human life conferred by the diffusion of scientific
knowledge, as all members of this great association must do—
to find that our work, and the objects of our meeting are so
highly appreciated by such a wide circle as that represented
by the signatories to this address. It is, moreover, particularly satisfactory to find such a recognition—where, perhaps,
at first sight, we might hardly have expected it—of the
advantages of pure or abstract science apart from its practical
application to the material welfare of mankind. You have
recognised what is certainly known to those who have
followed most closely the history of science, but what is not
so generally known or appreciated by others, that nearly all
the marvellous benefits which have been conferred on man by
the application of scientific knowledge have been the results
of the discoveries of philosophers who are pursuing knowledge
solely for its own sake—without any hope of reward, without
any hope of benefit to themselves or to others, and, very often,
amid the indifference, the neglect, and even the scorn of their

---

[1] Reply to an Address presented to the President of the British Association
by representatives of the various Trade Societies constituting the Newcastle-on-
Tyne, Gateshead, and District Trades-Council at the Working Men's Meeting,
Newcastle-on-Tyne, 14th September 1889.

contemporaries. The particular branch of science which it is my lot to represent is, at first sight, very little specially connected with the general welfare of man, and is looked upon by many as little more than idle speculation or mere curiosity. I well remember—though it is certainly many years ago now —one who, more than any one else living in this country, has advanced that branch of science—Professor Huxley—in a lecture delivered at the Royal Institution, saying that the common idea of a naturalist was " a wet, dirty man poking about the sea-shore with a net in one hand and a bottle in the other—an innocent and perhaps harmless individual, but a very useless one." Well, I may say that the description was made some thirty years ago, before Professor Huxley himself had done so much to raise the character of naturalists and natural history in this country, yet it still holds good in the opinion of many at the present time. You must recollect, however, that the researches of naturalists of that class, men who have occupied themselves in closely observing the ways and habits, and studying the structure of animals of a low type of organisation, have produced already marvellous results upon the happiness and welfare of mankind. Through researches of this kind we are obtaining knowledge of the causes and prevention of disease, which, when further advanced, for they are only beginning now, will, I have no doubt, lead to an enormous saving of health and of life. Moreover, through the researches of such naturalists greater results still have been produced. They have produced effects upon our mode of thinking on many subjects—on our relations to each other and to the universe—effects the end of which we hardly see at present. And they have taught us one great lesson, one that I alluded to in the address which I had the honour of giving here last Wednesday—namely, that progress in living nature has been due in great measure to the principle which Darwin most popularised, if he did not first enunciate, which he, at all events, brought into the condition in which we now know it—that of the survival of the fittest in the struggle for existence. Now, it is a law in nature that there should be a certain amount of individual differences or variations in the different animals and plants inhabiting the earth, and that the progress from

the lower to the higher forms of animals and plants has been
due to the opportunity of those individuals who are a little
superior in some respects to their fellows of asserting that
superiority, of continuing to live, and of propagating that
superiority as an inheritance to their descendants.   That law,
established in nature, is, I believe, equally applicable to
ourselves ; and this is the message which pure and abstract
biological research has sent to help us on in some of the
commonest problems of human life.   The lesson is this—that
there is always a certain amount of variability, that there is
no such thing as equality—equality in powers of work,
equality in powers of endurance, or equality in the powers of
men for doing great things in the world ; and that progress
depends on giving full liberty to that inequality, wherever it
asserts itself, having full play.   Now, supposing this law did
not exist in the animal kingdom, instead of the world being
filled with all the diversity and beauty which it now possesses,
every living thing would have been in the condition of slimy
polyps at the bottom of the sea, and if it did not exist in mankind, we should all still be in the condition of flint-chipping
savages.   I will leave it to you to apply to yourselves, in
your own social condition and social life, the application of
that law.   I will not press it any further at present, but
leave you to think it out afterwards.   You will, however, easily
see that it means that, as there is no such thing as equality
among yourselves, if you are to have progress there must be no
attempt whatever to keep down the capacities of the superior
to the level of the inferior.   Any man who gets a little rise
above his fellows helps on the progress of the world, and
brings all the others on with him.   Now, having merely
mentioned an application of abstract science from a quarter
where, perhaps, you little expect it, I will conclude by once
more thanking you, Mr. Girling, and those who have forwarded
the address in the name of the Association.   I am sure it will
give the Association great pleasure to hear what you have
done, and how you have sympathised with their work, and to
assure you that the hopes and wishes expressed at the end of
the address are, as far as possible, most fully realised.

XI

ON PALÆONTOLOGICAL EVIDENCE OF GRADUAL MODIFICATION OF ANIMAL FORMS[1]

I NEED scarcely say that one of the greatest, if not absolutely the greatest problem which has ever exercised the minds of naturalists is that of the fixity or the mutability of species.

Are the various specific forms under which animal and vegetable life exist upon earth, now and in all times past, fixed within certain narrow limits of variation, and did each originally appear upon the earth without genetic connection with any previously existing forms, having been created *de novo* in fact? or have these different species been produced by gradual modification from pre-existing living forms, under the influence of certain laws, at present very imperfectly understood, acting through vast and indefinite periods of time?

It is clear that these two views are strongly opposed to each other. Both have been held and still are held by men who are justly considered masters in the branch of knowledge to which they relate; and the solution of the question will exercise so important an influence on the progress of zoology that any real contribution towards it should be one of the most welcome additions to science that a naturalist of the present day can make.

[1] Lecture at the Royal Institution of Great Britain, 25th April 1873.

This essay represents the state of our knowledge twenty-five years ago. The accumulation of facts since that date, especially through the labours of the able palæontologists working in the rich fields of discovery in America has been enormous. If the essay were recast now, the scheme of classification at p. 144 would be crowded with additional names, but the main outlines would remain the same. Many gaps would be filled up, which would strengthen the argument, but it is perhaps more easily followed in the comparatively simple form in which it is here presented.

The question is, indeed, so far-reaching, so all-pervading, that it meets us everywhere in the study of every group of animal or vegetable life, and in almost every aspect in which the study can be carried out.

It bears largely upon, and is greatly illustrated by, descriptive zoology or botany. It adds vastly to the interest of the pursuit of anatomy, by calling out the meaning of rudimentary structures and so-called typical resemblances; it elucidates obscure questions relating to the habits and instincts of animals; it brings into prominence the signification of various facts of geographical distribution, and the life it throws into the study of palæontology is too obvious to need remark.

Evidences bearing upon, either for or against, the theory of *evolution* or *descent* can be collected from all these sources. I need only refer to Mr. Darwin's works, which must be familiar to you all, in illustration of the great variety and number of the branches of science which can be brought to throw light upon it. Indeed, in a subject like this, where direct observation can count for little, in consequence of the extreme shortness of the observing time of any individual compared with the enormous period required for the assumed changes, it is only by the accumulation of a vast number of facts from various sources, and observing the direction in which they all point, that anything like proof can be obtained.

Leaving aside, for the present occasion, all other sources of evidence in favour of either of these views, I propose this evening to enter only upon one which is in some respects, as all must admit, the most important, as it comes nearer than any other to show what actually has been the history of our existing species in times past; for as the most natural and conclusive way of ascertaining the method by which a nation has arrived at its present condition of society, customs, laws, etc., would certainly be to examine into the preserved records of its past history, so it must be with the present condition of animal and vegetable life.

We all know that such records have been preserved, that the solid rocks beneath our feet in many places teem with the actual remains of creatures which lived and died thousands or millions of years ago.

Why should they not yield to us the knowledge we are all so eager to acquire?

If species are and ever have been immutable, shall we not find the same hard and fast lines surrounding each as we do now? Shall we not find long series of similar forms following without change on an abrupt commencement? If the other alternative be correct, ought we not to find specimens of all the various stages through which the wonderful variety we meet with now has been brought about? Every gap which now so widely separates group from group ought to be filled up, and the various phases of modification should follow through the successive eras of geological time.

Now, there can be no hesitation in saying that the evidence of palæontology, in the present state of the science, does not reveal the last-described condition of things. Notwithstanding the vast increase of our knowledge in recent years, very many large groups of animals stand completely isolated, and the more nearly allied forms are mostly separated from each other by tolerably definite intervals.

Is, then, the question decisively answered against evolution or derivation by palæontology?

We must pause before we can join in the assertion that it is. The subject is far more complex than it may seem at first.

Before going further, a proper estimate must be arrived at of the nature and value of our evidence, and in doing this we must give full weight to the considerations derived from the " imperfection of the geological record " so strikingly elucidated in Mr. Darwin's chapter on the subject.

To those who have not fully considered this question, it is difficult to conceive how immense is the interval between our excessively fragmentary knowledge of extinct animals, and that perfect palæontological record which would imply evidence, first, of every form of life that has ever existed, and, secondly, of the period at which it existed.

If there were time, I might dwell long upon this part of the subject, but I must leave you to imagine (1) What the chances are against the fossilisation of any animal that dies. (2) What the chances of the stratum in which some fossil

PROVISIONAL CLASSIFICATION OF UNGULATA.

Phacochærus
Babirusa
Sus
Dicotyles
Palæochærus
Chærotherium
Merychyus
Anthracotherium
Merycochærus
Oreodon
Elotherium
Tragulus
Hyopotamus
Anoplotherium
Dichobune
Xiphodon
Oreodon
Caenotherium
Cebochærus
Choeropotamus
Hyæopotamus
Helladotherium
Bos
Capra
Dromotherium
Camelopardalis
Antilope
Sivatherium
Cervus
Ovis
Dicrocerus
Hippopotamus
Merycopotamus
Anthotherium
Poebrotherium
Procamelus
Auchenia
Camelus

Coryphodon
Lophiodon
Hyracotherium
Hyrachyus
Tapirus
Coryphodon
Palæotherium
Palæplotherium
Hyopotamus
Macrauchenia
Acerotherium
Hyracodon
Anchitherium
Hipparion
Rhinoceros
Equus
Protohippus
Merychippus

ARTIODACTYLA

PERISSODACTYLA

remains have been embedded being itself preserved during the constant changes going on on the earth's surface, and ultimately appearing in a situation accessible to man's research. (3) What the further chances against their being so found, even if they should have been preserved in an accessible locality.

I might refer you to the exceedingly minute portion of the earth's surface which has yet been really explored palæontologically; to the cases that are occurring every day of new and most unexpected forms and of whole species or orders, known only by an isolated individual, as the *Archæopteryx* of the Solenhofen oolite; to say nothing of more recondite speculations in the work above referred to, on the improbability of preservation of intermediate forms, owing to variation having usually been most rife during periods of elevation, when fossilisation is less likely to occur.

All these show in such a striking manner the extremely small value of negative evidence in palæontology, that I am quite justified in asking you to leave it altogether out of consideration in thinking of, or reasoning on, what is to follow.

Such being the material with which we have to deal, it will be seen that we must go to work upon it in a most careful and circumspect manner. We cannot rush at conclusions, but must be content cautiously, and often with much labour and anxiety, to piece together our facts, scrupulously observing the minutest hints, and following out the direction indicated by often very obscure signs, before we can reconstruct even an outline of the fabric from which we hope to gain an idea of the past history of the beings of which we treat.

I have selected for illustration of the subject this evening the division or order of Mammals called by naturalists UNGULATA, or hoofed animals, chiefly because it is the one of which the palæontological history—at least in the tertiary period (for beyond that we cannot trace it)—is better known than any other, and as that of which the classification,—that is, the relations of its various sub-groups to each other,—is on the whole better understood than in most other zoological divisions.

The order includes the most familiar of our domestic animals, and with the general appearance of the rest we are most of us well acquainted, thanks to the Zoological Gardens. They are the various forms of horses, asses, and zebras, the rhinoceroses and the tapirs, the pigs, hippopotamus, camels, deer, antelopes, sheep, oxen, and goats.

They are essentially herbivorous (though some few may be more or less omnivorous), and their teeth are modified accordingly. Their limbs are adapted for carrying the body in ordinary terrestrial progression, and are of very little use for any other purpose, such as climbing, seizing prey, or carrying food to the mouth. They never have clavicles or collar-bones, and their toes never exceed four in number (the digit which corresponds to the first of the complete pentadactyle foot being always wanting), and have the ends encased in hoofs instead of nails or claws. The species at present existing are very numerous, and widely diffused over the earth's surface, being wanting only in the Australian province. These Ungulate animals are divided into two natural groups, each having very many characters in common, the establishment of which, though contrary to the views of the great naturalists of the beginning of this century, has been a great gain to zoological science, especially as this division pervades all the known extinct as well as recent forms; and although some forms of either group may present some partial approximation to the other, no directly intermediate species are known. It is important, therefore, to apprehend thoroughly the distinction between these groups, which have received from Professor Owen the names of *Perissodactyle*, or odd-toed, and *Artiodactyle*, or even-toed, from one of their most striking external characteristics. The first have the toes of both feet arranged symmetrically to a line drawn through the middle of what would be the third toe of the typical pentadactyle foot, which toe is always the largest, and in some cases the only one fully developed. In the second, the toes are arranged symmetrically to a line drawn between the third and fourth toes, so that these two toes are equally developed, and may be alone present, or may be supplemented by an outer pair (the second and fifth), often in a more or less rudimentary condition. Besides these distinctions in the limbs, there are so many others correlated with them in the number of the vertebræ, the structure of the cranial bones, of the teeth, of the digestive organs, etc., that there can be no question about their forming natural divisions, very important to palæontologists, as it often happens that the position of an extinct and little-known form can be determined from a very small fragment of bone.

Each of these groups is further divided into genera, the names of which in what appears, in the present state of knowledge, to be their natural position and relation to each other are indicated on the diagram.   From this it will be seen that the existing Perissodactyles (excluding Hyrax, the position of which is doubtful, though I am inclined to consider it as an aberrant member of this group) consist of three groups, the tapirs, the rhinoceroses, and the horses, each represented by but few species, and (except in the case of the horse, through the agency of man) of rather restricted geographical distribution.

These groups at present are separated by very decided intervals, so much so, that one of them, containing the horses, has been considered by many naturalists as forming an order apart, the *Solidungula*.

The existing Artiodactyles range themselves around two principal types, the tubercular-toothed, or bunodont, and the crescentic-toothed, or selenodont.   To the former belong the pigs in all their modifications, including the babirussa and wart-hogs, and the hippopotamus and peccary.   To the latter the ruminants, *i.e.* vast numbers of species of animals included under the general designation of sheep, oxen, goats, antelopes, deer, musks, giraffes, and the two allied though aberrant forms, the camels, and the *Tragulidæ* or chevrotains, an interesting little group long confounded with the musk deer.

The two extremes of this division, represented by the pigs and the hollow-horned ruminants, seem to have very little in common at first sight, and if we were acquainted with the organisation only of the existing species, we might be justified in treating them as belonging to very distinct groups.   But even among existing forms there are some examples, which may almost be called intercalary types, so widely do they depart from the group to which they are most nearly relative in the direction of the other.

These are among the bunodonts, the little South American peccaries (*Dicotyles*), and among the selenodonts, in a far greater degree, the chevrotains (*Tragulus*). The latter in many remarkable characters deviate strongly from the ruminants, and approach the pigs, or rather, as will be shown presently, to the generalised type of the entire group.

Such being the present condition of the order, what does palæontology reveal of its past history ?

In the first place, it is most necessary to bear in mind the provisional character of all classifications of extinct animals, because of our imperfect knowledge of their structure ; but endeavouring to make the best use of what little we possess, I have added in the scheme opposite all the best-known extinct forms somewhere near the position, in relation to the existing forms and each other, in which their affinities would place them, and have shown by the different shading their relation as regards time.

The tertiary period, with which we are now alone concerned, has here been divided for convenience into six epochs. Of course, it were possible to have gone into minute details and made many more divisions, but it would have made the diagram less clear, and it is best, perhaps, not to attempt to refine too much in this somewhat tentative exposition of a biological history, especially as there is still much uncertainty as to the exact relative age of many of our fossiliferous strata.

The epochs chosen are the recent (including the pleistocene), the pliocene, late and early miocene, and late and early eocene, each represented by a different shading. It is not meant that if a genus or group is here assigned to one of these epochs, that some of its members may not have extended in some degree beyond its limits (as it must be always remembered that the boundaries of these epochs are quite artificial), either before or after, but that the period assigned to it was that in which it most chiefly flourished. When two shadings are represented, one within another, it signifies that the group existed in both, and of course in all intervening periods.

To begin with the Perissodactyles.   The earliest known
forms constitute a family called *Lophiodontidæ*, composed of
the genera *Lophiodon*, *Coryphodon*, and *Hyracotherium*.   Of
these animals little is known except the teeth, which, however,
indicate rather a primitive or root form, from which, by
modification, all the other teeth of Perissodactyles can be
derived.   The elevations and depressions of the molar teeth of
Lophiodon, for instance, are arranged on a pattern which is
the best key to that of all others of the sub-order; and it is
by going back, as it were, to it that we can understand and
compare all the other variously modified, and often more complicated, forms.   Moreover, these Lophiodonts possess a dental
character which distinguishes them from all other *Perissodactyles*, and brings them into a more generalised ungulate type,
for which reason I place them nearest to the earlier forms
of artiodactyles—that is, that all the premolars are smaller
and of a simpler form than the true molars.   Whether they
possessed any modification of the limbs or other structures which
bear them out in this position, we unfortunately cannot say.

At a somewhat later epoch in the earth's history appeared
on the scene the *Palæotheriidæ*, an important group, containing
animals the osseous structure and dentition of which are
completely known, chiefly through the famous researches of
Cuvier into the fossils found in the gypsum quarries at
Montmartre.   These were animals something like existing
tapirs, with three toes on each foot, complete and distinct
radius and ulna and tibia and fibula, complete typical number
of teeth, *i.e.* $i \frac{3}{3} c \frac{1}{1} p \frac{4}{4} m \frac{3}{3} = 44$; but the molar teeth
modified in pattern from that of the Lophiodonts.   They
flourished in the later eocene, after which period they are no
longer met with.   They have been divided into several genera,
but Gaudry has shown that these are united by transitional
forms, and present a gradual series of modifications, corresponding with successive geological epochs.   Another offset,
from the ancient Lophiodont stock (with which it appears
to be connected through the American eocene *Hyracahyus*),
constitutes the family *Tapiridæ*, first known in the miocene
and continued with scarcely any modification to our own day,
and therefore a most interesting form to contemplate in its

L

living state, as it brings back, in the most striking way, the
general facies of the fauna of those ancient times. In one
respect the tapir is remarkable among Perissodactyles, as it has
on its fore-feet as many as four toes, thus retaining a primitive
or generalised character. The other two existing forms, the
rhinoceros and the horse, appear to be more direct modifications
of the Palæotherium type, though in different directions. The
existing rhinoceros closely resembles the Palæotherium in the
general structure of its skeleton, limbs, number of toes, etc.,
and in the general pattern of the molar teeth: it differs,
however, in the greatly reduced number of front teeth, incisors
and canines, which in the African two-horned species are often
absolutely wanting; and also in the possession of those
singular epidermal appendages to the face, the well-known
horns, either one or two in number. Now palæontology
points out with tolerable precision the intermediate steps by
which these modifications have been brought about. A small
ancient rhinoceros has been found in the early miocene of
North America, to which Leidy has given the name of
Hyracodon, which had no horn, and had the complete number
of incisor and canine teeth, and was in many ways, at least as
far as the skull and teeth are concerned, intermediate between
Palæotherium and *Rhinoceros* proper. The earlier known
European rhinoceroses have had the name *Aceratherium* given
to them, the small size of the nasal bones being apparently
quite unfitted to support such a weapon as a horn. The
resemblance of their skull to *Palæotherium* has been pointed
out by H. v. Meyer.

The more recent fossil rhinoceroses present wonderfully
intermediate forms between some of the existing species, as
R. pachygnathus of Pikermi, as Gaudry has shown, is about
equally related to the two species of modern African rhinoceros, and might have been (upon the derivative hypothesis)
the ancestor of both. In the same way the Himalayan *R.
sivalensis* appears to be related to the modern *R. indicus* and
sondiacus, and the *R. schleirmacheri* to the Asiatic two-horned
species. One special line of variation indicated chiefly by the
ossification of the nasal septum culminated in the *R. ticho-
rhinus*, which became extinct only in the most recent geological

epoch. The history of this small group alone in its bearings
upon evolution might occupy many lectures; I must content
myself now only with one observation, borrowed from Mr.
Boyd Dawkins, that, in all modern rhinoceroses the molar
teeth have deeper crowns than in those which existed prior
to a certain epoch, so that the height of these teeth alone will
serve to distinguish a pleistocene from a pliocene form, in
other respects closely allied. The value of this observation
will be illustrated in the sequel.

The next line of modification from *Palæotherium*, is that
which culminates in the most specialised of mammals, the
modern horse, an animal we are so accustomed to look at
that we scarcely ever notice the most remarkably adaptive
character of its structure for its special mode of life. If
we were not acquainted with the horse (and here of course I
include its immediate allies, the asses and zebras) we could
scarcely conceive of an animal whose only support was the
tip of a single toe on each extremity, to say nothing of the
singular conformation of its teeth and other organs. So
striking have these characters appeared to many zoologists
that the animals possessing them have been reckoned as an
order apart called *Solidungula*; but palæontology has revealed
that in the structure of its skull, its teeth, its limbs, the horse
is nothing more than a modified *Palæotherium*; and though
still with gaps in certain places, many of the intermediate
stages of these modifications are already known to us, being
the *Paloplotherium, Anchitherium, Merychippus,* and *Hipparion.*
On this very interesting point, which looks more like a real
genealogical history than any other known, however, I need
not dwell, as it was so fully treated of in a lecture delivered
in this theatre three years ago by Professor Huxley—a lecture
entitled the " Pedigree of the Horse."

Lastly, there is *Macrauchenia,* a curiously modified Perissodactyle found in pleistocene times in South America, apparently another derivative of the palæotherium type, presenting
resemblances (though perhaps only analogical) to some of the
artiodactyles, especially the camels.

Directly intermediate forms between *Macrauchenia* and
the other animals of its group are not yet known; but

considering how little evidence we have of the animal life of
the middle or older tertiaries of South America, this is not to
be wondered at.

On the whole it will be seen that, taking actual anatomical
characters alone, palæontological research, even so far as it
has yet been carried, bridges over most of the gaps existing
between the modern forms of Perissodactyles, entirely abolish-
ing, for instance, the order *Solipedia*, as it is impossible to draw
a satisfactory line where the animal ceases to be equine and
becomes a palæotheroid; some drawing it between *Anchitherium* and *Palæotherium*, some between *Anchitherium* and
Hipparion. Moreover, and this is most important, the lines
from the modern more specialised forms converge towards the
ancient more generalised forms; so that if we could get a
side-view of what is shown in the diagram, the earliest forms
at the bottom and the latest at the top, we should have lines
(broken, it is true, here and there) diverging from a common,
or near a common centre, towards a circumference above—a
view, in fact, of the conventional genealogical tree.

We turn now to the *Artiodactyles*, represented at present
by the scattered groups before spoken of, clustering round two
type-forms so widely sundered in their structure and habits
as the pig and the ox; but the former history of this division
yields a totally different state of things. Of early eocene
Artiodactyles we know very little at present; but in the
later divisions of the same epoch forms appeared, such as
Anoplotherium, Dichobune, Chæropotamus, and *Hyopotamus,*
which were certainly neither pigs nor ruminants, but which
partook remarkably of the characters of both. They had
the complete number of teeth, *i.e.* incisors and canines, like
modern pigs, but molars with indications of the crescentic
pattern so characteristic of ruminants. They had two or four
toes; but the metacarpals and metatarsals were not united to
form a cannon bone as in ruminants, and they wanted the
horny appendages to the head, so usually met with in the
modern representatives of that group. From some of these
central forms, or more probably from a still earlier allied
group indicated by the genus *Acotherulum,* or by some other
still undescribed remains from Mauremont, transitions can be

traced with few breaks, through the successively modified miocene genera *Chœrotherium* and *Palæochœrus* to the genus *Sus*, or true pig, in which the dentition undergoes some remarkable specialisations, as the upturning of the upper canines, and great development and extremely tuberculated character of the posterior molars, which are both singularly exaggerated in some modern offsets of the pig family, the first in the babirussa, and the second in the wart-hog (*Phacochœrus*). More distantly related to the true pigs are the hippopotamus on the one hand, and the peccary on the other. In relation to the first, not found anteriorly to the latest miocene, it is significant that the earliest known forms had the more generalised number of incisor teeth (six) instead of four as in the modern hippopotamus, and hence has been made into a genus by itself, called *Hexaprotodon*.

The researches of Leidy into the ancient (miocene and early pliocene) fauna of Nebraska have furnished evidence of a remarkable group of animals now entirely extinct, the *Oreodontidæ*, the characters of which are perfectly inter-mediate between those of the pigs and the ruminants : animals with pig-like feet and complete number of incisors, canines, and molars, but with the latter important set of teeth, formed precisely on the same type as those of the deer. Within this particular group Leidy has noted a curious series of slight modifications coinciding with the successive age of the strata, in which the remains were found. *Agriochœrus*, the most ancient, approaches nearer to *Chœro-potamus*, has orbits open behind and very shallow-crowned teeth. Then follows *Oreodon* proper, and lastly *Merychyus*. more like the modern ruminants.

To return to the European forms, in the genus *Gelocus*, where the union of the two principal bones of the metapodium first occurs, Kowalevsky has noticed the gradual way in which this change seems to have been brought about in successive epochs of eocene and early miocene strata, at first free in the young, and only coalescing in old animals, afterwards coalescing at a much earlier age. The gradual perfecting of the foot by the development of the ridge round the lower articular end of the metapodium in later forms, the ridge

being quite wanting in early forms of the same group, has
been noticed by the same author in many different series of
Ungulates.

During the miocene period the peculiar dental characteristics of the modern ruminants, especially the loss of the upper
incisors, were developed, all selenodont artiodactyles henceforth
showing it.   Of this early race of imperfect ruminants,
still retaining many generalised characters, especially in the
skull, the cervical vertebræ, fibula, stomach, etc., the chevrotains
(*Tragulidæ*) are the survivors, especially the West African
Hyomoschus, which has existed almost unchanged since the
late miocene of Sansans and Steinheim.   Then for the first
time the appendages called antlers were introduced, but only
in a comparatively rudimentary condition, with long pedicles
and few branches, as in the modern Muntjaks.   It was
not till pliocene and especially pleistocene epochs that the
wonderful and luxuriant variety of cervine antlers reached
their full development.   As offsets of the deer group, the
giraffe, the gigantic Siwalik *Sivatherium*, and the *Helladotherium* of Greece may be mentioned, the two latter having
become extinct, apparently without descendants.

Later still, the yet more specialised forms of hollow-horned
ruminants appear—forms which now dominate the earth,
being of all Ungulates the most widely diffused and most
numerous in species, in individuals, and in outward variety,
though in essential structure all alike.   One of their principal
characteristics is the modification of their molar teeth in the
same way as in the modern horses, to which in some respects
they seem to form a parallel group.   The difference between
the molar tooth of a hollow-horned ruminant and that of a
deer consists in the great lengthening of the crown without
any change in the pattern of the enamel folds, and in the
addition of cement to support these folds.   This alteration
did not take place suddenly, and the crowns of teeth of the
artiodactyles before the time of deer were still shorter than in
those animals.

Among the deer themselves, as Lartet observed, the most
ancient have very short-crowned molars, and the depressions
on the surface are so shallow that the bottom is always

visible, while in the Cervidæ of the more recent tertiary periods, and especially the pleistocene and living species, these same cavities are so deep that, whatever be the state of dentition, the bottom cannot be seen.   This (he says) is a perfectly reliable rule for distinguishing the ancient from the more modern forms of deer, and can be applied to other animals as well as the Cervidæ.   From it he surmises that the duration of the life of modern is greater than that of ancient deer.   The same careful observer also remarks that in many natural groups a gradual progress is observed in the volume of the brain and complexity of its surface, as deduced from casts of the interior of the skull, from which fact he concludes that a gradual growth of vital energy and intelligence has occurred as the effect of the tendency towards improvement, of which the cause is always acting, and the limits indefinite.

Thus the history of the Even-toed Ungulates tells the same story as that of the Perissodactyles.   The modern forms are placed along lines which converge towards a common centre.   Moreover, the lines of both groups, to a certain extent, approximate; but within the limits of our knowledge they do not meet.   Both artiodactyles and perissodactyles existed low down in the eocene, just as Carnivores, Insectivores, bats, rodents, and other great groups then existed with boundary lines as distinctly marked as now.

Was the order according to which the introduction of new forms seems to have taken place since that epoch, then entirely changed? or did it continue as far back as the period when these lines would have been gradually fused into a common centre?

Here we are landed in the region of pure speculation; but bolder travellers than I have endeavoured to penetrate its mysteries, as may be seen by a perusal of Professor Huxley's presidential address to the Geological Society for 1870.

I have so far confined myself within the region of the known, and shown that at least in one group of animals the facts which we have as yet acquired point to the former existence of intermediate forms, so numerous that they go far to discredit the view of the sudden introduction of new species.

They also show many cases of gradual modification of particular organs, probably always to the benefit of the race, and also a general progress from lower to higher or more specialised types : though, as in all other cases of progress (human civilisation, for instance), attended with many exceptions, some local and temporary, some only apparent.

Whether the inferences which seem to me to follow from these facts are true or not may still be an open question ; for the sake of the stimulus that an open question of this sort lends to scientific research I am very glad that it is so ; but if true, if we are led by them to the conclusion that the world we live in is a world of gradual growth and progress, and orderly evolution, what grander view of the Creation and the history of that world can we have opened to us ?

A CENTURY'S PROGRESS IN ZOOLOGICAL KNOWLEDGE[1]

ON the 10th of January 1778 died the great Swedish naturalist, Charles Linné, more commonly known as Linnæus, a name which will ever be mentioned with respect and regard in an assembly devoted to the cultivation of the sciences of Zoology and Botany, as, whatever may be the future progress of those sciences, the numerous writings of Linnæus, and especially the publication of the *Systema Naturæ*, can never cease to be looked upon as marking an era in their development. That work contained a systematic exposition of all that was known on these subjects expressed in language the most terse and precise. The accumulated knowledge of all the workers at Zoology, Botany, and Mineralogy since the world began, was here collected together by patient industry, and welded into a complete and harmonious whole by penetrating genius.

Exactly a century has passed since Linnæus died. What of the progress of the subjects to which he devoted his long and laborious life? This one century is a brief space compared with the ages which have passed since man began to dwell upon the earth, surrounded by living objects, which have, more and more as time rolled on, awakened his curiosity, stimulated his faculties to observe, and impelled him to record the knowledge so gained for the benefit of those to come. How does it stand in comparison with those which preceded it, in the contributions it has thus acquired and recorded?

[1] Presidential address in the Department of Zoology and Botany. British Association. Dublin Meeting, 15th August 1878.

It may be not without interest in commencing our work at this meeting to cast our eyes back and take stock, as it were, of the knowledge of a hundred years ago, and of that of the present time, and see what advances have been made; to look at the living world as it was known to Linnæus and as it is known to ourselves. The *Systema Naturæ*, the last edition of which, revised by the author, was published in 1766, will be a convenient basis for the comparison; but as the subject is one which, even in a most superficial outline, might reach such lengths as would well tire out the most patient of audiences, and absorb time which will be more profitably occupied by the valuable contributions which are forthcoming from other members of the Association, I will merely take a small section of the work, about 100 pages out of the first of the four volumes, those devoted to the first class MAMMALIA. The comparison of this part is perhaps the easiest, as the contrast is the least striking, and the progress has been comparatively the slowest. The knowledge of large, accessible, and attractive-looking animals had naturally preceded that of minute and obscure organisms, and hence, while in many other departments the advance has altogether revolutionised the knowledge of Linnæus, in the Vertebrated Classes, especially the one of which I shall now speak, it has only extended and reformed it.

In taking the *Systema Naturæ* of Linnæus, the comparison is certainly carried back somewhat beyond the hundred years which have elapsed since his death, and the brilliant contributions to the knowledge of the Mammalia of Buffon and Daubenton, just then beginning to be known, and the systematic compilation of Erxleben (published in 1777), are ignored; but for the present purpose, especially considering the limited time at my disposal, it will be best not to go beyond the actual text of the work in question.

Before considering systematically the different groups into which Linnæus divides the class, I must remark in passing upon what is the greatest, and indeed most marvellous difference between the knowledge of Zoology of our time and that of Linnæus. Now we know that the animals at present existing upon the earth are merely the survivors of an

immensity of others, different in form, characters, and mode
of life, which have peopled the earth through vast ages of
time, and to which numerically our existing forms are but
infinitesimally small, and that the knowledge we already
possess of great numbers fully justifies the expectation of an
enormous further advance in this direction.   In the time of
Linnæus the existence in any past time of a species having
no longer living representatives on the earth, though perhaps
the speculation of a few philosophical minds, had not been
received among the certainties of science, and at all events
found no place in the great work we are now considering.

In the twelfth edition of the *Systema Naturæ* we find the
class MAMMALIA divided into seven orders: I. *Primates*,
II. *Bruta*, III. *Feræ*, IV. *Glires*, V. *Pecora*, VI. *Belluæ*,
VII. *Cete*.   These orders contain forty genera without any
intermediate subdivisions.   The genera are again divided into
species, of which the total number is 220.

The first order, PRIMATES, contains four genera: *Homo*,
Simia, *Lemur*, and *Vespertilio*.

The vexed question of man's place in the zoological system
was thus settled by Linnæus.   He belongs to the class
Mammalia, and the order *Primates*, the same order which
includes all known monkeys, lemurs, and bats: he differs only
generically from these animals.   But then we must remember
that the Linnæan genera were not our genera; they correspond
usually to what we call families, sometimes to entire orders.
So that practically man's position is much the same as that to
which, after several vicissitudes, as his separation as an order
by Blumenbach and Cuvier, or as a sub-class by Owen, he has
returned in the systems of nearly all the zoologists of the
present day who treat of him as a subject for classification upon
zoological and not metaphysical grounds.

Yet since the time of Linnæus the whole science of
Anthropology has been created.   There is certainly an
attempt at the division of the species *Homo sapiens* into six
varieties in the *Systema Naturæ*, but it has scarcely any
scientific basis.   Zoological Anthropology may be said to have
commenced with Blumenbach, who, it is interesting to recall
as an evidence of the rapid growth of the science, was a

contemporary with most of us in this room, for he died as lately as 1840, although his first work on the subject, *De generis humani varietate nativa*, was published three years before the death of Linnæus, too late, however, to influence the work we are now speaking of. The scientific study of the natural history of man is therefore, we may say, but one century old. To what it has grown during that time you are probably aware. Scarcely an important centre of civilisation in the world but has a special Society devoted to its cultivation. It forms by itself a special department of the Biological Section of our Association—a department of such importance that on this occasion no less distinguished a person than a former most eminent President of the whole Association was thought fit to take charge of it. From him you will doubtless hear what is its present scope, aim, and compass. I need only remind you that except the one cardinal point of the zoological relation of man to other forms of life, which Linnæus appears to have appreciated with intuitive perception, all else that you will now hear in that department was not dreamt of in his philosophy.

As might naturally be supposed, apes and monkeys have, for various reasons, attracted the attention of observers of nature from very early times, and consequently Linnæus was able to give rather a goodly list of species of these animals, amounting to thirty-three; but of their mutual affinities, and of the important structural differences which exist between many of them, he seems to have had no idea, his three divisions being simply regulated by the condition of the tail, whether absent, short, or long.

We now know that the so-called Anthropoid or man-like apes, the gorilla, chimpanzee, orang, and gibbons, form a group apart from all the others of such importance that everything related to their history, structure, and habits has been most assiduously studied, and there is now an immense literature devoted to this group alone. Nothing could better illustrate the advances we have made in a hundred years, than the contrast of our present knowledge of these forms with that of Linnæus. It is true that, as shown in the most interesting story of the gradual development of our knowledge relating to

them in the first chapter of Huxley's *Man's Place in Nature*,
the animal now called gorilla was, without doubt, the pongo,
well known to and clearly described by our countryman,
Andrew Battle, a contemporary of Shakespeare; and that a
really accurate and scientific account of the anatomy of the
chimpanzee had been published as far back as 1699 by Dr.
Edward Tyson, who, as the first English comparative anatomist,
I am proud to claim as in some sort a predecessor in the chair
I have the honour to hold in London, as he is described on
the title-page of his work as " Reader of Anatomy at Chirurgeons'
Hall."

Linnæus was, however, not acquainted with these, and his
second species of the genus *Homo, H. troglodytes*, and his first
of the genus *Simia, S. satyrus*, were both made up of vague
and semi - fabulous accounts of the animals now known as
chimpanzees and orangs, but hopelessly confounded together.
Of the gorilla, and what is stranger still, of any of the large
genus of gibbons, or long-armed apes of South-eastern Asia, he
had at the time he revised the *Systema* no idea.

The remaining monkeys, we now know, fall into three
very distinct sections : the *Cercopithecidæ* of the Old World,
and the *Cebidæ* and *Hapalidæ* of the New, or by whatever
other names we may like to designate them. Although
members of all three groups appear in the list in the *Systema*,
they are all confusedly mixed together. Even that the
American monkeys belong to a totally different stock from
those of the Old World does not seem to have been suspected.

The genus *Lemur* of Linnæus comprehends five species, of
which the first four were all the then known forms of a most
interesting section of the Mammalia. These animals, mostly
inhabitants of the great island of Madagascar, though some
are found in the African continent, and others in some of the
Southern and Eastern parts of Asia, constitute a well-defined
group, but one of which the relations are very uncertain. At
one time, as in the system of Linnæus, they were closely
associated with the monkeys. As more complete knowledge
of their organisation has been gradually attained, the interval
which separates them structurally from those animals has
become continually more evident, and since they cannot be

placed within the limits of any of the previously constituted orders, it has been considered advisable by some naturalists to increase the ordinal divisions in their behalf and to allow them to take rank as a distinct group, related to the *Primates* on the one hand, and to the *Carnivora* and *Insectivora* on the other. The knowledge of their relations, however, bids fair to be greatly increased by the discoveries of fossil forms lately made both in France and America, some of which seem to carry their affinities even to the *Ungulata*.

Existing upon the earth at present, besides the more ordinary lemurs to which the species known to Linnæus belong, there are two aberrant forms, each represented by a single species. These are the little *Tarsius* of Borneo and Celebes, and the singular *Chiromys*, or Aye-aye, which, though an inhabitant of the headquarters of the group, Madagascar, and living in the same forests and under the same conditions as the most typical lemurs, exhibits a most remarkable degree of specialisation in the structure both of limbs and teeth, the latter being modified so as to resemble, at least superficially, those of the Rodents, a group with which in fact it was once placed. It was discovered by Sonnerat in Madagascar in 1780, two years after the death of Linnæus. The specimen brought to Paris by this traveller was the only one known until 1860. Since that date, however, its native land has been more freely open than before to explorers, and many specimens have been obtained, one having lived for several years in the Gardens of the London Zoological Society.

The history of a name is often not a little curious. Linnæus applied the term *Lemures*, *i.e.* the departed spirits of men, to these animals on account of their nocturnal habits and ghost-like aspect. The hypothetical continent in the Indian Ocean, supposed to have connected Madagascar with the Malayan Archipelago is called by Mr. Sclater, *Lemuria*, as the presumed original home of the lemur-like animals. Although the steps are not numerous, it might puzzle a classical scholar, ignorant of Zoology, to explain the connection between this continent and the Roman festival of the same name.

The fifth animal which Linnæus places in his genus *Lemur*, under the name of *L. volans*, is the very singular creature to

which the generic term *Galeopithecus* has since been applied.
It is one of those completely aberrant forms which, having no
near existing relations, and none yet discovered among extinct
forms, are perfect puzzles to systematic zoologists.   It is
certainly not a lemur, and not a bat, as has been supposed by
some.   We shrink from multiplying the orders for the sake of
single genera containing only two closely-allied species ; so we
have generally allowed it to take refuge among the *Insectivora*,
though without being able to show to which of that somewhat
heterogeneous group it has any near affinities.

The fourth genus of the PRIMATES is *Vespertilio*, comprising
six species of bats.   This genus has now by universal consent
expanded into an order, and one of the best characterised and
distinctly circumscribed of any in the class : indeed, those
who have worked most at the details of the structure of bats
find so much diversity in the characters of the skull, teeth,
digestive organs, etc., associated with the modification of the
fore-limbs for flight common to all, as almost to entitle them
to be regarded rather as a sub-class.   Anatomical as well as
palæontological evidence shows that they must have diverged
from the ordinary mammalian type at a very far distant date,
as the earliest known forms, from the Eocene strata, are
quite as specialised as any now existing, and no trace has
hitherto been discovered of forms linking them to any of the
non-volant orders.   By the publication within the last few
weeks of a valuable monograph on the existing species of the
group, entitled " A Catalogue of the Chiroptera in the Collection of the British Museum," by G. E. Dobson, we are enabled
to contrast our present knowledge with that of the time of
Linnæus.   Although the author has suppressed a large number
of nominal species which formerly encumbered our catalogues,
and wisely abstained from the tendency of most monographists
to multiply genera, he describes four hundred species, arranged
in eighty genera ; nearly double the number of species, and
exactly double the number of genera, of the whole class
MAMMALIA in the *Systema Naturæ*, and these Dr. Günther
remarks in his Preface are probably only a portion of those
existing.   The small size, nocturnal habits, and difficulty of
capture of these animals, are sufficient reasons for the supposi-

tion that there are still large numbers unknown to science. In the list of Linnæus, the first primary group of Dobson, the *Megachiroptera*, now containing seventy species, is represented by a single one, *V. vampyrus*, obviously a *Pteropus*, to which the bloodthirsty habits of the fabulous Vampyre are attributed, but which is not absolutely identified with any one of the known species. The other species described by Linnæus can almost all be identified with bats at present well known.

A curious example of the results of basing classification upon a few, and those somewhat artificial characters, is afforded by one of the true bats, now called *Noctilio leporinus*, though admitted by Linnæus to be "*simillimus vespertilionibus, similiter pedibus alatus,* being separated from the others, not only generically, but even placed in another order, that of the GLIRES or Rodents, because it did not, or was supposed not to fall under the definition of the order PRIMATES, which begins "*Dentes primores incisores superiores IV. paralleli.*" In reality this bat has four upper incisors, but the outer ones are so small as to have been overlooked when first examined. But even if this were not so, no one would now dream of basing an animal's position upon such a trivial character when opposed to the totality of its organisation and habits.

The characters of the incisor teeth are placed in the first rank in the definitions of all the orders in the *Systema Naturæ*, and hence the next order called BRUTA, characterised by "*dentes primores nulli superius aut inferius,*" contains a curious mixture of heterogeneous animals, as the names of the genera *Elephas, Trichechus, Bradypus, Myrmecophaga, Manis,* and *Dasypus* will indicate. It contains, in fact, all the animals then known comprised in the modern orders of *Proboscidea, Sirenia,* and *Edentata,* together with the walrus, one of the *Carnivora.* The name BRUTA has been revived for one of these orders, that more generally called *Edentata,* but I think very inappropriately, for it was certainly not equivalent, and if retained at all, should rather belong to the *Proboscidea,* as *Elephas* stands first in the list of genera which Linnæus assigned to the group.

It is curious to find that the striking differences between the African and the Indian elephants, now so well understood

by every beginner in Zoology, and all the facts which have
already been accumulated relating to the numerous extinct
forms of Proboscideans, whether Mammoths, Mastodons, or
Dinotheria, were quite unknown to Linnæus.   One species
only, *Elephas maximus*, represented in the zoology of a hundred
years ago all that was known of the elephants or elephant-like
animals.

The genus *Trichechus* of this edition exhibits a very curious
phase of zoological knowledge.   It contains two species—1.
T. rosmarus, the walrus, now known to be a modified seal,
and therefore a member of the Linnæan order FERÆ, and 2.
T. manatus, a name under which were included all the known
forms of manatees and dugongs, in fact the whole of the
modern order *Sirenia*,—animals widely removed in all essential
points of their organisation from the walrus, with which they
are here generically united.   Their position, however, between
the elephant on the one hand and the sloths on the other,
is far better than their association with the *Cetacea*, as in
Cuvier's system, an association from which it has been most
difficult to disengage them, notwithstanding their total dissimilarity except in a few external characters.   Although the
discovery of many fossil forms has done much to link together
the few existing species and to show the essential unity of the
group, it has thrown no light upon their origin, or their
affinities to other mammals.   They still stand, both by their
structure and their habits, in a strangely isolated position, and
it baffles conjecture to say whence they have been derived, or
how they have attained their present singular organisation.

The remaining genera of the Linnæan order BRUTA constitute the group out of which Cuvier, following Blumenbach,
formed his order *Edentata*, a name certainly not happily
chosen for a division which includes species like the great
armadillo, having a larger number of teeth than any other
land mammal, but which, nevertheless, has been so generally
adopted, and is so well understood, that to attempt to change
it would only introduce an element of confusion.   Four out
of five of the principal modifications of form in the group at
present known are indicated by the four Linnæan genera,
Bradypus or sloth, *Myrmecophaga* or ant-eater, *Manis* or

pangolin, and *Dasypus* or armadillo.   The advances during
the century have consisted in the accumulation of a great
mass of details respecting these groups; the addition of a
fifth and very distinct existing form, the *Orycteropus* or Cape
ant-eater ; and the discovery of numerous and very remarkable
extinct forms, such as the Megatheriums and Glyptodons of
South America, now so fully known by their well-preserved
osseous remains.   There is, however, still much to be done in
working out the real relationship of the somewhat isolated
members of the order, if it be a natural order, both to each
other, and to the rest of the Mammalia, from which they
stand widely removed in many points of organisation.

The third order of Linnæus, FERÆ, contained all the then
known animals, which, with whatever diversities of general
structure, agreed in their predatory habits, and possessed
certain general characters of teeth and claws to correspond,
though the terse definition of " *Dentes primores superiores sex,
acutiusculi canini solitarii*," is by no means universally
applicable to them.   This order was broken up by Cuvier into
the orders *Carnivora* and *Insectivora*, and the genus *Didelphys*,
included in it by Linnæus, has been since by universal assent
removed to another group.

The first six genera belong to the very well-defined and
probably natural group now called *Carnivora*.   The one placed
at the head of the list, *Phoca*, is equivalent to the large and
important modern sub-order *Pinnipedia*,—the walrus, however,
though essentially a seal, having been, as before mentioned,
relegated by Linnæus to another order, on account of its
aberrant dentition.   But three species are recorded in the
genus; *P. ursina*, the sea-bear of the North Pacific (now
Otaria ursina); *P. leonina*, founded on Anson's sea-lion, now
commonly called the elephant seal, or sea-elephant (*Macrorhinus
proboscideus*, or more properly *leoninus*); and *P. vitulina*, the
common seal of our coasts.

The terrestrial sub-order of *Carnivora* is represented by
five genera: 1. *Canis*, including the dog, wolf, hyæna, fox,
arctic fox, jackal, etc.   2. *Felis*, with only six species, but
still one of the few Linnæan genera, which covers exactly the
same ground as at present in the opinion of the majority of

zoologists, although it may be mentioned as an example of the tendency towards excessive and unnecessary multiplication of generic names which exists in some quarters, that it has been divided into as many as fourteen.   3. *Viverra*, a heterogeneous group, containing ichneumons, coatis, and skunks, animals belonging to three very distinct families, according to modern ideas.   4. *Mustela*, a far more natural group, being nearly equivalent to the modern family *Mustelidæ;* and, lastly, a very comprehensive genus *Ursus*, consisting of *U. meles*, the badger, *U. lotor*, the raccoon, *U. luscus*, the wolverene, and all the true bears known, comprised in the single species *U. arctos*. Many interesting forms of *Carnivora*, as *Cryptoprocta*, *Proteles*, *Eupleres*, *Ailurus*, and *Ailuropus*, have no place in the Linnæan system, being comparatively modern discoveries. The very recent date (1869) at which the last-named remarkable animal was made known to science by the enterprising researches of the Abbé David into the Fauna of Eastern Thibet, gives hope that we may not yet be at the end of the discovery of even large and hitherto unsuspected forms of existing mammals.

Next in the Linnæan system comes the genus *Didelphys*, constituted for the reception of five species of American opossums.   This is a very interesting landmark in the history of the progress of the knowledge of the animal life of the world, as these five opossums, forming a genus in the midst of the order FERÆ, were all that was then known of the great sub-class *Marsupialia*, now constituting a group entirely apart from the ordinary members of the class.   It is difficult now to imagine an animal world without kangaroos, without wombats, without phalangers, without thylacines, without dasyures, and so many other familiar forms, and yet such was the animal world known to Linnæus.   It is true that a species of kangaroo from one of the islands of the Austro-Malayan Archipelago was described as long ago as 1714 by De Bruyn, who saw it alive at the house of the Dutch governor of Batavia, and that Captain Cook and Sir Joseph Banks saw and killed kangaroos on the east coast of Australia in 1770, and had published figures and descriptions of them in 1773, or five years before the death of Linnæus; but the work we are now considering contains no traces of knowledge of the existence of such a remarkable and now so well-known animal.

The three remaining genera of FERÆ, *Talpa, Sorex,* and *Erinaceus,* contained all the known species of the present order INSECTIVORA, which now embraces many and very varied forms, quite unsuspected a century ago, and to which it is probable that many others will be added by the time the exploration of the animal products of the world is completed.

The fourth order, GLIRES, has remained practically unchanged to our day, although the name *Rodentia* has generally superseded that bestowed upon it by Linnæus. The five genera of the *Systema Naturæ, Hystrix, Lepus, Castor, Mus,* and *Sciurus,* have been vastly increased, partly by subdivision and partly by the discovery of new forms. *Noctilio* is, as before mentioned, removed to the Chiroptera, but its loss is well compensated for by *Hydrochœrus,* the well-known Capybara, the largest existing member of the group, which in the Linnæan system is placed among the Belluæ, in the same genus with the pigs.

The fifth Linnæan order, PECORA, is a fairly natural group, equivalent to Cuvier's *Ruminantia;* but it is no longer considered of the value of an order, since the animals composing it have now been shown to be as closely related to certain of those belonging to the next order as they are to each other. The first genus, *Camelus,* contains both the American lamas and the Old World camels, the demonstration of the common origin and close affinities of which has been one of the important results of the recent discoveries in the palæontology of the Western continent. In the next genus, *Moschus,* were placed the well-known musk deer of the highlands of Central Asia, and two small African antelopes, which have no special affinity with it. The subsequent inclusion in the same genus of the small chevrotains (*Tragulinæ*), which was very natural at the time, as they agree perfectly with the musk in the absence of horns and the presence of large canine tusks, by which artificial characters the genus was defined by Linnæus, was one of those unfortunate associations which have greatly retarded the progress of knowledge of the true affinities of the

group. Judging by the popular works on Zoology, it is still
as difficult to apprehend that a chevrotain is not a musk deer,
as it is that a manatee is not a cetacean,—both errors of the
same kind, if not quite so gross, as that of regarding a whale
as a fish, or a bat as a bird. The Linnean genus *Cervus*
contains six species of true deer, including the moose, rein-
deer, red deer, fallow, and roe, associated with the giraffe.

The twenty-one species of the great group of hollow-horned
Ruminants, at that time recognised, are distributed quite
arbitrarily in three genera, *Capra*, *Ovis*, and *Bos*. Though
subsequent investigations have greatly increased the number
of species known, we are still in much uncertainty about their
mutual affinities and generic distinctions. Being a group of
comparatively modern origin, and only just attaining its
complete development, variation has chiefly affected the less
essential and superficial organs. Moreover, the process of
extinction of intermediate forms has not operated sufficiently
long to break it up into distinctly-separated natural minor
groups, as is the case with many of the older families, which
lend themselves, therefore, far more readily to systematic classification, especially as long as the extinct forms are unknown
or ignored.

The sixth order of land mammals, BELLUÆ, corresponding
to the *Pachydermata* of Cuvier, contains what is now known
to be a heterogeneous collection, viz. the horses, the hippopotamus, the pigs, rhinoceroses, and the rodent capybara. The
abolition of these two last orders and the entire rearrangement
of the ungulate mammals into two different natural groups,
now called *Artiodactyla* and *Perissodactyla*, first indicated by
Cuvier in the "Ossemens fossiles," from the structure of the
limbs alone, and afterwards confirmed by Owen from comparison
of every part of the organisation, has been one of the most
solid advances made in our knowledge of the classification of
the Mammalia during the present century.

The past history of this, as of so many other groups of
vertebrated animals, has been brought to light in an unexpected
manner by the wonderful discoveries of fossil remains made
during the last ten years in the Rocky Mountains of
America,—discoveries, the importance of which will only be

fully appreciated when the elaborate and beautifully illustrated work which Professor Marsh has now in progress is completed.

The last Linnæan order, CETE, is exactly conterminous with the order so named, or rather more generally modified to *Cetacea*, in the best modern systems, for Linnæus did not commit the error of Cuvier and others, of including the *Sirenia* among the whales. His knowledge of the animals composing the group was necessarily very imperfect; indeed it is only within the last few years, especially since the impulse given to their study by Eschricht of Copenhagen, that the great difficulties which surround the investigation of the structure and habits of these denizens of the open sea have been so far surmounted that we have begun to obtain clear views of their organisation, affinities, and geographical distribution.

Two most remarkable forms of mammals, so abnormal in their organisation as now to be generally considered deserving the rank of a distinct sub-class, the *Echidna* and *Ornitho-rhynchus*, were first made known to science in 1792 and 1799 respectively, and consequently have no place in the *Systema Naturæ*. The very recent discovery of a third form to this group, or at least a very striking modification of one of the forms, the large New Guinea echidna (*Acanthoglossus bruynii*), is the last important acquisition to our knowledge of the class.

In this brief review of the progress of one small section of one branch of zoological knowledge, it will be seen that it is chiefly of systems of arrangement, of classification, and of names, that I have been treating. By many biologists of the present day these are looked upon as the least attractive and least profitable branches of the subject. The interest of classification, though it has lost much in some senses by the modern advances of scientific biology, has, however, gained vastly in others. The idea that has now, chiefly in consequence of the writings of Darwin, taken such strong hold upon all working naturalists—the idea of a gradual growth and progressive evolution, and therefore genetic connection between all living things—breaks down the artificial barriers which zoologists raise around their groups, and shows that such names as *species, genera, families, orders*, etc., are merely more or less clumsy attempts to express various shades of differences among creatures connected by infinite gradations, and in this sense destroys the importance attached to them by our predecessors. On the other hand, it immensely increases the interest contained in the word "relationship," as it implies that the word is used in a real and not, as formerly, in a metaphorical sense. There is a kind of classification, such as we might apply to inanimate substances or manufactured articles. We may say, for instance, that a tumbler, a wine-glass, and a tea-cup are more closely related to each other than either one is to a chair or a table, and that they might be formed into one group, and the last-named objects be placed in a second. This kind of classification is certainly useful in its way for methodical arrangement and descriptive purposes. It is the kind of arrangement which Linnæus and his contemporaries applied to animals. It is, however, a very different classification from that which supposes that the members of a group having common essential characters are descended from a common ancestor, and have gradually, by whatever cause or means, become differentiated from other groups. On this view a true classification, if it could be obtained, would be a revelation of the whole secret of the evolution of animal life, and it is no wonder that many are willing to devote so large a share of their energies in an endeavour to attain it.

The right application of the principles of nomenclature, first clearly established by Linnæus, to the groups we form is, again, by no means to be despised, as laxity and carelessness in this respect are becoming more and more the greatest hindrances to the study of Zoology. The introduction of any new term, especially a generic name, and indeed the use of an old one by any person whose authority carries weight, has an appreciable effect upon the progress of science, and should never be made without a full sense of the responsibility incurred. All beginners are puzzled and often repelled by the confused state of zoological nomenclature to an extent to which those who have advanced so far as only to care for the things, and to whom the actual names by which they are called are comparatively indifferent, have little idea. Those whose special gift or inclination leads them to the pursuit of other branches of Biology, as Morphology, Physiology, or Embryology, must have definite names for the objects they observe, depict, or describe, and are dependent upon the researches of the systematic zoologist for supplying them, and should not neglect to take his counsel, otherwise much of their work will lose its value.

Several times has the British Association thought this a worthy subject for the consideration of its members, and through the instrumentality of a committee of working naturalists an excellent code of regulations and suggestions on the subject of zoological nomenclature was drawn up in 1842. These rules were revised and reprinted in 1865 ; and in accordance with a resolution adopted at the last annual meeting at Plymouth they have been again republished at the cost of the Association during the present year. The mere issue of such rules must have had a beneficial effect, as they have undoubtedly been a guide to many careful and conscientious workers. Unfortunately no means exist of enforcing them upon those of a different class, but there is still something wanting, short of enforcing them, which possibly may be within the power of the Association to effect. In the administration of the judicial affairs of a nation, besides the makers of the laws, we have an equally essential body to interpret or apply the law to particular cases—the judges. However carefully compiled or excellent a code of regulations may be, dubious and difficult cases will arise, to which the application of the law is not always clear, and about which individual opinions will differ. The necessary permission given in the Association rules to change names which are either "glaringly false," or "not clearly defined," opens the door to considerable latitude of private interpretation. As what we are aiming at is simply convenience and general accord, and not absolute justice or truth, there are also cases in which the rigid law of priority, even if it can be ascertained, requires qualification, and others in which it may be advisable to put up with a small error or inconvenience to avoid falling into a larger one. I may name such cases as the propriety or the reverse of reviving an obsolete or almost unknown name for one which, if not strictly legitimate, has been universally accepted, or the retention of a name when already applied to a different genus, instead of the institution of another in its place. For instance, should the name *Echidna*, by which the well-known Monotrematous Mammal is known in every text-book and catalogue in every language, be superseded by *Tachyglossus*, because the former name had previously been applied to a genus of snakes? or should the chimpanzee be no longer called *Troglodytes* lest it should be confounded with a wren? Should *Chiromys* be discarded for *Daubentonia*, *Trichechus* for *Odobenus*, and *Tapirus* for *Hydrochœrus*? Should the Java slow lemur be called *Loris*, *Stenops*, or *Nycticebus*? Should Sowerby's whale be placed in the genus *Physeter*, *Delphinus*, *Delphinorhynchus*, *Heterodon*, *Diodon*, *Aodon*, *Nodus*, *Ziphius*, *Micropterus*, *Micropteron*, *Mesodiodon*, *Dioplodon*, or *Mesoplodon*, in all of which it may be found in various systematic lists? Should one of the largest and best known of the Cetaceans of our seas be called *Balœnoptera musculus*, *Physalus antiquorum*, or *Pterobalœna communis*, all names used for it by authors of high authority? Should the smallest British seal be called *Phoca hispida*, *fœtida*, or *anellata*?

I might go on indefinitely multiplying instances which will be answered differently by different naturalists, the arguments for one or the other name being often nicely balanced. What 'is wanted, therefore, is some kind of judicial authority for deciding which should in future be used. If a committee of eminent naturalists, selected from various nations, and divided into several sections, according to the subjects with which each member is most familiar, could be prevailed upon to take up the task of revising the whole of our existing nomenclature upon the basis of the laws issued by the Association in 1842, occasionally tempering their strictly legal decisions with a little discretion and common sense, and with a view, as much as possible, of avoiding confusion and promoting general convenience,—and if the working zoologists of the world generally would agree to accept the decisions of such a committee as final,—we should dispose of many of the difficulties with which we are now

troubled. There seems to me no more reason why the nomenclature of such a committee, if it were composed of men in
whose judgment their fellow-workers would have confidence,
should not be as universally accepted as is the nomenclature
of the last edition of the *Systema Naturæ* of Linnæus. We
have agreed not to look beyond that work for evidence of
priority, and why should we not agree in the same way to
accept decisions which would probably be arrived at with even
fuller knowledge and greater sense of responsibility ?

Whether this suggestion will be received with favour or
not, it appeared to me that it was one not inappropriate for
the consideration of this Section, which has already dealt with
the question in a manner so advantageous to science, and also
for this year, which has witnessed the hundredth anniversary of
the death of the great teacher of systematic zoology.

Our knowledge of the living inhabitants of the earth has
indeed changed since that time. Our views of their relations
to the universe, to each other, and to ourselves, have undergone
great revolutions. The knowledge of Linnæus far surpassed
that of any of his contemporaries; but yet of what we now
know he knew but an infinitesimal amount. Much that he
thought he knew we now deem false. Nevertheless, some of
the oldest words to be found in all his writings contain
sentiments which still claim a response in the hearts of many.
Although we are less accustomed to see such words in works
of science, that is no proof that their significance has been
impaired by the marvellous progress of knowledge. With the
words which Linnæus selected to place at the head of his great
work I will conclude—

> *O Jehova,*
> *Quam ampla sunt tua opera !*
> *Quam sapienter ea fecisti !*
> *Quam plena est terra possessione tua !*

THE ZOOLOGICAL SOCIETY OF LONDON[1]

NOWHERE has the progress which the world has made during the fifty years of Her Majesty's reign, the completion of which we are now happily celebrating, been more strikingly manifested than in the advance of that so-called "natural knowledge" for the improvement of which our Royal Society was instituted more than two centuries ago. Although there have been, without doubt, immense strides in other directions —in morals, in art, in historical and literary criticism—I venture to say that none of these can be compared with the marvellous progress that has been made in scientific knowledge and scientific methods.

The tangible results that have followed the practical applications of mechanics, physics, and chemistry have so deeply affected the material interests of mankind, that the progress of these branches of knowledge may seem to put into the shade the wonderful changes that have taken place in the kindred sciences. Nevertheless, I think we may safely say that Zoology, in a certain sense one of the oldest of human studies, has in these latter times undergone a new birth, which has not only changed the standpoint from which we view the special objects of our studies, but has also spread its influence far and wide, and profoundly modified our conceptions on many questions at first sight entirely remote from its sphere. The universal abandonment of the doctrine of fixity of species, which was an article of faith with almost every zoologist in 1837, has introduced new interests, as well, it

[1] Address to the General Meeting of the Society, held in the Zoological Gardens in celebration of the Fiftieth Anniversary of Her Majesty's reign, 16th June 1887.

must be confessed, as new difficulties, the extent of which we are only beginning to appreciate. The definite systems of classification and methods of nomenclature on which our fathers relied utterly fail before the wider field of vision which it is the privilege, as well as the embarrassment, of the present generation of zoologists to realise.

But it is no part of my intention, in the brief space of time for which I shall ask your patience, to attempt to give a history of the recent advances of zoological science in general, but only, as requested by your Council, to say a few words on the progress of the particular Institution established for its cultivation in which we are personally interested, and the duration of which is so nearly contemporaneous with that of Her Majesty's reign.

Before this Society was founded there was no distinct organisation in the country devoted solely to collecting, recording, and discussing the facts upon which zoological science rests. The dignified parent of all our scientific societies, the *Royal*, certainly undertook, as it does still, the discussion of many zoological subjects; but it could not be expected to treat them in any detail. The Linnæan was a society of great respectability, devoted solely to biological research, both zoological and botanical, already nearly forty years of age, and possessed of all the usual appurtenances of a scientific organisation—meetings, library, and collections for reference. I cannot help thinking that if its leading Fellows had, at that time, displayed more energy, it might have kept in its hands the principal direction of the biological studies of the country, instead of allowing what has since proved so formidable a rival to spring up, and to absorb so large a portion of its useful functions. However, for reasons which it is perhaps not worth while to inquire into now, it did not supply all the needs of the lovers of Zoology; and in the year 1826 an active and zealous band united together, and, as the Charter tells us, "subscribed and expended considerable sums of money for the purpose" of founding the *Zoological Society of London*.

The leading spirit of this band was Sir Stamford Raffles, then just returned from the administration of those Eastern islands of which the history, both natural and political, will ever be intimately associated with his name. He was chosen for the office of President, but his death, on the 5th of July 1826, deprived the Society, while yet in its infancy, of his valuable services even some years before it acquired its Charter of Incorporation. In this deed, dated 27th March 1829, Henry, Marquis of Lansdowne, is named as the first President of the chartered Society, Joseph Sabine as the first Treasurer, and Nicholas Aylward Vigors the first Secretary.

The Society appears to have acquired great popularity in a surprisingly short time. The first printed list of members that I can discover (dated 1st January 1829) contains the names of 1294 ordinary Fellows and 40 honorary and corresponding members. The list is an interesting one, from the number of names it includes of persons eminent either in science, art, literature, politics, or social life ; indeed, there were not many people of distinction in the country at that time who are not to be found in it.

A piece of ground in the Regent's Park having been obtained from the Government at little more than a nominal rent, the Gardens were laid out, and opened in 1828, during which year 98,605 visitors are recorded as having entered. In the following (the first complete) year there were as many as 189,913 visitors, and this number was increased in 1831 to 262,193.

While the menagerie of living animals was being formed in the Regent's Park, the Officers and Fellows of the Society were also engaged in establishing a museum of preserved specimens, which soon assumed very considerable dimensions. A catalogue printed as early as the year 1828 contains a classified list of 450 specimens of Mammalia alone; and it continued for many years to attract donations from travellers and collectors in all parts of the world, and became of great scientific importance, inasmuch as it contained very many types of species described for the first time in the publications of the Society. It was at first lodged in rooms in the society's house in Bruton Street; but these becoming so crowded as to present the " confused air of a store rather than the appearance of an arranged museum," premises were taken in 1836 in Leicester Square, the same which were formerly occupied by the museum of John Hunter before its removal to the College of Surgeons. At this time the museum is reported to have contained as many as 6720 specimens of vertebrated animals, and numerous additions were still being made both by donations and by purchase. The rooms in Leicester Square being found inconvenient for the purpose, it was finally resolved, after considerable discussion of various sites, to transfer the collection to the Gardens in the Regent's Park; and in 1843 the building which is now occupied as a lecture-room on the upper floor and a storeroom below was constructed and fitted up for its reception.

Although the museum was at one time looked upon as a very important part of the Society's operations, being spoken of as "the centre of the Society's scientific usefulness" (Report of Council, 1837), and one upon which considerable sums of money were spent, it was afterwards a cause of embarrassment, from the difficulty and expense of keeping it up in a state of efficiency; and when the Zoological Department of the British Museum acquired such a development as to fulfil all the objects proposed by the Society's collection, the uselessness of endeavouring to maintain a second and inferior zoological museum in the same city became apparent, and in 1856 it was, as I think very wisely, determined to part with the collection, the whole of the types being transferred to the National Museum, and the remaining specimens to other institutions, where it was thought their value would be most appreciated.

Another enterprise in which the Fellows of the Society were much interested in its early days was the farm at Kingston, the special object of which was thus defined: "It will be useful in receiving animals which may require a greater range and more quiet than the Gardens at the Regent's Park can afford. It is absolutely necessary for the purpose of breeding and rearing young animals and giving facilities for observations on matters of physiological interest and research, and, above all, in making attempts to naturalise such species as are hitherto rare or unknown in this country." The farm, however, apparently not fulfilling the objects expected of it, and being a source of expense which the Society could not then well afford, was gradually allowed to fall into neglect, and finally abandoned in 1834.

The mention of this establishment, however, causes me to allude to one of the objects on which the Society laid considerable stress at its foundation, and which is defined in the Charter as "the introduction of new and curious subjects of the Animal Kingdom," but which, as may be gathered from the Annual Reports of the Council and from other documents, meant not only the temporary introduction of individuals for the purpose of satisfying curiosity about their external characters and structure, but also the permanent domestication of foreign animals which might become of value to man, either for their utility in adding to our food-supplies or for the pleasure they afford by their beauty.

Abundant illustrations of the vanity of human expectations are afforded by the details of the hopes and disappointments recorded in the reports of the Society relating to this subject. It is mentioned in the report of the year 1832 that "the armadillo has three times produced young, and hopes are entertained of this animal, so valuable as an article of food, being naturalised in this country." More than fifty years have passed, and British-grown armadillo has not yet appeared upon the menu-cards of our dinner-tables. At one time the South American curassows and guans were confidently looked upon as future rivals to our barn-door fowls and turkeys. Various species of pheasants and other game birds from Northern India, collected and imported at great expense, were to add zest and variety to the battue of the English sportsman. The great success which for many years attended the breeding of giraffes in the Gardens not unnaturally led to the expectation that these beautiful creatures might become denizens of our parks, or at all events a source of continued profit to the Society ; and it is possible that some who are here now may have been present at the feast, for which an eland was sacrificed, amid loudly-uttered prognostications that the ready acclimatisation of these animals would result, if not in superseding, at least in providing a change from our monotonous round of mutton, beef, and pork. Unfortunately for these anticipations, no giraffe has been born in the Gardens during the last twenty years, and elands are still far too scarce to be killed for food of man in England.

It is well that these experiments should have been tried; it may be well, perhaps, that some of them should be tried again when favourable opportunities occur; but it is also well that we should recognise the almost insuperable difficulties that must attend the attempt to introduce a new animal able to compete in useful qualities with those which, as is the case of all of our limited number of domestic animals, have gradually acquired the peculiarities making them valuable to man, by the accumulation of slight improvements through countless generations of ancestors. While all our pressing wants are so well supplied by the animals we already possess, it can no longer pay to begin again at the beginning with a new species. This appears to be the solution of the singular fact, scarcely sufficiently appreciated, that no addition of any practical importance has been made to our stock of truly domestic animals since the commencement of the historic period of man's life upon earth.

I now turn to the history of one of the most important features of the Society, the scientific meetings. In the early days of the Society there was only one class of general meetings for business of all kinds; and the exhibition of specimens and the communication of notices on subjects of zoological interest formed part of the ordinary proceedings at those meetings. The great extent, however, of the general business was soon found to interfere with such an arrangement. The number of the elections and of the recommendations of candidates, the reports on the progress of the Society in its several branches during each month, and other business, were found to require so much time as to leave little for scientific communications, and the Council saw with regret that these were frequently and necessarily postponed to matters of more pressing but less permanent interest. To obviate this inconvenience and to afford opportunities for the reception and discussion of communications upon zoological subjects, the Council had recourse to the institution of a " Committee of Science and Correspondence," composed of such members of the Society as had principally applied themselves to science,—at the meetings of which communications upon zoological subjects might be received and discussed, and occasional selections made for the purpose of publication.

The first meeting of the Committee took place on the evening of Tuesday, 9th November 1830, at the Society's house in Bruton Street, when a communication was received upon the anatomy of the uran-utan by a young and then unknown naturalist, Richard Owen by name. This was the first of that long series of memoirs, extending over a period of more than fifty years, the publication of which in our *Transactions* has done so much to advance the knowledge of comparative anatomy and to give an illustrious place to their author in the annals of science.

Among the names of others who are mentioned as having taken part in the business of the Committee during the first year of its existence, either by their actual presence or by forwarding communications, are N. A. Vigors, W. Yarrell, J. E. Gray, J. Gould, E. T. Bennett, Andrew Smith, Bryan H. Hodgson,[1] T. Bell, W. Martin, Joshua Brookes, W. Kirby, W. H. Sykes, Marshall Hall, W. Ogilby, and John Richardson.

The Committee continued in existence for two years, having met for the last time on 11th December 1832. The success of its meetings was so great that it was thought desirable to make an alteration in the by-laws, by which the meetings of the Committee were replaced by the " General Meetings of the Society for Scientific Business." The first of these meetings took place on Tuesday, the 8th of January 1833, and they have continued to be held on two Tuesdays in each month during the season to the present time. As long as the Society retained its house in Bruton Street, the meetings were held there. In 1843 the Society took another house, which it occupied for forty-one years, No. 11 Hanover Square ; but its needs having outgrown the accommodation afforded there, it removed in 1884 to the far more spacious and commodious premises in No. 3 of the same square, which we at present occupy. These meetings of the Society, which are open to all the Fellows and to friends introduced

---

[1] Who was present at the reading of this address.

by them, have exercised a considerable influence upon the
progress of zoological knowledge, not only by the reading and
discussing of communications formally brought before them,
but also by the interchange of ideas at the informal social
gatherings over the coffee-table in the Library afterwards,
which have great value as affording a common meeting-ground
and bond of union for all the working zoologists of the
country, as well as for many visitors from foreign lands.

The more important scientific communications to these
meetings have from the commencement been published in
the form of quarto *Transactions* and octavo *Proceedings,*
which constitute a series of inestimable importance both for
the value of the material contained in them and for the
excellence of the illustrations of new or rare forms of animal
life with which they are embellished.   In later times they
have also formed a vehicle for communicating to the world the
important results obtained from the dissection of animals
which have died at the Gardens, and which, since the
establishment of the office of Prosector in 1865, have been
systematically used for this purpose.

In connection with the scientific meetings must be mentioned the Library, the first formation of which is described
in the report of the Council for the year 1837, and which
has been steadily growing ever since by donations of books,
by exchange of publications with other learned Societies, and
by judicious annual expenditure of money, until it has become
one of the best-selected, well-arranged, and most accessible collection of works of reference that it is possible for the zoological
student to enjoy.   Its value has been greatly increased by the
publication within the past month of an excellent catalogue,
which contains the titles of about 6560 works.

The most recent addition to the functions that the Society
has undertaken with a view to carry out the purposes of
its foundation is the publication of an Annual Record of
Zoological Literature, containing a summary of the work done
by British and Foreign naturalists in the various branches of
Zoology in each year, a publication of the utmost value to the
working zoologist.   Such a Record has been carried on for
some years past by a voluntary association of naturalists, but,

owing to the difficulties met with in obtaining sufficient
support, it was in danger of being abandoned, until the
Council, after the full consideration which the importance of
the subject deserved, resolved to take it in hand as part of the
operations of the Society.

The Society has, however, not only been mindful of
advancing scientific knowledge—it has also endeavoured to
spread some of this knowledge in a popular manner by means
of lectures. In former years these were only given in an
occasional manner; but the liberal bequest of Mr. Alfred
Davis to the Society in 1870 has enabled the Council to
undertake a more regular and systematic method of instruction; and the Fellows and others have had every summer
for several years past the opportunity of hearing many of our
most eminent naturalists and able expositors upon subjects
which they have made especially their own. I regret, however,
to add that the interest taken by the Society generally in
these lectures has not quite equalled the expectations that
were raised when the question of establishing them was first
brought before the notice of the Council.

Although, as will be seen by a consideration of the various
subjects which I have already referred to, the Society has a
wide sphere of operation and many methods by which the
objects of its founders are carried out, it is undoubtedly the
maintenance of the menagerie of living animals in the Gardens
where we are now assembled by which it is best known both
to the public as well as to a large number of our Fellows. It
will be well, therefore, before concluding, to add a few words
upon some points of interest connected with the past history
and present condition of this branch of the Society's operations,
the one which is at the same time the largest source of its
revenue and cause of expenditure.

The collection and exhibition of rare and little known
living animals has long been a subject of interest and instruction in civilised communities, and in many countries either
the State or the Sovereign has considered it as part of their
duty or privilege to maintain a more or less perfect establishment of the kind.

Before the Zoological Society was formed the " lions " at

the Tower had been for centuries a national institution; and
it may be interesting to those who derive pleasure in tracing
the links between the present and the past, to be reminded
that our collection is in some measure a lineal continuation of
that time-honoured institution, as it appears from the Reports
of the Council that in the year 1831 His Majesty King
William the Fourth " was graciously pleased to present to the
Society all the animals belonging to the Crown lately maintained at the Tower."   It is also recorded that in the previous
year His Majesty had made a munificent donation of the
whole of the animals belonging to the Royal Menagerie kept
in Windsor Park.   This may perhaps be the place to mention
that in the Report read April 1837 the Council " had the
gratification to call the special attention of the members to
a donation from Her Royal Highness the Princess Victoria,"
consisting of a pair of those pretty and interesting little
animals, the Stanley musk-deer.   During the fifty years that
have elapsed since this first-recorded mark of interest in the
society on the part of her present Majesty, the Queen and
her family have never failed to show their regard for its
welfare whenever any opportunity has arisen, of which the
acceptance of the Presidency by the late Prince Consort, on
the death of the Earl of Derby in 1851, was one of the
most signal instances.   The advantages which the Society has
received from the numerous donations to the Menagerie, and
the constant kindly interest shown in its general progress by
H.R.H. the Prince of Wales, are so continually before the
observation of the Fellows, that I need scarcely do more than
allude to them here, beyond stating that in no year of the
Society's existence has the number of visitors to the Gardens
or the Society's income been so great as in 1876, when the
large collection of animals brought from India by His Royal
Highness formed the special object of attraction.

Except for the collection, necessarily of limited extent,
exhibited in the Tower, and a few others having their origin
in commercial enterprise, as Mr. Crosse's menagerie at Exeter
Change and the various itinerant wild-beast shows, there were,
before the foundation of the Society's Gardens, little means in
the country of gaining knowledge of the strange forms of

exotic animal life with which the world abounds. An extensive, well-arranged, and well-kept collection, where the circumstances of exhibition were more favourable than in the institutions just referred to, seemed then to fulfil a national need, as the rapidly-acquired popularity of the Society already alluded to testifies. Indeed, when we consider the amount of enjoyment and instruction which has been afforded to the 24,572,405 visitors who are registered as having entered our Gardens from their first opening in 1828 to the end of last year, it is easy to realise what a loss the country would have sustained if they had not existed. There was a period, it is true, in which they fell rather low in popular favour, the record of 1847 showing both the smallest number of visitors and the lowest income of any year in the Society's existence. A new era of activity in the management of the Society's affairs was then happily inaugurated, which resulted in a prosperity which has continued ever since, with only slight fluctuations, arising from causes easy to be understood—a prosperity to which the scientific knowledge, zeal, and devotion to the affairs of the Society of our present Secretary, ably seconded in all matters of detail by the Resident Superintendent, have greatly contributed.

Among the great improvements which have been gradually effected in the Gardens in recent years is the erection of larger, more commodious, and more substantial buildings for the accommodation of the animals than those that existed before. A few examples will suffice to illustrate the successive steps that have been taken in this direction. The primary habitation of the lions and other large feline animals was the building near the north-east corner of the Gardens, which many of us may remember as a Reptile-house, and which has been lately restored as a dwelling-place for the smaller Carnivora. The Council Reports of the period frequently speak of the bad accommodation it afforded to the inmates, the consequent injury to their health, and the disagreeable effects on visitors from the closeness of the atmosphere. In September 1843 the terrace, with its double row of cages beneath, was completed; and the report of the following spring, speaking of this as " one of the most important works ever undertaken at the

Gardens," congratulates the Society upon the fact that the
anticipations of the increased health of this interesting portion
of the collection, resulting from a free exposure to the external
air and total absence of artificial heat, have been fully realised.
The effects of more air and greater exercise were indeed said
to have become visible almost immediately.   Animals which
were emaciated and sickly before their removal became plump
and sleek in a fortnight after, and the appetites of all were
so materially increased that they began to kill and eat each
other.   This, however, led to an immediate increase in their
allowance of food, since which time, it is stated, no further
accidents of the kind have occurred.   As this structure,
looked upon at that period as so great an improvement upon
its predecessors, still remains, though adapted for other
inmates, we all have an opportunity of contrasting the size
of its dens and the provision it affords generally for the
health and comfort of the animals and the convenience of
visitors, with those of the magnificent building which superseded it in 1876.

In the report of the year 1840 it is stated that the only
work of considerable magnitude undertaken since the last
anniversary was the erection of the " New Monkey-house,"
and the Council speak with great satisfaction of the substantial nature of the structure and the superior accommodation
which its internal arrangements were calculated to afford to
its inmates.

Many of us may remember this building, which stood on
the space now cleared in the centre of the Gardens.   Twenty-
four years after its erection, in their report, dated April
1864, we find the Council speaking of it as "what is at
present perhaps the most defective portion of the Society's
Garden establishment," and the erection of a second " New
Monkey-house " was determined upon.   This is the present
light and comparatively airy and spacious building, the
superiority of which over the old one in every respect is
incontestable.

Up to the year 1848 the only attempt which had been
made to familiarise the visitors with the structure and habits
of animals of the class Reptilia was by the occasional display

of a pair of pythons, which were kept closely covered in a box of limited dimensions in one of the smaller Carnivora-houses. In 1849 the building which had been rendered vacant by the removal of the lions to the new terrace was fitted up with cases with plate-glass fronts on a plan entirely novel in this country, and which for many years afforded an instructive exhibition of the forms, colours, and movements of many species of serpents, lizards, and crocodiles. This house was a vast improvement upon anything of the kind ever seen before; but the contrast between it and the present handsome and spacious building so recently erected in the south-eastern corner of the grounds affords another illustration of the great progress we are making.

If time allowed I might also refer to the Elephant-house, completed in 1870, to the Insect-house, opened in 1881, and to various others of less importance.

The erection of these houses has necessarily been a very costly undertaking; in fact, since what may be called the reconstruction of the permanent buildings of the Gardens, which commenced in the year 1860, more than £50,000 has been expended upon them. It is only in years of great prosperity, when the Society's income has considerably exceeded its necessarily large permanent expenditure, that works such as these can be undertaken.

Much as has been done in this direction, we must all admit that there is still more required. The buildings of to-day will, we may even hope, some day seem to our successors what the former ones appear to us. The old idea of keeping animals in small cramped cages and dens, inherited from the Tower and travelling wild-beast shows, still lingers in many places. We have a responsibility to our captive animals, brought from their native wilds to minister to our pleasure and instruction, beyond that of merely supplying them with food and shelter. The more their comfort can be studied, the roomier their place of captivity, the more they are surrounded by conditions reproducing those of their native haunts, the happier they will be, and the more enjoyment and instruction we shall obtain when looking at them. Many of our newest improvements are markedly in this direction. I may especially mention the new inclosure for wild sheep near the Lion-house in the South Garden, with its picturesque rockwork and fall of water, and the large Aviary for herons and similar birds just completed on what used to be called the Water-Fowls' Lawn.

All such improvements can, however, only be carried out by the continued aid of the public, either by becoming permanently attached to the Society as Fellows or by visiting the Gardens. I trust that this brief record of the principal events of the Society's history will show that such support is not undeserved by those who have had the management of its affairs.

WHALES, AND BRITISH AND COLONIAL WHALE
FISHERIES [1]

WHEN asked by your Council to lecture upon some subject connected with natural history, it occurred to me that the great link between Britain and all her Colonies was the ocean, and that, therefore, something concerning its animal inhabitants might be interesting to those whose avocations and situation in life call them to traverse its pathless ways. Even in an ordinary voyage, such as is necessitated by any intercourse between a Colony and the Mother Country, some familiarity may be acquired with the gigantic denizens of the deep; but the knowledge of them so gained is generally so slight and superficial that I venture to think that some more accurate and definite information about them would be welcome. The subject I have chosen is, however, so large that I can only, in the limited time allowed on one evening, select a few of the more interesting features in the history of these remarkable animals, both from the point of view of the naturalist and in their relation to human civilisation and commerce.

In the admirable sketch, by our greatest living naturalist, of the history of the science of comparative anatomy during the present century, appended to the recently published *Life of Sir Richard Owen*, Professor Huxley says: "Take, for example, the question whether a whale is a fish or not, which, I observe, is not yet quite settled for some people. As the whale is not a little like a fish outside, and lives permanently in the sea, why should it not be classed with the fishes?" Before proceeding to answer this question, I must ask another. In

---

[1] Lecture at the Royal Colonial Institute, 8th January 1895.

what sense do you use the term " fish "?   It happened to me
a few years ago to receive a semi-official inquiry from the
Colonial Office, as to whether a lobster was a fish, because
an important point in the dispute between the French and
English about the Newfoundland fisheries depended upon the
interpretation of an old treaty in which the word " fish "
occurs.   After giving the modern naturalists' definition of a
fish, by which a lobster is clearly excluded from the class, of
course I felt it necessary to remind my correspondents that in
such a case the real answer to the question lay in the sense in
which the word was used at the time of the treaty, and by
those who were parties to drawing it up, and if that could be
ascertained it would be more to the point than the strictest of
scientific definitions.   Now, on turning to what was, in the
beginning of the present century, our greatest authority on
the meaning of words, I find in *Johnson's Dictionary* (I now
quote from Todd's edition, 1818) "*fish*" defined as " an animal
that inhabits the water."   Without doubt this was the general
and popular view, as the universally used expressions *shell-fish*,
lobster and oyster *fisheries*, whale *fisheries*, and even seal
fisheries, abundantly testify.   I therefore cannot say that in
a certain vague and antiquated sense of the word, " fish " may
not be applied to the animals of which I propose to speak
to you this evening.   This must not, however, cause us to
forget that, tested by the light of modern scientific knowledge,
a whale is in everything essential in its structure entirely
removed from the class of animals to which zoologists now
restrict the term " fish," a very clearly defined group of cold-
blooded creatures, breathing by means of gills the air which
is dissolved in the water in which they swim, with lowly
organised brain, and producing their young from eggs, and
after they are born not nourishing them by the mother's milk
—in all of which, and many other important characters,
whales are entirely removed from them.   In fact, as Professor
Huxley continues, in response to the question with which I
stopped the quotation :—

The answer, of course is, that the moment one compares a whale with
any one of the thousands of ordinary fishes, the two are seen to differ in
almost every particular of structure and, moreover, in all those points

in which the whale differs from the fish, it agrees with ordinary mammals. Therefore the zoologists put the whales into the same class with the mammals, and not into that of the fishes. But this conclusion implies the assumption that animals should be arranged according to the totality of their resemblances. It means that the likenesses in structure of whales and mammals are greatly more numerous and more close than the likenesses between whales and fishes.

It also means, if the derivative hypothesis of animal species is true, that the whale is far more nearly related to, say, a horse or a cow than it is to a cod or a shark.

It is decided, then, that, from the point of view of a zoologist at all events (whatever the fisherman and the man of business may continue to say), the whales and their allies belong to the class *Mammalia*, and not to that of *Pisces*. We can easily fix their place in that class as constituting a distinct and clearly defined order, the *Cetacea*, derived from the Latin word *cetus* = a whale. Although the term "whale" is generally, if somewhat vaguely, restricted to the larger and middle-sized members of the order, the smaller ones, commonly called "dolphins" and "porpoises," to all intents and purposes belong to it, and no line can be drawn to separate them except size; and even in this respect there is a regular gradation between the colossal rorqual of 80 feet in length to the pontoporia, or dolphin of the estuary of the La Plata, which scarcely exceeds a yard from snout to tail. On this occasion, after a few general observations on the group, I propose to limit myself almost entirely to the larger species, to which the term "whale" is most especially appropriate, and which have the greatest interest to man, on account of the industries to which the commercial value of their products gives rise.

Taken altogether, as I have mentioned, the *Cetacea* constitute a perfectly distinct and natural order of mammals, characterised by their purely aquatic mode of life and external fish-like form. Their body passes anteriorly into the head without any distinct constriction or neck, and posteriorly tapers off gradually to the tail, which is provided with a pair of lateral pointed expansions of skin, supported by dense fibrous tissue, called "flukes," forming together a horizontally placed triangular propelling organ, quite different from the vertically-placed tail-fin of a fish. The fore limbs are reduced to the condition of flattened paddles, encased in a continuous skin, showing no external sign of division into arm, forearm, hand, or fingers, and without any trace of nails. There are no vestiges of hind limbs visible externally, although in many species rudiments of the hip and thigh bones, and of the muscles and joints connecting them, are found buried far away below the surface. The general surface of the body is smooth and glistening and devoid of hair, the absence of which, as a preserver of the animal heat, is compensated for by the remarkable layer of dense fat or " blubber " immediately beneath the skin. The whole organisation necessitates their life being passed entirely in the water, as on land they are absolutely helpless; but they have to rise very frequently to the surface for the purpose of respiration. The position of the respiratory orifice, nostril, or " blowhole " on the highest part of the head is very important for this mode of life, since it is the only part of the body of which the exposure above the surface is absolutely necessary. Of the numerous erroneous ideas connected with natural history, few are so widespread and still so firmly believed, notwithstanding repeated ex-positions of its falsity, as that whales spout out through their blowholes water taken in at their mouth. The fact is, the " spouting," or more properly " blowing," of the whale is nothing more than the ordinary act of breathing, which, taking place at longer intervals than in land animals, is performed with a greater amount of emphasis. The moment the animal rises to the surface it forcibly expels from its lungs the air taken in at the last inspiration, which of course is highly charged with watery vapour in consequence of the natural respiratory changes. This, rapidly condensing in the cold atmosphere in which the phenomenon is generally observed, forms a column of steam or spray, which has been erroneously taken for water.

It also often happens, especially when the surface of the ocean is agitated into waves, that the animal commences its expiratory puff before the orifice has quite cleared the top of the water, some of which may thus be driven upwards with the blast, tending to complete the illusion. In hunting whales

the harpoon often pierces the lungs or air passages of the unfortunate victim, and then fountains of blood may be forced
high in the air through the blow-holes, as commonly depicted
in scenes of Arctic adventure; but this is nothing more
(allowance being made for the whale's peculiar mode of
breathing) than what always follows severe wounds of the
respiratory organs of other warm-blooded animals.

The *Cetacea* all subsist on animal food of some kind. One
genus alone (the killers, *Orca*) eat other warm-blooded animals,
as seals, and even members of their own order, both large and
small. Some feed on fish, others on small floating crustacea,
pteropods, and medusæ; while the staple food of many is
constituted of the various species of cephalopods (squid and
cuttlefish), which abound in some seas in vast quantities.
With some exceptions they are generally timid, inoffensive
animals, active in their movements, sociable and gregarious
in their habits. They are remarkable for the great care and
affection with which they treat their young.

Among the existing members of the order there are two
very distinct types—the toothed whales (*Odontoceti*) and the
whalebone or baleen whales (*Mystacoceti*), which present
throughout their organisation markedly distinct structural
characters, and have no transitional forms between them.
The giants of the order, of which I am about to speak,
contain representatives of both groups.

The whales that have teeth and no whalebone are far the
most numerous, and include all the smaller members of the
order, the various kinds of dolphins and porpoises, among
which are the fresh-water dolphins of the great rivers of
India and South America. Some of the moderate-sized
animals of this group, especially those spoken of by sailors
under the vague designations of " grampuses," " bottle-noses," and
" black fish " (*Orca, Hyperoodon*, and *Globicephalus* of zoologists),
may be classed as whales, and some of them, as well as the
narwhal and beluga or white whale of the Arctic Seas, are
objects of pursuit by man, and when captured yield products,
mainly oil, of commercial value. But time will only allow
me now to speak in any detail of one species, which
greatly surpasses all the others, not only in size, but in value

and interest, as having long afforded material for a regular and important branch of human industry.   This is the animal commonly called the "sperm whale," known in books by its French name of *cachalot*, or its scientific designation of *Physeter macrocephalus*, which, taken altogether—not in length, but in bulk and weight—is the most colossal of all animals.

Although a contrary opinion prevailed at one time, it is now fairly well established that there is but one species of sperm whale, which has a remarkably wide geographical distribution, being met with, usually in herds, or "schools" as they are termed, in almost all tropical or subtropical seas, but not occurring, except accidentally, in the Polar regions. Not unfrequently specimens appear on the coasts of Great

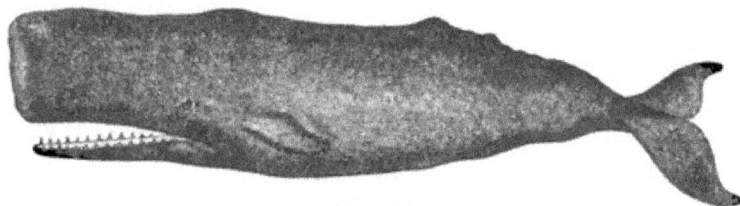

Fig. 8.[1]

Britain, but only as solitary stragglers, or as dead carcasses floated northwards by the Gulf Stream.   It is remarkable that every case of these of which we have an accurate record has been an old male.   The females and young appear never to wander so far from their usual haunts, although they have been met with in the Mediterranean, and even on the Atlantic coast of France.   The sperm whale (Fig. 8) is a strange-looking animal, and cannot be mistaken for any other cetacean.   The head is about one-third of the whole length of the animal, very massive, high, and truncated in front, and owes its huge size and remarkable form mainly to the great accumulation of a peculiar form of oily matter, contained in great cells, connected with the nasal passages, and filling the

[1] All the figures which illustrate this paper are taken, with the kind permission of the publishers, Messrs. A. & C. Black, from Flower & Lydekker's *Introduction to the Study of Mammals.*

large hollow on the upper surface of the skull. This oily matter, liquid at the natural temperature of the body, crystallises when cold, and yields when refined the spermaceti of commerce, so valuable in the manufacture of surgical ointments and candles. The nostril or " blowhole " is single, in the form of a longitudinal slit, and placed, not near the top of the head, as in most other cetaceans, but near the front end of the great snout, and rather to the left of the middle line. Consequently the " blowing " of the sperm whale is so different from that of all other species that the whalers can recognise it at any distance. The steamy jet, instead of being double and projected directly upwards, as in an ordinary fountain, which is the case with all the large whalebone whales, is single and directed obliquely forwards. The opening of the mouth is on the under side of the head, considerably behind the end of the snout. The lower jaw is extremely narrow, and has on each side from twenty to twenty-five stout conical teeth, which furnish ivory of good quality, though not in sufficient bulk for most of the purposes for which that article is required. The upper teeth are quite rudimentary and buried in the gum. The pectoral fin, or flipper, is short and broad, and on the back, where many whales have a dorsal fin, there is a series of low, rounded protuberances, scarcely to be called fins. The general colour of the surface is black above and gray below, the colours gradually shading into each other. The food of the sperm whale consists mainly of various species of cephalopods (squid and cuttlefish), but they also eat fish of considerable size.

The length of the sperm whale has been, as is always the case in which size is the most striking characteristic, greatly exaggerated. Giants are always said to be much larger than they really are, as tested by rigid measurements. To say nothing of the fabulous dimensions given by older writers, even Beale, who had immense opportunities of actual observation, says that one captured in the Japan seas measured 84 feet in length. Such statements, however intended in good faith, can never be relied upon; the difficulties and sources of fallacy in making such measurements are very great, and we are not assured whether the length is taken, as it should be, in a straight line between the front end of the head and the middle of the end of the tail, or following the curves of the surface of the body, which of course would give a considerably greater length. I have taken pains to obtain careful measurements of all the skeletons available of perfectly adult or even aged animals in various museums which I have visited, and it is curious how nearly alike they are. Allowing for the distance between the vertebræ and for the soft parts at either end, about 55 feet seems to be the usual length of the male sperm whale, and I have never been able to find any substantial proof that any one has even attained the length of 60 feet, fairly measured. The skeleton at Burton Constable, prepared from a whale which came ashore on the Yorkshire coast in 1825, is that of a very aged animal, and now measures 48 feet 4 inches in length, the vertebræ being articulated in close apposition, but even if 10 feet are allowed in addition, this would not bring it up to 60 feet. In the museum of the Royal College of Surgeons is a lower jaw presented by the late Mr. W. L. Crowther, of Hobart Town, which was considered as that of the largest sperm whale ever killed in the Tasmanian seas, and quite unique on account of its size. It only measures one inch more than the jaw of the Yorkshire specimen. The female of this species, contrary to what occurs among the whalebone whales, is very much smaller than the male.

The products of the sperm whale which render it commercially valuable when killed are (1) sperm oil, obtained by boiling the thick coating of blubber which everywhere envelops the body of the animal; (2) spermaceti, contained in the great cavity on the top of the head: (3) ambergris, formerly used in medicine, and now in perfumery—a concretion formed in the intestine of this whale, and often found floating on the surface of the seas it inhabits. Its genuineness is proved by the presence of the horny beaks of the cephalopods on which the whale feeds.

The capture or "fishery" of the sperm whale will be spoken of later on in conjunction with that of the other species of whales.

All the other large whales which are of importance to man

belong to the group called "whalebone whales," because they
are provided with a remarkable apparatus in the mouth for
the purpose of obtaining their food, to which the rather
misleading name of "whalebone" has been given,—a name for
which naturalists have substituted "baleen"; but the former
is so completely engrafted into our language, especially in all
commercial transactions connected with the subject, that it
will probably last as long as the material itself, although
it is now well known that it has nothing to do with "bone"
in the ordinary acceptation of the word. What whalebone
actually is was apparently a mystery to our forefathers.
Belon in 1551 hazarded the conjecture that it was the eyebrows of the whale; but others thought that it was the
apparatus by which it steered itself through the water. This
notion is probably connected with the old feudal law cited
by Blackstone (vol. i. p. 233), that the tails of all whales
belonged to the Queen as a perquisite to furnish her Majesty's
wardrobe with whalebone. The whalebone whales have no
teeth in either jaw (except some of a most rudimentary
nature, which disappear even before birth), but in the place
teeth usually occupy in the upper jaw, or rather upon each
side of the palate, the whalebone grows. This consists of
flattened horny plates, several hundred in number on each
side, separated by a narrow bare interval along the middle
line. The chemical composition and general character of
these plates resemble those of hair, horn, or hoof. In minute
structure they more nearly resemble the horn of the rhinoceros than any other similar growth. They are placed transversely to the long axis of the palate, with very short spaces
between them. Each plate or blade is somewhat triangular
in form, with the base attached to the palate and the apex
hanging downwards. The outer edge of the blade is hard and
smooth, but the inner edge and apex fray out into long bristly
fibres, so that the roof of the whale's mouth looks as if
covered with hair, as described by Aristotle. As the bony
palate is more or less arched from before backwards, the
blades are longest near the middle of the series, and gradually
diminish towards the front and back of the mouth.

The use of the whalebone to the whale is to strain the

water from the small marine molluscs, crustaceans, or fish upon which the whales subsist. In feeding they fill the immense mouth with water containing shoals of these small creatures, and then on their closing the jaws and raising the tongue, so as to diminish the cavity of the mouth, the water streams out through the narrow intervals between the hairy fringe of the whalebone blades, and escapes through the lips, leaving the living prey to be swallowed. In the different kinds of whales, which I shall now speak of, there are great differences in the character of the baleen. In the Californian grey whale (*Rachianectes glaucus*), an animal which attains a length of from 30 to 40 feet, the baleen blades are fewer than two hundred on each side, and far apart, very short (the longest being from 14 to 16 inches in length), coarse, and inelastic, light brown or nearly white in colour. From this there is a gradual transition, through the rorquals or finners, the humpbacks, the southern right whales, up to the Greenland whale, which exhibits this structure in its greatest perfection, both for the purposes it serves in the animal economy, and for the uses to which it has been applied by man.

All the known whalebone whales may be divided into five different groups or *genera*, as they are called by naturalists, the first of which (genus *Balæna*) have long been distinguished by practical whalers as " right whales," as they are, compared to all the others, the right whales to catch, being of far the greater commercial value. They are readily distinguished externally by the perfectly smooth back, without any trace of a dorsal fin, and by the skin of the throat and chest being also smooth, whereas in most of the other forms this region presents a number of deep longitudinal plaits or furrows. Of the right whales there are two perfectly distinct forms, though whether each of these represents a single species, or can be subdivided into several, is still a matter of uncertainty, and for our present purpose of little importance, as if minute investigation can prove that they are separable, they are most closely allied and perfectly similar to all ordinary observation. The two forms, which I shall speak of as species or kinds, are the Greenland or rather Arctic right whale (*Balæna mysticetus*) and the southern right whale (*Balæna australis*).

The Arctic right whale (Fig. 9), when full grown, attains the length of from 45 to 50 feet, a size which, as in the case of the sperm whale, has generally been greatly exaggerated in old descriptions.   As is apparently the case in all whalebone whales, but contrary to what occurs in the sperm whale, the female is rather larger than the male.   Its external form is shown in Fig. 9, from a careful drawing by Mr. Robert Gray. In this species all the peculiarities which distinguish the head and mouth of the whales from those of other mammals have attained their greatest development.   The head is of enormous size, exceeding one-third of the whole length of the creature. The cavity of the mouth is fully as large as that of the body, chest and abdomen together.   The upper jaw is very narrow

FIG. 9.

but greatly arched from before backwards, to increase the height of the cavity and allow for the great length of the baleen or "whalebone" blades.   The enormous rami of the lower jaw are widely separated behind, and have a still farther outward sweep before they meet in front, giving the floor of the mouth the shape of an immense spoon.   The baleen blades attain the number of 380 or more on each side, and those in the middle of the series have a length of 10 or sometimes 12 feet.   They are black in colour, fine, and highly elastic in texture, and fray out at the inner edge and ends into long, delicate, soft, almost silky, but very tough hairs.   The remarkable development of the mouth, and of the structures in connection with it, which distinguishes the right whale among its allies is entirely in relation to the nature of its food.   It is by this apparatus that it is enabled to avail itself of the minute but highly nutritious crustaceans and pteropods which swarm in immense shoals in the seas it frequents.   The large mouth enables it to take in at one time a sufficient quantity of water filled with these small organisms, and the length and delicate structure of the baleen provide a sufficient strainer or hair-sieve by which the water can be drained off.   If the baleen were rigid, and only as long as is the aperture between the upper and lower jaws when the mouth is shut, a space would be left beneath it when the jaws are separated, through which the water and the minute particles of food would escape together.   But instead of this, the long, slender, brush-like, elastic ends of the whalebone blades fold back when the mouth is closed, the front ones passing below the hinder ones in a channel lying between the tongue and the lower jaw. When the mouth is opened their elasticity causes them to straighten out like a bow unbent, so that at whatever distance the jaws are separated the strainer remains in perfect action, filling the whole of the interval.   The mechanical perfection of the arrangement is completed by the great development of the lower lip, which rises stiffly above the jawbone, and prevents the long, slender, flexible ends of the baleen from being carried outwards by the rush of water from the mouth, when its cavity is being diminished by the closure of the jaws and raising of the tongue.

If, as appears highly probable, the "bowhead" or right whale of the Okhotsk Sea and Behring Strait belongs to this species, its range is circumpolar, but it is strictly limited to the icy seas of the north.   "Though," as Scammon says, "it is true that these animals are pursued in the open water during the summer months, in no instance have we learned of their being captured south of where winter ice-fields are occasionally met with."   In the Behring Sea it is seldom seen south of the 55th parallel, and the southern limit of its range in the North Sea has been ascertained by Eschricht and Reinhardt to be from the east coast of Greenland at 64° N. lat. along the north of Iceland towards Spitzbergen.   Though found in the seas on both sides of Greenland, and passing freely from one to the other, it is never seen so far south as Cape Farewell; but on the Labrador coast, where a cold

stream sets down from the north, its range is somewhat
further.   There is no authentic instance of its having been
seen or captured upon any European coast.

The southern right whale, or "black whale" (Fig. 10), as it
is often called by whalers, attains about the same length as
the last, but differs in being more slender in form, in possessing
a smaller head in proportion to the body, shorter baleen
(scarcely more than half the length), a differently shaped
contour of the upper margin of the lower lip, and a greater
number of vertebræ.   Animals of this group closely resembling
each other have been found abundantly in the temperate seas
of both hemispheres, North Atlantic and North Pacific (where
they are regularly hunted by the Japanese), and in the

FIG. 10.

neighbourhood of the Cape of Good Hope, Kerguelen's Island,
Australia, and New Zealand, but according to Captain Maury's
charts they are never or rarely seen in the tropical seas.   It is
chiefly this supposed isolation of distribution rather than any
constant distinctive characters which has given rise to the
idea that the North Atlantic whale (called *Balæna biscayensis*),
the Japanese (*B. japonica*), the Cape whale (*B. australis*), and
the New Zealand whale (*B. antipodarum*), must be of different
species.   Until more numerous specimens of skeletons are
procured for our museums, or more accurate descriptions can
be obtained, the question cannot be satisfactorily determined.

The whalebone whales, not called right whales, are (1) the
humpback (*Megaptera*), or "hunchback" (Fig. 11), so called by
whalers on account of the low hump-like form of the dorsal
fin.   In Dudley's account of the whales of the New England
coast (*Phil. Trans.* 1725), the fourth species is "The bunch

or humpback whale, distinguished from the right whale by having a bunch standing in the place where the fin does in the finback. This bunch is as big as a man's head, and a foot high, shaped like a plug pointing backwards." A better distinction from all other whalebone whales is the immense length of the pectoral fins or flippers, which are indented or

Fig. 11.

scalloped along their margins. The usual length of the adult ranges from 45 to 50 feet. The baleen plates are short and broad, and of a deep black colour. This whale has a very wide range, being found, with no important differences, in both the North and South Atlantic and Pacific Oceans, and from Greenland to South Georgia. When caught it yields a fair supply of oil, but much less than the right whales, and its whalebone is of very inferior quality.

Fig. 12.

(2) The rorquals, or finners (*Balænoptera*) (Fig. 12). These have the plicated skin of the throat like that of the humpback, the furrows being more numerous and close set, but the pectoral fins are comparatively small, and the dorsal fin distinct, compressed, and triangular. The head is comparatively small and flat, and pointed in front, the whalebone short and coarse, the body long and slender, and the tail very much compressed before it expands into the "flukes." The rorquals are perhaps the most abundant and widely distributed

of all the whales, being found in all seas, except the extreme
Arctic and probably Antarctic regions.   Owing to the small
quantity and inferior quality of their whalebone, the comparatively limited amount of blubber or subcutaneous fat, their
great activity, and the difficulty of capturing them by the old
methods, these whales were not until recently an object of
pursuit by whale-fishers ; but since the introduction of steam-
vessels, and especially of explosive harpoons fired from guns in
the place of those hurled by the human hand, a regular fishery
of " finbacks " has been established on the coast of Finmark,
where many hundreds are killed every year, their bodies being
towed to shore for the purpose of flensing.   Some of the
rorquals attain the largest size of any of the *Cetacea*, the blue
whale (*Balænoptera sibbaldii*), not unfrequent in the British
Seas, reaching a length of 80 or perhaps 85 feet when fully
adult.   On the other hand, one common species (*B. rostrata*)
never exceeds 30 feet in length.

(3) Besides these there are other two forms of whalebone
whales not reckoned as right whales, the grey whale of the
North Pacific (*Rhachianectes glaucus*), and a small and very
peculiar species from the seas around New Zealand and
Australia (*Neobalæna marginata*); but these being very local
in distribution, and of little value except for museums, are
more interesting from a scientific than from a commercial
point of view.

I now proceed to the next part of my subject—the pursuit
and capture of whales by man for the sake of the materials
they yield, which are of value in commerce.   The method
universally followed is attacking the animal when it comes to
the surface of the water to breathe by means of a weapon called
a harpoon, with an iron sharp-pointed barbed head, and to
the shaft of which is attached a strong line.   In former days
this was always thrown by the hand, and to do so effectually
required great skill on the part of the man who wielded it.
Now the harpoon is generally projected from a gun, which
carries a considerably greater distance than could be traversed
by a hand-thrown weapon, and the necessity of approaching
so closely to the animal is avoided.   Various methods to
increase the efficacy of harpoons have also been introduced

including devices by which they explode within the body of
the victim.   Long straight spears or lances are also used to
despatch the animal when it has been secured by the line
attached to the harpoon.   The whale is always approached
by rowing in a boat as closely to it as necessary.   These boats
may be directly connected with a station on the coast or with
a ship out at sea.   This gives rise to the primary division of
whale fishing into two principal methods : shore fishing and
open-sea fishing, both of which are extensively practised in
various parts of the world.   In the first, a look out is kept
from a station on some projecting headland, and when a whale
appears within sight signals are given on which the boats go
out in pursuit, and when a capture takes place the body is
towed to shore for the purpose of obtaining its valuable
products.   In the second, the ships sail to some distant part
of the open sea, where it is supposed that whales are likely to
be met with.   The look out is kept from the " crow's nest "
on the masthead, and the boats being all in readiness row out
in pursuit the moment a whale is sighted, and if successful
tow their prey to the side of the vessel.

As I mentioned before, all cetaceans have immediately
beneath their skin, and closely connected with it, a very
dense layer of what is called " blubber," in large whales as
much as a foot in thickness, composed of a network of cellular
tissue, the interspaces of which are filled with oil.   This
layer, though so adherent to the outer skin as to be separated
from it with difficulty, is only connected with the flesh or
muscles which lie below by loose tissue, and so is easily
stripped off.   With a large whale this process, called " flensing," is effected as follows.   If the animal is caught at sea,
the carcass is lashed alongside the ship, and men with spikes
in their shoes, descend upon the slippery surface, and with
large sharp-edged spades perform the cutting part of the
operation, having first fixed, by means of a hook, a strong rope
into the blubber at the junction of the head and the body.
This rope runs over pulleys fixed to the rigging of the ship,
and the blubber, separated by the spades into strips about
two or three feet broad, is gradually hauled up on to the
deck of the ship.   The cuts being made in a spiral direction

round the body of the whale, the blubber is stripped off from head to tail, much as a spiral roller or bandage might be, the body of the whale meanwhile performing a rotatory motion.   When the blubber is brought on board, it is cut up into smaller pieces, and either stowed in casks or tanks to be brought home to undergo the next process, that of "trying out," or if the voyage is of lengthened duration, as in the case of the South Sea whalers hailing from European or American ports, this is done on board the ship.   It simply consists of boiling the blubber in large iron pots until the oil is separated from the mesh of cellular tissue which contained it, the latter being generally used for fuel in subsequent boilings.   In the case of the sperm whale the upper surface of the great head is opened, and the liquid spermaceti is baled out of the cavities which contain it, and in the case of the whalebone whales the whalebone is removed from the mouth.   All the rest of the animal being useless is turned adrift into the sea, and speedily becomes the prey of voracious sharks and other fish and sea birds.   When whales are caught near the shore, as in many of the "fisheries," from boats without the intervention of sea-going vessels, they are towed into shallow water for the purpose of flensing and removing the whalebone.

The earliest known regular whale fishery is that which took place from the Basque towns of France and Spain, Bayonne, Biarritz, St. Jean de Luz, Fuenterrabia, St. Sebastian, Guetaria, Ondarroa, and many others.   From the tenth century onwards the hardy fishermen of the towns and villages of this coast pursued the Atlantic right whales in the Bay of Biscay, at first only catching them from open boats near the shore, but afterwards, as the whales became more scarce and the whalers more adventurous, following them in ships across the Atlantic to the Bermudas, Newfoundland, and Iceland.   From this source all the whale oil and all the whalebone used by our forefathers down to the year 1600 was derived.   Queen Elizabeth and all her court depended upon the Basque fishermen for the most prominent characteristics of their costume. The supply was, however, diminishing when the attempt to discover the North-East route to China, about the close of the sixteenth century, led to the opening up of the sea between

Greenland and Spitzbergen, and the discovery of the Arctic right whale, an animal up to that time practically unknown to man. This being much more valuable, both on account of the larger quantity and finer quality of the whalebone it produced, and also the larger amount of oil, for many years attracted the principal attention of the whaling ships of Europe. The English entered into the business at a very early period, but, being unacquainted with the methods of capturing whales, engaged Basque harpooners for all their earliest voyages, and closely followed their methods. The very word "harpoon" is said to be Basque. The Dutch also took the fishing up on a very extensive scale, and established a permanent settlement upon the northern shore of Spitzbergen, which they named "Smeeremberg," which was the rendezvous of the whaling fleet during the summer, and to which the blubber was brought for boiling. In its most flourishing period, about the year 1680, the Dutch whale fishery employed as many as 260 ships and 14,000 men. When, however, the whales became scarcer in the neighbourhood of the coast, and the ships had to seek them further in the open sea, it was found more economical to bring the blubber direct to Holland, and Smeeremberg was deserted. The great war at the end of the last century, in which England kept possession of the North Sea, put an end to the whale fishery, not only of Holland, but of France and of all other countries which had engaged in it, and henceforth we maintained a monopoly of the trade. From the year 1732 to 1824, our government paid bounties, amounting altogether, it is calculated, to £2,500,000, to vessels engaged in the northern whaling business, with a view to encourage the enterprise. The ships at first sailed from London, then Hull, Yarmouth, and Whitby entered into the field. In 1819 as many as sixty-five ships went to the north from Hull. Since 1836 no ship has gone from London, and now Dundee and Peterhead are the only ports in the British Islands which keep up the northern whale fishery, though on a much more limited scale than formerly.

The fishery between Greenland and Spitzbergen, which in the last century proved so productive, is almost played out, but that of Davis Straits and Lancaster Sound is still remunerative, owing to the very high price that whalebone has lately been fetching.   At the beginning of the century the average value was from £70 to £90 a ton, but a few years ago a sale was effected at the enormous sum of £2650 per ton ; this is the highest price which has ever been given for it, and recently it has somewhat declined.   In 1893 four Dundee vessels secured between them twenty-seven whales. An average-sized Greenland whale will produce about fifteen hundredweight of whalebone and about fifteen tons of oil. The Greenland fishery begins early in May, and goes on to the end of September.   A few vessels remain all winter in Cumberland Inlet, ready to take advantage of the opening of the ice in the following spring.   The Arctic right whale, called locally by the American whalers the " bowhead," has since 1848 been regularly hunted in the neighbourhood of Behring Strait and the Okhotsk Sea, where its southern limit, according to Scammon, is about 54°.

Although doubtless individual sperm whales approaching near the shore, especially in the neighbourhood of the right whale fisheries, had often fallen a prey to man, the systematic capture of this species is of recent date compared to that of some other kinds.   It began about the end of the seventeenth century, from the Atlantic coasts of North America, especially the part then called New England, at first only from the shore, but afterwards in sea-going vessels from Nantucket, New Bedford, and other ports, which gradually extended their voyages into the Indian and Pacific Oceans.   From the year 1775 vessels engaged in this trade (assisted for a time by Government bounties) regularly left the mouth of the Thames for the South Seas, making voyages of three or four years' duration ; but since 1853 the business has been abandoned by the English, and what little remains of it has almost entirely reverted into the hands of the Americans.   At one time our Australian Colonies had a considerable number of ships engaged in the sperm whale fishery, and a few still sail every year from Hobart Town.   Sperm oil has fallen so greatly in price that its production is now hardly a remunerative industry, and it has found a rival, possessing all the qualities which render it of special value, in the oil of an allied but much smaller species of whale, the bottlenose (*Hyperoodon*), which has consequently become the object of a regular fishery in the North Sea, especially by the Norwegians.

Let me now return to the whales of the Basques, the North Atlantic right whale. It is a singular fact that its existence was quite overlooked by naturalists till lately, all accounts of it which are to be found in the numerous records of European whale fishing having been attributed to the Greenland whale, which was supposed by Cuvier, for instance, to have had formerly a much wider distribution than now, and to have been driven by the persecution of man to its present circumpolar haunts. To the two Danish naturalists Eschricht and Reinhardt is due the credit of having proved its existence as a distinct species from a careful collation of numerous historical notices of its structure, distribution, and habits, and, although they were at one time disposed to think that the species had become extinct, they were able to show that this was not the case, an actual specimen having been captured in the harbour of San Sebastian in January 1854, the skeleton of which Eschricht was fortunate enough to secure for the Copenhagen Museum. More recently other specimens have been captured on the Spanish coast, the Mediterranean, North America, and Norway. A skeleton has fortunately been secured for the British Museum, the exhibition of which is only delayed by the want of a proper room in which it can be mounted. In the North Pacific a very similar if not identical whale is regularly hunted by the Japanese, who tow the carcasses ashore for the purpose of flensing and extracting the whalebone. In the tropical seas, according to Captain Maury's whale charts, right whales are never or rarely seen, but when the southern temperate seas were explored, they were found to be abundantly inhabited by right whales called "black whales," so closely similar in character to the Atlantic and Japanese species that, although described and named as if distinct, at present no satisfactory and constant characters have been pointed out by which they can be separated. Of course this may arise from our very imperfect knowledge of these animals, very few specimens [1] having been preserved in museums, and still fewer accurate descriptions and drawings have been made of recently killed individuals, notwithstanding the hundreds of thousands which have been slaughtered from British and American ships during the present century. Just as its northern representative approached the coasts during the winter, and left for the open seas in the summer months, so these southern whales resorted to the bays and inlets of the Cape of Good Hope, Australia, and New Zealand during the southern winter (May to October), and departed for higher latitudes during the remainder of the year.

Though certain numbers of the southern right whales were caught in the open sea by American or Colonial ships engaged mainly in the sperm whale fishery, the principal fishery at one time, as remunerative to those who pursued it as it was destructive to the whales, was carried on from the shore, at first at the Cape, then in Australia and Tasmania, and more recently in New Zealand. Of the latter we have very detailed accounts in many contemporary works, Wakefield, Dieffenbach, and others, and a good epitome of its history will be found in Sherrin's *Handbook of the Fishery of New Zealand* (1886). Whaling vessels from America and England were in the habit of visiting the New Zealand seas for the purposes of their trade ever since the beginning of the century, but the first shore station was established in 1827 at Preservation Inlet, near the south end of the Middle Island, and in a few years there were twelve stations between that place and Banks Peninsula. In 1833 Messrs. G. and E. Weller, merchants, of Sydney, founded a whaling establishment at Otago, which was for a short time the most successful and important of any on the coast. In 1834 the whales caught yielded 310 tons of oil, besides bone, and for several years there were on this station from seventy-five to eighty Europeans constantly employed. In 1840 the oil fell off to fourteen tons, and the fishery was abandoned. The value of "black oil" at this time was £8 to £12 a ton

---

[1] The skeleton is almost the only part of a whale which can be satisfactorily preserved, but life-sized models of the principal species will soon be exhibited in the British Natural History Museum.

in New Zealand, and as much as £30 in London. In other parts of the islands the fishery was continued for a longer period. It is stated that in 1843 the whale fishery on the whole of the coast employed eighty-five boats and some 730 men, and the oil taken amounted to 1290 tons, valued at £20,000, the whalebone being valued at £12,000. After this period the decline in the take became very rapid, and the stations with their great shears for hoisting the bodies of the whales on shore for the purpose of flensing, the furnaces and boiling pots for " trying out " the oil, and the bold and hardy, but unhappily rough and dissolute, inhabitants entirely disappeared from the scene. The cause of this is not difficult to divine. The result was fully anticipated by all who carefully observed what was going on. Almost literally the goose that laid the golden egg was being killed. The whales appeared in the month of May and remained till October. Those that approached nearest to the coast and were the easiest prey to the fishermen were females about to bring forth their young. It was the regular habit of this species of whale to seek at this season some quiet, sheltered harbour, bay, or inlet, and there to remain with the new-born young, until it acquired strength and vigour enough to take care of itself in the open sea. The very affection of the whale for her young thus became the principal cause of its destruction. The whalers soon discovered that if the calf (as they called it) was wounded or caught the mother would never leave it, and they found that the calf, though of no value in itself, being inexperienced and slow, was easily captured, and then the mother became a sure prey. To the old code, regulating the northern whale fisheries, which assigned the whale to the boat which first fixed a harpoon securely in it, they added : " The boat making fast to a calf has a right to the cow, because the cow will not desert her young." It should be added that strong protests were made against this cruel and in the end unprofitable mode of capture, but they all passed unheeded. The result has been to the southern right whale much the same as that which happened to its Atlantic ally after its persecution by the Basques, although it was brought about in a much shorter space of time. The whale not only became scarce on the New Zealand coasts, but in all parts of its range. To destroy it in its last remaining breeding haunts was to destroy it everywhere. Although we have at present, unfortunately, very little accurate information about the habits and migrations of whales, there is every reason to believe that in the Antarctic summer this species retired nearer the South Pole. Sir James C. Ross in 1840, in lat. 64°, nearly due south of New Zealand, and again in 1842 in nearly the same latitude, south of the Falkland Islands, found right whales very abundant in the month of December. On the strength of this observation it has been thought that a whale corresponding to the Arctic right whale might be a permanent inhabitant of the Antarctic icy seas. Two years ago some ships sailed from Dundee in the hope of meeting with it, but they were completely disappointed. No trace of such whales was found, for doubtless Sir James Ross had only come across the summer haunts of the same whales which were then undergoing the process of ruthless extermination in their winter breeding places on the Australian and New Zealand coasts. Such having been the fate of this species, and the sperm whale not being habitually found in icy seas, the probability of any large whale being again met with in the Antarctic regions is very remote.

Our Colonial whale fisheries are practically extinct, probably never to be revived, at all events with anything like the success they met with formerly. Not that any species of whale is likely to be completely exterminated by man. It is only too easy to exterminate an animal whose habitat is confined to land, especially if that land is of limited extent, as in the case of an island; but the ocean is vast, and the possibilities of escape from pursuers in it are great. When the numbers of any species become so small that it no longer pays to hunt them, they have a chance, as has been most strikingly shown in the case of the North Atlantic or Biscay whale. Without doubt, they would increase again in the southern seas, and if means could be taken to give the whale an effectual close time, the Australian and New Zealand black-whale fishery could be revived, as it has been partially in Tasmania; but the difficulty and expense of establishing any sufficient protection would probably be greater than the value of the produce. Although the better qualities of whalebone still maintain a very high price, owing to their great scarcity, substitutes for it are being gradually invented, and the competition of mineral oils, now found in such abundance in so many parts of the world, and the rapid advances of other methods of lighting have greatly reduced the value of both sperm and train oil. Unless whales can be caught easily and cheaply, they will not be worth catching at all; and this indicates their best chance of maintaining their place in the world, as experience shows that if any profit can be made out of them, the cupidity of man will give them no quarter.

XV

ON WHALES, PAST AND PRESENT, AND THEIR PROBABLE ORIGIN [1]

FEW natural groups present so many remarkable, very obvious, and easily appreciated illustrations of several of the most important general laws which appear to have determined the structure of animal bodies, as that selected for my lecture this evening. We shall find the effects of the two opposing forces—that of heredity or conformation to ancestral characters, and that of adaptation to changed environment, whether brought about by the method of natural selection or otherwise—distinctly written in almost every part of their structure. Scarcely anywhere in the animal kingdom do we see so many cases of the persistence of rudimentary and apparently useless organs, those marvellous and suggestive phenomena which at one time seemed hopeless enigmas, causing despair to those who tried to unravel their meaning, looked upon as mere will-of-the-wisps, but now eagerly welcomed as beacons of true light, casting illuminating beams upon the dark and otherwise impenetrable paths through which the organism has travelled on its way to reach the goal of its present condition of existence.

It is chiefly to these rudimentary organs of the Cetacea and to what we may learn from them that I propose to call your attention. In each case the question may well be asked, granted that they are, as they appear to be, useless, or nearly so, to their present possessors, insignificant, imperfect, in fact *rudimentary*, as compared with the corresponding or homologous parts of other animals, are they survivals,

[1] Lecture at the Royal Institution of Great Britain, 25th May 1883.

P

or vestiges of a past condition, become useless owing to change of circumstances and environment, and undergoing the process of gradual degeneration, preparatory to their final removal from an organism to which they are only, in however small a degree, an incumbrance, or are they incipient structures, beginnings of what may in future become functional and important parts of the economy? These questions will call for an attempt at least at solution in each case as we proceed.

Before entering upon details, it will be necessary to give some general idea of the position, limits, and principal modifications of the group of animals from which the special illustrations will be drawn. The term "whale" is commonly but vaguely applied to all the larger and middle-sized Cetacea, and though such smaller species as the dolphins and porpoises are not usually spoken of as whales, they may for all intents and purposes of zoological science be included in the term, and will come within the range of the present subject. Taken altogether the *Cetacea* constitute a perfectly distinct and natural order of mammals, characterised by their purely aquatic mode of life and external fishlike form. The body is fusiform, passing anteriorly into the head without any distinct constriction or neck, and posteriorly tapering off gradually towards the extremity of the tail, which is provided with a pair of lateral pointed expansions of skin supported by dense fibrous tissue, called "flukes," forming together a horizontally-placed, triangular propelling organ. The fore-limbs are reduced to the condition of flattened ovoid paddles, incased in a continuous integument, showing no external sign of division into arm, forearm, and hand, or of separate digits, and without any trace of nails. There are no vestiges of hind-limbs visible externally. The general surface of the body is smooth and glistening, and devoid of hair. In nearly all species a compressed median dorsal fin is present. The nostrils open separately or by a single crescentic valvular aperture, not at the extremity of the snout, but near the vertex.

Animals of the order *Cetacea* abound in all known seas, and some species are inhabitants of the larger rivers of South America and Asia. Their organisation necessitates their life being passed entirely in the water, as on the land they are

absolutely helpless; but they have to rise very frequently to
the surface for the purpose of respiration. They are all predaceous, subsisting on living animal food of some kind. The
members of one genus alone (*Orca*) eat other warm-blooded
animals, as seals and even animals of their own order, both
large and small. Some feed on fish, others on small floating
crustacea, pteropods, and medusæ, while the staple food of many
is constituted of the various species of Cephalopods, chiefly
Loligo and other *Teuthidæ*, which must abound in some seas in
vast numbers, as they form almost the entire support of some
of the largest members of the order. The Cetacea are, with
some exceptions, timid, inoffensive animals. They are active in
their movements, and sociable and gregarious in their habits.

Among the existing members of the order there are two
very distinct types—the toothed whales, or *Odontoceti*, and
the baleen whales, or *Mystacoceti*, which present throughout
their organisation most markedly distinct structural characters,
and have in the existing state of nature no transitional forms.
The extinct *Zeuglodon*, so far as its characters are known,
does not fall into either of these groups as now constituted,
but is in some respects intermediate, and in others more
resembles the generalised mammalian type.

The important and interesting problems of the origin of
the Cetacea and their relations to other forms of life are at
present involved in the greatest obscurity. They present no
more signs of affinity with any of the lower classes of
vertebrated animals than do many of the members of their
own class. Indeed, in all that essentially distinguishes a
mammal from one of the oviparous vertebrates, whether in
the osseous, nervous, vascular, or reproductive systems, they
are as truly mammalian as any, even the highest, members of
the class. Any supposed signs of inferiority are, as we shall
see, simply modifications in adaptation to their peculiar mode
of life. Similar modifications are met with in another quite
distinct group of mammalia, the *Sirenia* (Dugongs and
Manatees), and also, though in a less complete degree, in the
aquatic Carnivora or seals. But these do not indicate any
community of origin between these groups and the Cetacea.
In fact, in the present state of our knowledge, the Cetacea are

absolutely isolated, and little satisfactory reason has ever been given for deriving them from any one of the existing divisions of the class rather than from any other. The question has indeed often been mooted whether they have been derived from land mammals at all, or whether they may not be the survivors of a primitive aquatic form which was the ancestor not only of the whales, but of all the other members of the class. The materials for—I will not say solving—but for throwing some light upon this problem, must be sought for in two directions—in the structure of the existing members of the order, and in its past history, as revealed by the discovery of fossil remains. In the present state of science it is chiefly on the former that we have to rely, and this therefore will first occupy our attention.

One of the most obvious external characteristics by which the mammalia are distinguished from other classes of vertebrates is the more or less complete clothing of the surface by the peculiar modification of epidermic tissue called hair. The Cetacea alone appear to be exceptions to this generalisation. Their smooth, glistening exterior is, in the greater number of species, at all events in adult life, absolutely bare, though the want of a hairy covering is compensated for functionally by peculiar modifications of the structure of the skin itself, the epidermis being greatly thickened, and a remarkable layer of dense fat being closely incorporated with the tissue of the derm or true skin; modifications admirably adapted for retaining the warmth of the body, without any roughness of surface which might occasion friction and so interfere with perfect facility of gliding through the water. Close examination, however, shows that the mammalian character of hairiness is not entirely wanting in the Cetacea, although it is reduced to a most rudimentary and apparently functionless condition. Scattered, small, and generally delicate hairs have been detected in many species, both of the toothed and of the whalebone whales, but never in any situation but on the face, either in a row along the upper lip, around the blowholes or on the chin, apparently representing the large, stiff " vibrissæ " or " whiskers " found in corresponding situations in many land mammals. In some cases these seem to persist throughout the life of the animal; more often they are only found in the young or even the fœtal state. In some species they have not been detected at any age.

Eschricht and Reinhardt counted in a new-born Greenland right whale (*Balæna mysticetus*) sixty-six hairs near the extremity of the upper jaw, and about fifty on each side of the lower lip, as well as a few around the blowholes, where they have also been seen in *Megaptera longimana* and *Balænoptera rostrata*. In a large rorqual (*Balænoptera musculus*), quite adult and sixty-seven feet in length, stranded in Pevensey Bay in 1865, there were twenty-five white, straight, stiff hairs about half an inch in length, scattered somewhat irregularly on each side of the vertical ridge in which the chin terminated, extending over a space of nine inches in height and two and a half inches in breadth. The existence of these rudimentary hairs must have some significance beyond any possible utility they may be to the animal. Perhaps some better explanation may ultimately be found for them, but it must be admitted that they are extremely suggestive that we have here a case of heredity or conformation to a type of ancestor with a full hairy clothing, just on the point of yielding to complete adaptation to the conditions in which whales now dwell.

In the organs of the senses the Cetacea exhibit some remarkable adaptive modifications of structures essentially formed on the Mammalian type, and not on that characteristic of the truly aquatic Vertebrates, the fishes, which, if function were the only factor in the production of structure, they might be supposed to resemble.

The modifications of the organs of sight do not so much affect the eyeball as the accessory apparatus. To an animal whose surface is always bathed with fluid, the complex arrangement which mammals generally possess for keeping the surface of the transparent cornea moist and protected, the movable lids, the nictitating membrane, the lachrymal gland, and the arrangements for collecting and removing the superfluous tears when they have served their function, cannot be needed, and hence we find these parts in a most rudimentary condition or altogether absent. In the same way the organ of hearing in its essential structure is entirely mammalian, having not only the sacculi and semicircular canals common to all but the lowest vertebrates, but the cochlea, and tympanic cavity with its ossicles and membrane, all, however, buried deep in the solid substance of the head; while the parts specially belonging to terrestrial mammals,—those which collect the vibrations of the sound travelling through air, the pinna and the tube which conveys it to the sentient structures within,—are entirely or practically wanting. Of the pinna or external ear there is no trace. The meatus auditorius is certainly there, reduced to a minute aperture in the skin like a hole made by the prick of a pin, and leading to a tube so fine and long that it cannot be a passage for either air or water, and therefore can have no appreciable function in connection with the organ of hearing, and must be classed with the other numerous rudimentary structures that whales exhibit.

The organ of smell, when it exists, offers still more remarkable evidence of the origin of the Cetacea. In fishes this organ is specially adapted for the perception of odorous substances permeating the water; the terminations of the olfactory nerves are spread over the inner plicated surface of a cavity near the front part of the nose, to which the fluid in which the animals swim has free access, although it is quite unconnected with the respiratory passages. Mammals, on the other hand, smell substances with which the atmosphere they breathe is impregnated; their olfactory nerve is distributed over the more or less complex foldings of the lining of a cavity placed more deeply in the head, but in immediate relation to the passages through which air is continually driven to and fro on its way to the lungs in respiration, and therefore in a most favourable position for receiving impressions from substances floating in that air. The whalebone whales have an organ of smell exactly on the mammalian type, but in a rudimentary condition. The perception of odorous substances diffused in the air, upon which many land mammals depend so much for obtaining their food, or for protection from danger, can be of little importance to them. In the more completely modified

Odontocetes the olfactory apparatus, as well as that part of the brain specially related to the function of smell, is entirely wanting, but in neither group is there the slightest trace of the specially aquatic olfactory organ of fishes. Its complete absence and the presence of vestiges of the aerial organ of land mammals in the Mystacocetes are the clearest possible indications of the origin of the Cetacea from air-breathing and air-smelling terrestrial mammalia. With their adaptation to an aquatic mode of existence, organs fitted only for smelling in air became useless, and so have dwindled or completely disappeared. Time and circumstances do not seem to have permitted the acquisition of anything analogous to the specially aquatic smelling apparatus of fishes, the result being that whales are practically deprived of whatever advantage this sense may be to other animals.

It is characteristic of the greater number of mammalia to have their jaws furnished with teeth having a definite structure and mode of development. In all the most typical forms these teeth are limited in number, not exceeding eleven on each side of each jaw, or forty-four in all, and are differentiated in shape in different parts of the series, being more simple in front, broader and more complex behind. Such a dentition is described as " heterodont." In most cases also there are two distinct sets of teeth during the lifetime of the animal, constituting a condition technically called "diphyodont."

All the Cetacea present some traces of teeth, which in structure and mode of development resemble those of mammals, and not those of the lower vertebrated classes, but they are always found in a more or less imperfect state. In the first place, at all events in existing species, they are never truly heterodont, all the teeth of the series resembling each other more or less, or belonging to the condition called " homodont," and not obeying the usual numerical rule, often falling short of, but in many cases greatly exceeding it. The most typical Odontocetes, or toothed whales, have a large number of similar, simple, conical, recurved, pointed teeth, alike on both sides and in the upper and under jaws, admirably adapted for catching slippery, living prey, such as fish, which are swallowed whole without mastication. In one genus

(*Pontoporia*) there may be as many as sixty of such teeth on each side of each jaw, making 240 in all. The more usual number is from twenty to thirty. These teeth are never changed, being "monophyodont," and they are, moreover, less firmly implanted in the jaws than in land mammals, having never more than one root, which is set in an alveolar socket, generally wide and loosely fitting, though perfectly sufficient for the simple purpose which the teeth have to serve.

Most singular modifications of this condition of dentition are met with in different genera of toothed whales, chiefly the result of suppression—sometimes of suppression of the greater number, combined with excessive development of a single pair. In one large group, the Ziphioids, although minute rudimentary teeth are occasionally found in young individuals, and sometimes throughout life, in both jaws, in the adults the upper teeth are usually entirely absent, and those of the lower jaw reduced to two, which may be very large and projecting like tusks from the mouth, as in *Mesoplodon*, or minute and entirely concealed beneath the gums, as in *Hyperoodon*,—an animal which is for all practical purposes toothless, yet in which a pair of perfectly formed though buried teeth remain throughout life, wonderful examples of the persistence of rudimentary and to all appearance absolutely useless organs. Among the *Delphinidæ* similar cases are met with. In the genus *Grampus* the teeth are entirely absent in the upper, and few and early deciduous in the lower jaw. But the narwhal exceeds all other Cetaceans, perhaps all other vertebrated animals, in the specialisation of its dentition. Besides some irregular rudimentary teeth found in the young state, the entire dentition is reduced to a single pair, which lie horizontally in the upper jaw, and both of which in the female remain permanently concealed within the bone, so that this sex is practically toothless, while in the male the right tooth usually remains similarly concealed and abortive, and the left is immensely developed, attaining a length equal to more than half that of the entire animal, projecting horizontally from the head in the form of a cylindrical or slightly tapering pointed tusk, with the surface marked by spiral grooves and ridges.

The meaning and utility of some of these strange modifications it is impossible, in the imperfect state of our knowledge of the habits of the Cetacea, to explain, but the fact that in almost every case a more full number of rudimentary teeth is present in early stages of existence, which either disappear, or remain as concealed and functionless organs, points to the present condition in the aberrant and specialised forms as being one derived from the more generalised type, in which the teeth were numerous and equal.

The Mystacocetes, or whalebone whales, are distinguished by entire absence of teeth, at all events after birth. But it is a remarkable fact, first demonstrated by Geoffroy St. Hilaire, and since amply confirmed by Cuvier, Eschricht, Julin, and others, that in the fœtal state they have numerous minute calcified teeth lying in the dental groove of both upper and lower jaws. These attain their fullest development about the middle of fœtal life, after which period they are absorbed, no trace of them remaining at the time of birth. Their structure and mode of development have been shown to be exactly that characteristic of ordinary mammalian teeth, and it has also been observed that those at the posterior part of the series are larger, and have a bilobed form of crown, while those in front are simple and conical, a fact of considerable interest in connection with speculations as to the history of the group.

It is not until after the disappearance of these teeth that the baleen, or whalebone, makes its appearance. This remarkable structure, though, as will be presently shown, only a modification of a part existing in all mammals, is, in its specially developed condition as baleen, peculiar to one group of whales. It is therefore perfectly in accord with what might have been expected, that it is comparatively late in making its appearance. Characters that are common to a large number of species appear early—those that are special to a few, at a late period, alike both in the history of the race and of the individual.

Baleen consists of a series of flattened, horny plates, several hundred in number, on each side of the palate, separated by a bare interval along the middle line. They are placed transversely to the long axis of the palate, with very short spaces between them. Each plate or blade is somewhat triangular in form, with the base attached to the palate, and the apex hanging downwards. The outer edge of the blade is hard and smooth, but the inner edge and apex fray out into long, bristly fibres, so that the roof of the whale's mouth looks as if covered with hair, as described by Aristotle. The blades are longest near the middle of the series, and gradually diminish towards the front and back of the mouth. The horny plates grow from a dense fibrous and highly vascular matrix, which covers the palatal surface of the maxillæ, and which sends out lamellar processes, one of which penetrates the base of each blade. Moreover, the free edge of each of these processes is covered with very long vascular thread-like papillæ, one of which forms the central axis of each of the hair-like epidermic fibres of which the blade is mainly composed. A transverse section of fresh whalebone shows that it is made up of numbers of these soft vascular papillæ, circular in outline, each surrounded by concentrically arranged epidermic cells, the whole bound together by other epidermic cells, which constitute the smooth cortical (so-called "enamel") surface of the blade, and which, disintegrating at the free edge, allows the individual fibres to become loose and to assume the hair-like appearance spoken of before. These fibres differ from hairs in not being formed in depressed follicles in the enderon, but rather resemble those of which the horn of the rhinoceros is composed. The blades are supported and bound together for a certain distance from their base, by a mass of less hardened epithelium, secreted by the surface of the palatal membrane or matrix of the whalebone in the intervals of the lamellar processes. This is the "intermediate substance" of Hunter, the "gum" of the whalers.

The function of the whalebone is to strain the water from the small marine molluscs, crustaceans, or fish upon which the whales subsist. In feeding they fill the immense mouth with water containing shoals of these small creatures, and then, on their closing the jaws and raising the tongue, so as to diminish the cavity of the mouth, the water streams out through the narrow intervals between the hairy fringe of the whalebone blades, and escapes through the lips, leaving the living prey to be swallowed. Almost all the other structures to which I am specially directing your attention, are, as I have mentioned, in a more or less rudimentary state in the Cetacea; the baleen, on the other hand, is an example of an exactly contrary condition, but an equally instructive one, as illustrating the mode in which nature works in producing the infinite variety we see in animal structures. Although appearing at first sight an entirely distinct and special formation, it evidently consists of nothing more than the highly modified papillæ of the lining membrane of the mouth, with an excessive and cornified epithelial development.

The bony palate of all mammals is covered with a closely adhering layer of fibro-vascular tissue, the surface of which is protected by a coat of non-vascular epithelium, the former exactly corresponding to the derm or true skin, and the latter to the epiderm of the external surface of the body. Sometimes this membrane is perfectly smooth, but it is more often raised into ridges, which run in a direction transverse to the axis of the head, and are curved with the concavity backwards; the ridges, moreover, do not extend across the middle line, being interrupted by a median depression or *raphé*. Indications of these ridges are clearly seen in the human palate, but they attain their greatest development in the Ungulata. In oxen, and especially in the giraffe, they form distinct laminæ, and their free edges develop a row of pointed papillæ, giving them a pectinated appearance. Their epithelium is thick, hard, and white, though not horny. Although the interval between the structure of the ridges in the giraffe's palate and the most rudimentary form of baleen at present known is great, there is no difficulty in seeing that the latter is essentially a modification of the former, just as the hoof of the horse, with its basis of highly developed vascular laminæ and papillæ, and the resultant complex arrangement of the epidermic cells, is a modification of the simple nail or claw of other mammals, or as the horn of the rhinoceros is only a modification of the ordinary derm and epiderm covering the animal's body differentiated by a local exuberance of growth.

Though the early stages by which whalebone has been modified from more simple palate structures are entirely lost to our sight, probably for ever, the conditions in which it now exists in different species of whales, show very marked varieties of progress, from a simple comparatively rudimental and imperfect condition, to what is perhaps the most wonderful example of mechanical adaptation to purpose known in any organic structure. These variations are worth dwelling upon for a few minutes, as they illustrate in an excellent manner the gradual modifications that may take place in an organ, evidently in adaptation to particular requirements, the causation of which can be perfectly explained upon Darwin's principle of natural selection.

In the rorquals or fin-whales (genus *Balœnoptera*, Fig. 12, p. 198), found in almost all seas, and so well known off our own coasts, the largest blades in an animal 70 feet long do not exceed 2 feet in length, including their hairy terminations; they are in most species of a pale horn colour, and their structure is coarse and inelastic, separating into thick, stiff fibres, so that they are of no value for the ordinary purposes to which whalebone is applied in the arts. These animals feed on fish of considerable size, from herrings up to cod, and for foraging among shoals of these creatures the construction of their mouth and the structure of their baleen is evidently sufficient. This is the type of the earliest known extinct forms of whales, and it has continued to exist, with several slight modifications, to this day, because it has fulfilled one purpose in the economy of nature. Other purposes for which it was not sufficient have been supplied by gradual changes taking place, some of the stages of which are seen in the intermediate conditions still exhibited in the Megaptera and in the Atlantic and southern right whales. Before describing the extreme modifications in the direction of complexity, I may mention, to show the range at present presented in the development of baleen, that there has lately been discovered in the North Pacific a species called by the whalers the Californian grey whale (*Rachianectes glaucus*), which shows the opposite extreme of simplicity. The animal is from 30 to 40 feet in length; the baleen blades are only 182 on each side (according to Scammon) and far apart, very short (the longest being from 14 to 16 inches in length), light brown or nearly white in colour, and still more coarse in grain and inelastic than that of the rorquals. The food of these whales is not yet known with certainty. They have been seen apparently seeking for it along soft bottoms of the sea, and fuci and mussels have been found in their stomachs.

In the Greenland right whale of the circumpolar seas, the bowhead of the American whalers (*Balæna mysticetus*, Fig. 9, p. 195), all the peculiarities which distinguish the head and mouth of the whales from other mammals have attained their greatest development. The head is of enormous size, exceeding one-third of the whole length of the creature. The cavity of the mouth is actually larger than that of the body, thorax and abdomen together. The upper jaw is very narrow, but greatly arched from before backwards, to increase the height of the cavity and allow for the great length of the baleen, the enormous rami of the mandibles are widely separated posteriorly, and have a still further outward sweep before they meet at the symphysis in front, giving the floor of the mouth the shape of an immense spoon. The baleen blades attain the number of 350 or more on each side, and those in the middle of the series have a length of ten or even twelve feet. They are black in colour, fine and highly elastic in texture, and fray out at the inner edge and ends into long, delicate, soft, almost silky, but very tough hairs.

How these immensely long blades depending vertically from the palate were packed into a mouth the height of which was scarcely more than half their length was a mystery not solved until a few years ago. Captain David Gray, of Peterhead, at my request, first gave us a clear idea of the arrangement of the baleen in the Greenland whale, and showed that the purpose of its wonderful elasticity was not primarily at least the benefit of the corset and umbrella makers, but that it was essential for the correct performance of its functions. It may here be mentioned that the modification of the mouth structure of the right whale is entirely in relation to the nature of its food. It is by this apparatus that it is enabled to avail itself of the minute but highly nutritious crustaceans and pteropods which swarm in immense shoals in the seas it frequents. The large mouth enables it to take in at one time a sufficient quantity of water filled with these small organisms, and the length and delicate structure of the baleen provides an efficient strainer or hair sieve by which the water can be drained off.   If the baleen were, as in the rorquals, short and rigid, and only of the length of the aperture between the upper and lower jaws when the mouth was shut, when the jaws were separated a space would be left beneath it through which the water and the minute particles of food would escape together.   But instead of this, the long, slender, brush-like ends of the whalebone blades, when the mouth is closed, fold back, the front ones passing below the hinder ones in a channel lying between the tongue and the bone of the lower jaw.   When the mouth is opened their elasticity causes them to straighten out like a bow that is unbent, so that at whatever distance the jaws are separated, the strainer remains in perfect action, filling the whole of the interval.   The mechanical perfection of the arrangement is completed by the great development of the lower lip, which rises stiffly above the jawbone, and prevents the long, slender, flexible ends of the baleen being carried outwards by the rush of water from the mouth, when its cavity is being diminished by the closure of the jaws and raising of the tongue.   The interest and admiration excited by the contemplation of such a beautifully adjusted piece of mechanism is certainly heightened by the knowledge that it has been brought about by the gradual adaptation and perfection of structures common to the whole class of animals to which the whale belongs.

Few points of the structure of whales offer so great a departure from the ordinary mammalian type as the limbs. The fore-limbs are reduced to the condition of simple paddles or oars, variously shaped, but always flattened and more or less oval in outline.   They are freely movable at the shoulder-joint, where the humerus or upper-arm bone articulates with the shoulder-blade in the usual manner, but beyond this point, except a slight flexibility and elasticity, there is no motion between the different segments.   The bones are all there, corresponding in number and general relations with those of the human or any other mammalian arm, but they are flattened out, and their contiguous ends, instead of presenting hinge-like joints, come in contact by flat surfaces, united together by strong ligamentous bands, and all wrapped up in an

undivided covering of skin, which allows externally of no sign
of the separate and many-jointed fingers seen in the skeleton.

Up to the year 1865 it was generally thought that there
was nothing to be found between this bony framework and the
covering skin, with its inner layer of blubber, except dense
fibrous tissue, with blood-vessels and nerves sufficient to
maintain its vitality. Dissecting a large rorqual, 67 feet in
length, upon the beach of Pevensey Bay in that year, I was
surprised to find lying upon the bones of the forearm well-
developed muscles, the red fibres of which reached nearly to
the lower end of these bones, ending in strong tendons,
passing to, and radiating out on, the palmar surface of the
hand. Circumstances then prevented me following out the
details of their arrangement and distribution, but not long
afterwards Professor Struthers, of Aberdeen, had an opportunity
of carefully dissecting the fore-limb of another whale of the
same species, and he has recorded and figured his observations
in the *Journal of Anatomy* for November 1871. He found
on the internal or palmar aspect of the limb three distinct
muscles corresponding in attachments to the flexor carpi
ulnaris, the flexor profundus digitorum, and the flexor longus
pollicis of man, and on the opposite side but one, the extensor
communis digitorum.[1] Large as these muscles actually are,
yet, compared with the size of the animal, they cannot but be
regarded as rudimentary, and being attached to bones without
regular joints and firmly held together by unyielding tissues,
their functions must be reduced almost to nothing. But
rudimentary as the muscles of the fin-whales are, lower stages
of degradation of the same structures are found in other
members of the group. In some they are indeed present in
form, but their muscular structure is gone, and they are
reduced in most of the toothed whales to mere fibrous bands,
scarcely distinguishable from the surrounding tissue which
connects the inner surface of the skin with the bone. It is
impossible to contemplate these structures without having the
conviction forced home that here are the remains of parts once
of use to their possessor, now, owing to the complete change of

---

[1] The muscles of the forearm of an allied species, *Balænoptera rostrata*, were
described by Macalister in 1868, and Perrin in 1870.

purpose and mode of action of the limb, reduced to a condition of atrophy verging on complete disappearance.

The changes that have taken place in the hind-limbs are even more remarkable. In all known Cetacea (unless *Platanista* be really an exception) a pair of slender bones are found suspended a short distance below the vertebral column, but not attached to it, about the part where the body and the tail join. In museum skeletons these bones are often not seen, as, unless special care has been taken in the preparation, they are apt to get lost. They are, however, of much importance and interest, as their relations to surrounding parts show that they are the rudimentary representatives of the pelvic or hip bones, which in other mammals play such an important part in connecting the hind-limbs with the rest of the skeleton. The pelvic arch is thus almost universally present, but of the limb proper there is, as far as is yet known, not a vestige in any of the large group of toothed whales, not even in the great cachalot or sperm whale, although it should be mentioned that it has never been looked for in that animal with any sort of care. With the whalebone whales, however, at least in some of the species, the case is different. In these animals there are found, attached to the outer and lower side of the pelvic bone, other elements, bony or only cartilaginous as the case may be, clearly representing rudiments of the first, and in some cases the second segment of the limb, the thigh or femur, and the leg or tibia. In the small *Balænoptera rostrata* a few thin fragments of cartilage, imbedded in fibrous tissue attached to the side of the pelvic bone, constitute the most rudimentary possible condition of a hind-limb, and could not be recognised as such but for their analogy with other allied cases. In the large rorqual, *Balænoptera musculus*, 67 feet long, previously spoken of, I was fortunate enough in 1865 to find attached by fibrous tissue to the side of the pelvic bone (which was sixteen inches in length) a distinct femur, consisting of a nodule of cartilage of a slightly compressed, irregularly oval form, and not quite one inch and a half in length. Other specimens of the same animal dissected by Van Beneden and Professor Struthers have shown the same; in one case, partial ossification had taken place. In the genus

Megaptera a similar femur has been described by Eschricht; and the observations of Reinhardt have shown that the Greenland right whale (*Balæna mysticetus*) has not only a representative of the femur developed far more completely than in the rorqual, being from six to eight inches in length and completely ossified, but also a second smaller and more irregularly formed bone, representing the tibia. Our knowledge of these parts in this species has recently been greatly extended by the researches of Dr. Struthers, who has published in the *Journal of Anatomy* for 1881 a most careful and detailed account of the dissection of several specimens, showing the amount of variation to which these bones (as with most rudimentary structures) are liable in different individuals, and describing for the first time their distinct articulation one with the other by synovial joints and capsular ligaments, and also the most remarkable and unlooked-for presence of muscles passing from one bone to the other, representing the adductors and flexors of mammals with completely developed limbs, but so situated that it is almost impossible to conceive that they can be of any use; the whole limb, such as it is, being buried deep below the surface, where any movement, except of the most limited kind, must be impossible. Indeed, that the movement is very limited and of no particular importance to the animal was shown by the fact that in two out of eleven whales dissected the hip-joint was firmly anchylosed (or fixed by bony union), though without any trace of disease. In the words of Dr. Struthers, " Nothing can be imagined more useless to the animal than rudiments of hind-legs entirely buried beneath the skin of a whale, so that one is inclined to suspect that these structures must admit of some other interpretation. Yet, approaching the inquiry with the most sceptical determination, one cannot help being convinced, as the dissection goes on, that these rudiments really are femur and tibia. The functional point of view fails to account for their presence. Altogether they present for contemplation a most interesting instance of those significant parts, rudimentary structures."

We have here a case in which it is not difficult to answer the question before alluded to, often asked with regard to rudimentary parts, Are they disappearing or are they incipient organs ?   We can have no hesitation in saying that they are the former.   All we know of the origin of limbs shows that they commence as outgrowths upon the surface of the body, and that the first-formed portions are the most distal segments. The limb, as proved by its permanent state in the lowest Vertebrates, and by its embryological condition in higher forms, is at first a mere projection or outward fold of the skin, which, in the course of development, as it becomes of use in moving or supporting the animal, acquires the internal frame-work which strengthens it and perfects its functions.   It would be impossible, on any theory of causation yet known, to conceive of a limb gradually developed from within outwards. On the other hand, its disappearance would naturally take place in the opposite direction ; projecting parts which had become useless, being in the way, would, like all the other prominences on the surface of the whales,—hair, ears, etc.,—be removed, while the most internal, offering far less interference with successful carrying on the purposes of life, would be the last to disappear, lingering, as in the case of the Greenland whale, long enough to reveal their wonderful history to the anatomist who has been fortunate enough to possess the skill and the insight to interpret it.

Time will not allow of more illustrations drawn from the structure of existing Cetacea ; we turn next to what the researches of palæontology teach of the past history of the order.   Unfortunately this does not at present amount to very much.   As is the case with nearly all other orders of mammals, we know nothing of their condition, if they existed, in the mesozoic age.   Even in the Cretaceous seas, the deposits at the bottom of which are so well adapted to preserve the remains of the creatures which swam in them, not a fragment of any whale or whale-like animal has been found.   The earliest Cetaceans of whose organisation we have any good evidence are the Zeuglodons of the Eocene formations of North America.   These were creatures whose structure, as far as we know it, was intermediate between that of the existing sub-orders of whales, having the elongated nasal bones and anterior position of the nostrils of the Mystacocetes, with the teeth of the Odontocetes, and with some characters more like those of the generalised mammalian type than of any of the existing forms. In fact *Zeuglodon* is precisely what we might have expected *a priori* an ancestral form of whale to have been. The remarkable smallness of its cerebral cavity, compared with the jaws and the rest of the skull, so different from that of modern Cetaceans, is exactly paralleled in the primitive types of other groups of mammals. The teeth are markedly differentiated in different parts of the series. In the anterior part of both jaws they are simple, conical, or slightly compressed and sharp pointed. The first three of the upper jaw are distinctly implanted in the premaxillary bone, and so may be reckoned as incisors. The tooth which succeeds, or the canine, is also simple and conical, but it does not greatly exceed the others in size. This is followed by five teeth with two distinct roots and compressed pointed crowns, with denticulated cutting edges. It has been thought that there was evidence of a vertical succession of the molar teeth, as in diphyodont mammals, but the proof of this is not quite satisfactory. Unfortunately the structure of the limbs is most imperfectly known. A mutilated humerus has given rise to many conjectures; to some anatomists it appears to indicate freedom of motion at the elbow-joint, while to others its characters seem to be those of the ordinary Cetacea. Of the structure of the pelvis and hind-limb we are at present in ignorance.

From the middle Miocene period fossil Cetacea are abundant, and distinctly divided into the two groups now existing. The Mystacocetes, or whalebone whales, of the Miocene seas were, as far as we know now, only *Balænopteræ*, some of which (as the genus *Cetotherium*) were, in the elongated flattened form of the nasal bones, the greater distance between the occipital and frontal bones at the top of the head, and the greater length of the cervical vertebræ, more generalised than any now existing. In the shape of the mandible also, Van Beneden, to whose researches we are chiefly indebted for a knowledge of these forms, discerns some approximation to the Odontocetes. Right whales (*Balæna*) have not been found earlier than the Pliocene period, and it is interesting to note that instead of the individuals diminishing in bulk as we approach the times we live in, as with many other groups of animals, the contrary has been the case, no known extinct species of whales equalling in size those that are now to be met with in the ocean.   The size of whales, as of all other things whose most striking attribute is magnitude, has been greatly exaggerated; but when reduced to the limits of sober fact, the Greenland right whale of 50 feet long, the sperm whale of 60, and the great northern rorqual (*Balœnoptera sibbaldii*) of 80, exceed all other organic structures known, past or present.   Instead of living in an age of degeneracy of physical growth, we are in an age of giants, but it may be at the end of that age.   For countless centuries impulses from within and the forces of circumstances from without have been gradually shaping the whales into their present wonderful form and gigantic size; but the very perfection of their structure and their magnitude combined, the rich supply of oil protecting their internal parts from cold, the beautiful apparatus of whalebone by which their nutrition is provided for, have been fatal gifts, which, under the sudden revolution produced on the surface of the globe by the development of the wants and arts of civilised man, cannot but lead in a few years to their partial if not complete extinction.

It does not need much foresight to divine the future history of whales, but let us return to the question with which we started, What was their probable origin ?

In the first place, the evidence is absolutely conclusive that they were not originally aquatic in habit, but are derived from terrestrial mammals of fairly high organisation, belonging to the placental division of the class,—animals in which a hairy covering was developed, and with sense organs, especially that of smell, adapted for living on land; animals, moreover, with four completely-developed pairs of limbs on the type of the higher vertebrata, and not of that of fishes. Although their teeth are now of the simple homodont and monophyodont type, there is much evidence to show that this has taken place by the process of degradation from a more perfect type, even the fœtal teeth of whalebone whales showing signs of differentiation into molars and incisors,

and many extinct forms, not only the Zeuglodons, but also
true dolphins, as the Squalodons, having a distinct heterodont
dentition, the loss of which, though technically called a
"degradation," has been a change in conformity to the habits
and needs of the individuals. So much may be considered
very nearly, if not quite, within the range of demonstrated
facts, but it is in determining the particular group of
mammals from which the Cetacea arose that greater difficulties
are met with.

One of the methods by which a land mammal may have
been changed into an aquatic one is clearly shown in the
stages which still survive among the Carnivora. The seals
are obviously modifications of the land Carnivora, the Otariæ,
or sea-lions and sea-bears, being curiously intermediate. Many
naturalists have been tempted to think that the whales
represent a still further stage of the same kind of modification. So firmly has this idea taken root that in most
popular works on zoology in which an attempt has been
made to trace the pedigree of existing mammals, the Cetacea
are definitely placed as offshoots of the Pinnipedia, which in
their turn are derived from the Carnivora. But there is to
my mind a fatal objection to this view. The seal of course
has much in common with the whale, inasmuch as it is a
mammal adapted for an aquatic life, but it has been converted
to its general fish-like form by the peculiar development of
its hind-limbs into instruments of propulsion through the
water; for though the thighs and legs are small, the feet are
large and are the special organs of locomotion, the tail being
quite rudimentary. The two feet applied together form an
organ very like the tail of a fish or whale, and functionally
representing it, but only functionally, for the time has, I trust,
quite gone by when the Cetacea were defined as animals with
the "hinder limbs united, forming a forked horizontal tail."
In the whales, as we have seen, the hind-limbs are aborted
and the tail developed into a powerful swimming organ. Now
it is very difficult to suppose that when the hind-limbs had
once become so well adapted to a function so essential to the
welfare of the animal as that of swimming, they could ever
have become reduced and their action transferred to the tail;

—the animal must have been in a too helpless condition to maintain its existence during the transference, if it took place, as we must believe, gradually. It is far more reasonable to suppose that whales were derived from animals with large tails, which were used in swimming, eventually with such effect that the hind-limbs became no longer necessary, and so gradually disappeared. The powerful tail, with lateral cutaneous flanges, of an American species of otter (*Pteronura sandbachii*), or the still more familiar tail of the beaver, may give some idea of this member in the primitive *Cetacea*. I think that this consideration disposes of the principal argument in favour of the whales being related to the seals, as most of the other resemblances, such as those in the characters of their teeth, are evidently resemblances of analogy related to similarity of habit.

As pointed out long ago by Hunter, there are numerous points in the visceral organs of the Cetacea which far more resemble those of the Ungulata than of the Carnivora. These are the complex stomach, the simple liver, the respiratory organs, and especially the reproductive organs and structures relating to the development of the young. Even the skull of Zeuglodon, which has been cited as presenting a great resemblance to that of a seal, has quite as much likeness to one of the primitive pig-like Ungulates, except in the purely adaptive character of the form of the teeth.

Though there is, perhaps, generally more error than truth in popular ideas on natural history, I cannot help thinking that some insight has been shown in the common names attached to one of the most familiar of Cetaceans by those whose opportunities of knowing its nature have been greatest —" Sea-Hog," " Sea-Pig," or " Herring-Hog " of our fishermen, *Meerschwein* of the Germans, corrupted into the French " Marsouin," and also " Porcpoisson," shortened into " Porpoise."

The difficulty that might be suggested in the derivation of the Cetacea from the Ungulata, arising from the latter being at the present day mainly vegetable feeders, is not great, as the primitive Ungulates were probably omnivorous, as their least modified descendants, the pigs, are still; and the aquatic branch might easily have gradually become more and more piscivorous, as we know from the structure of their bones and teeth, the purely terrestrial members have become by degrees more exclusively graminivorous.

One other consideration may remove some of the difficulties that may arise in contemplating the transition of land mammals into whales. The gangetic dolphin (*Platanista*) and the somewhat related *Inia* of South America, which retain several rather generalised mammalian characters, and are related to some of the earliest known European Miocene dolphins, are both to the present day exclusively fluviatile, being found in the rivers they inhabit almost up to their very sources, more than a thousand miles from the sea. May this not point to the freshwater origin of the whole group, and thus account for their otherwise inexplicable absence from the Cretaceous seas?

We may conclude by picturing to ourselves some primitive generalised, marsh-haunting animals with scanty covering of hair like the modern hippopotamus, but with broad, swimming tails and short limbs, omnivorous in their mode of feeding, probably combining water-plants with mussels, worms, and freshwater crustaceans, gradually becoming more and more adapted to fill the void place ready for them on the aquatic side of the borderland on which they dwelt, and so by degrees being modified into dolphin-like creatures inhabiting lakes and rivers, and ultimately finding their way into the ocean. Here the disappearance of the huge Enaliosaurians, the *Ichthyosauri* and *Plesiosauri*, which formerly played the part the Cetacea do now, had left them ample scope. Favoured by various conditions of temperature and climate, wealth of food supply, almost complete immunity from deadly enemies, and illimitable expanses in which to roam, they have undergone the various modifications to which the Cetacean type has now arrived, and gradually attained that colossal magnitude which we have seen was not always an attribute of the animals of this group.

Please to recollect, however, that this is a mere speculation, which may or may not be confirmed by subsequent palæontological discovery. Such speculations are, I trust, not without their use and interest, especially when it is distinctly understood that they are offered only as speculations and not as demonstrated facts.

ANTHROPOLOGY

XVI

PRESIDENTIAL ADDRESS TO THE DEPARTMENT OF ANTHROPOLOGY[1]

It is impossible for us to commence the work of this section of the Association without having vividly brought to our minds the loss which has befallen us since our last meeting—the loss of one who was our most characteristic representative of the complex science of Anthropology—one who had for many years conducted with extraordinary energy, amidst multifarious other avocations, a series of researches into the history, customs, and physical characters of the early inhabitants of our island, for which he was so especially fitted by his archæological, historical, and literary as well as his anatomical knowledge, and who was also the most popular and brilliant expositor, to assemblies such as meet together on these occasions, of the results of those researches. I need scarcely say that I refer to Professor Rolleston.

Within the last few months the study of our subject in this country has received an impulse from the publication of a book —small in size, it is true, but full of materials for thought and instruction—the *Anthropology* of Mr. E. B. Tylor, the first work published in English with that title, and one very different in its scope and method from the older ethnological treatises.

The immense array of facts brought together in a small compass, the terseness and elegance of the style, the good taste and feeling with which difficult and often burning questions are treated, should give this book a wide circulation among all classes, and thoroughly familiarise both the word and the subject to English readers.

[1] British Association for the Advancement of Science (York Meeting), 1st September 1881.

The origin and early history of man's civilisation, his language, his arts of life, his religion, science, and social customs in the primitive conditions of society, are subjects in which, in consequence of their direct continuity with the arts and sciences, religious, political, and social customs among which we all live, by which we are all influenced, and about which we all have opinions, every person of ordinary education can and should take an interest.   In fact, really to understand all these problems in the complex condition in which they are presented to us now, we ought to study them in their more simple forms, and trace them as far as may be to their origins.

But, as the author remarks, this book is only an introduction to anthropology, rather than a summary of all that it teaches; and some, even those that many consider the most important branches of the subject, are but lightly touched upon, or wholly passed over.

In one of the estimates of the character and opinions of the very remarkable man and eminent statesman, whose death the country was mourning last spring, it was stated: "Lord Beaconsfield had a deep-rooted conviction of the vast importance of race, as determining the relative dominance both of societies and of individuals;"[1] and with regard to the question of what he meant by "race," we have a key in the last published work of the same acute observer of mankind: "Language and religion do not make a race—there is only one thing which makes a race, and that is blood."[2]   Now "blood" used in this sense is defined as "kindred; relation by natural descent from a common ancestor; consanguinity."[3]   The study of the true relationship of the different races of men is then not only interesting from a scientific point of view, but of great importance to statesmanship in such a country as this, embracing subjects representing almost every known modification of the human species whose varied and often conflicting interests have to be regulated and provided for.   It is to want of appreciation of its importance that many of the inconsistencies and shortcomings of the government of

[1] *Spectator*, 23rd April 1881.   [2] *Endymion*, vol. ii. p. 205.
[3] Webster's *Dictionary*.

our dependencies and colonies are due, especially the great inconsistency between a favourite English theory and a too common English practice—the former being that all men are morally and intellectually alike, the latter being that all are equally inferior to himself in all respects: both propositions egregiously fallacious. The study of race is at a low ebb indeed when we hear the same contemptuous epithet of "nigger" applied indiscriminately by the Englishmen abroad to the blacks of the West Coast of Africa, the Kaffirs of Natal, the Lascars of Bombay, the Hindoos of Calcutta, the aborigines of Australia, and even the Maoris of New Zealand!

But how is he to know better? Where in this country is any instruction to be had? Where are the books to which he may turn for trustworthy information? The subject, as I have said, is but slightly touched upon in the last published treatise on anthropology in our language. The great work of Pritchard, a compendium of all that was known at the time it was written, is now almost entirely out of date. In not a single university or public institution throughout the three kingdoms is there any kind of systematic teaching, either of physical or of any other branch of anthropology, except so far as comparative philology may be considered as bearing upon the subject. The one Society of which it is the special business to promote the study of these questions, the Anthropological Institute of Great Britain and Ireland, is, I regret to say, far from flourishing. An anthropological museum, in the proper sense of the word, either public or private, does not exist in this country.

What a contrast is this to what we see in almost every other nation in Europe! At Paris there is, first, the Museum d'Histoire Naturelle, where man, as a zoological subject—almost entirely neglected in our British Museum—has a magnificent gallery allotted to him, abounding not only in illustrations of osteology, but also in models, casts, drawings, and anatomical preparations, showing various points in his physical or natural history, which are expounded to the public in the free lectures of the venerable Professor Quatrefages and his able coadjutor, Dr. Hamy. There is also the vigorous Society of Anthropology, which is stated in the last annual report to

number 720 members, showing an increase of 44 during
the year 1880, and which is forming a museum on a most
extensive scale; and, finally, the School of Anthropology,
founded by the illustrious Broca, whose untimely death last
year, instead of paralysing, seems to have stimulated the
energies of colleagues and pupils into increased activity.
In this school, supported partly by private subscriptions, partly
by the public liberality of the Municipality of Paris, and of
the Department of the Seine, are laboratories in which all the
processes of anthropological manipulation are practised by
students and taught to travellers.   Here all the bodies of
persons of outlandish nationalities dying in any of the hospitals
of Paris are dissected by competent and zealous observers, who
carefully record every peculiarity of structure discovered, and
are thus laying the foundation for an exhaustive and trustworthy collection of materials for the comparative anatomy
of the races of man.   Here, furthermore, are lectureships on
all the different branches.   Biological and anatomical anthropology, ethnology, prehistoric, linguistic, social, and medical
anthropology are all treated of separately by eminent professors
who have made these departments their special study.   The
influence of so much activity is spreading beyond the capital.
The foundation of an Anthropological Society at Lyons has
been announced within the present year.

In Germany, although there is not at present any institution organised like the school at Paris, the flourishing
state of the Berlin Ethnological Society, which also reports
a large increase in the number of its members, the various
other societies and journals, and the important contributions
which are continually being made from the numerous intellectual centres of that land of learning, all attest the
interest which the study of man excites there.   In Italy, in
the Scandinavian kingdoms, in Russia, and even in Spain,
there are signs of similar activity.   A glance at the recent
periodical literature of America, especially the publications of
the Smithsonian Institution, will show how strongly the
scientific work of that country is setting   in   the   same
direction.

It is true that a very great proportion of the energies of

the societies, institutions, and individuals who cultivate this vast subject are, in all these lands, as it is indeed to so great an extent in our own, devoted to that branch which borders upon the old and favourite studies of archæology and geology. The fascinating power of the pursuit of the earliest traces of man's existence upon the earth, with the possibilities of obtaining some glimpses of his mode of origin, is attested in the devotion seen everywhere in museums, in separate publications, and in journals, to prehistoric anthropology.

But, though the study of man's origin and earliest appearances upon the earth, and that of the structural modifications to which in course of time he has arrived, or the study of races, are intimately related, and will ultimately throw light upon one another, I venture to think that the latter is the more pressing of the two, as it is certainly the more practically important; and hence the necessity for greater attention to physical anthropology. In seeking for a criterion upon which to base our study of races, in looking for essential proofs of consanguinity of descent from common ancestors in different groups of men, I have no hesitation in saying that we must first look to their physical or anatomical characters, next to their moral and intellectual characters—for our purpose more difficult of apprehension and comparison—and, lastly, as affording hints, often valuable in aid of our researches, but rarely to be depended upon, unless corroborated from other sources, to language, religion, and social customs.

The study of the physical or anatomical characters of the races of man is unfortunately a subject beset with innumerable difficulties. It can only be approached with full advantage by one already acquainted with the ordinary facts of human anatomy, and with a certain amount of zoological training. The methods used by the zoologist in discriminating species and varieties of animals, and the practice acquired in detecting minute resemblances and differences that an ordinary observer might overlook, are just what are required in the physical anthropologist.

As the great problem which is at the root of all zoology is to discover a natural classification of animals, so the aim of zoological anthropology is to discover a natural classification of man. A natural classification is an expression of our knowledge of real relationship, of consanguinity—of "blood," as the author of *Endymion* expresses it. When we can satisfactorily prove that any two of the known groups of mankind are descended from the same common stock, a point is gained. The more such points we have acquired, the more nearly shall we be able to picture to ourselves, not only the present, but the past distribution of the races of man upon the earth, and the mode and order in which they have been derived from one another.

The difficulties in the way of applying zoological principles to the classification of man are vastly greater than in the case of most animals; the problem being, as we shall see, one of much greater complexity. When groups of animals become so far differentiated from each other as to represent separate species they remain isolated; they may break up into further subdivisions—in fact, it is only by further subdivision that new species can be formed; but it is of the very essence of species, as now universally understood by naturalists, that they cannot recombine, and so give rise to new forms. With the varieties of man it is otherwise. They have never so far separated as to answer to the physio-logical definition of species. All races are fertile one with another, though perhaps in different degrees. Hence new varieties have constantly been formed, not only by the segmentation, as it were, of a portion of one of the old stocks, but also by various combinations of those already established.

Neither of the old conceptions of the history of man, which pervaded the thought, and form the foundation of the works of all ethnological writers up to the last few years, rest on any solid basis, or account for the phenomena of the present condition and distribution of the species.

The one view—that of the monogenist—was that all races, as we see them now, are the descendants of a single pair, who, in a comparatively short period of time spread over the world from one common centre of origin, and became modified by degrees in consequence of changes of climate and other external conditions. The other—that of the polygenist—is that a certain number of varieties or species (no agreement has been arrived at as to the number, which is estimated by different authorities at from three to twenty or more) have been independently created in different parts of the world, and have perpetuated the distinctive characters as well as the geographical position with which they were originally endowed.

The view which appears best to accord with what is now known of the characters and distribution of the races of man, and with the general phenomena of nature, may be described as a modification of the former of these hypotheses.

Without entering into the difficult question of the method of man's first appearance upon the world, we must assume for it a vast antiquity—at all events as measured by any historical standard.   Of this there is now ample proof. During the long time he existed in the savage state—a time compared to which the dawn of our historical period was as yesterday—he was influenced by the operation of those natural laws which have produced the variations seen in other regions of organic nature.   The first men may very probably have been all alike ; but, when spread over the face of the earth, and become subject to very diverse external conditions— climate, food, competition with members of his own species or with wild animals—racial differences began slowly to be developed through the potency of various kinds of selection acting upon the slight variations which appeared in individuals in obedience to the tendency implanted in all living things.

Geographical position must have been one of the main elements in determining the formation and the permanence of races.   Groups of men isolated from their fellows for long periods, such as those living on small islands, to which their ancestors may have been accidentally drifted, would naturally, in course of time, develop a new type of features, of skull, of complexion or hair.   A slight set in one direction, in any of these characters, would constantly tend to intensify itself, and so new races would be formed.   In the same way different intellectual or moral qualities would be gradually developed and transmitted in different groups of men.   The longer a race thus formed remained isolated, the more strongly impressed and the more permanent would its characteristics

R

become, and less liable to be changed or lost, when the surrounding circumstances were altered, or under a moderate
amount of intermixture from other races—the more " true,"
in fact, would it be.   On the other hand, on large continental
tracts, where no " mountains interposed make enemies of
nations," or other natural barriers form obstacles to free
intercourse between tribe and tribe, there would always be
a tendency towards uniformity from the amalgamation of races
brought into close relation by war or by commerce.   Smaller
or feebler races have been destroyed or absorbed by others impelled by superabundant population or other causes to spread
beyond their original limits ; or sometimes the conquering
race has itself disappeared by absorption into the conquered.

Thus, for untold ages, the history of man has presented a
shifting, kaleidoscopic scene ; new races gradually becoming
differentiated out of the old elements, and, after dwelling
awhile upon the earth, either becoming suddenly annihilated
or gradually merged into new combinations ; a constant
destruction and reconstruction ; a constant tendency to
separation and differentiation, and a tendency to combine
again into a common uniformity—the two tendencies acting
against and modifying each other.   The history of these
processes in former times, except in so far as they may be
inferred from the present state of things, is a difficult study,
owing to the scarcity of evidence.   If we had any approach
to a complete palæontological record, the history of man
could be reconstructed ; but ˙nothing of the kind is forthcoming.   Evidences of the anatomical characters of man, as
he lived on the earth during the time when the great racial
characteristics were being developed, during the long ante-
historic period in which the Negro, the Mongolian, and the
Caucasian were being gradually fashioned into their respective
types, is entirely wanting, or, if any exists, it is at present
safely buried in the earth, perhaps to be revealed at some
unexpected time, and in some unforeseen manner.

It will be observed, and perhaps observed with perplexity
by some, that no definition has as yet been given of the oft-
recurring word " race."   The sketch just drawn of the past
history of man must be sufficient to show that any theory

implying that the different individuals composing the human
species can be parcelled out into certain definite groups, each
with its well-marked and permanent limits separating it
from all others, has no scientific foundation; but that, in
reality, these individuals are aggregated into a number of
groups of very different value in a zoological sense, with
characters more or less strongly marked and permanent,
and often passing insensibly into one another. The great
groups are split up into minor subdivisions, and filling up
the gaps between them are intermediate or intercalary forms,
derived either from the survival of individuals retaining the
generalised or ancestral characters of a race from which two
branches have separated and taken opposite lines of modification, or from the reunion of members of such branches in
recent times. If we could follow those authors who can
classify mankind into such divisions as trunks, branches,
races, and sub-races, each having its definite and equivalent
meaning, our work would appear to be greatly simplified,
although perhaps we should not be so near the truth we are
seeking. But being not yet in a position to define what
amount of modification is necessary to constitute distinction
of race, I am compelled to use the word vaguely for any
considerable group of men who resemble each other in certain
common characters transmitted from generation to generation.

In approaching the question of the classification of the
races of man from a physical point of view, we must bestow
great care upon the characters upon which we rely in
distinguishing one group from another. It is well known in
zoology that the modifications of a single organ or system
may be of great value, or may be quite useless, according as
such modifications are correlated with others in different
organs or systems, or are mere isolated examples of variation
in the economy of the animal without structural changes elsewhere. The older ornithologists associated in one order all
the birds with webbed feet, and the order thus constituted,
Natatores or *Palmipedes*, which received the great sanction of
Cuvier, still stands in many zoological compilations. Recent
investigations into the anatomy of birds have shown that the
species thus associated together show no other sign of natural

affinity, and no evidence of being derived from the same stock. In fact, there is tolerably good proof that the webbing of the feet is a merely adaptive character, developed or lost, present or absent, irrespective of other structural modifications. In the same way, when anthropology was less advanced than it is now, it was thought that the distinction between long and short headed, dolichocephalic and brachycephalic people, pointed out by Retzius, indicated a primary division of the human species; but it was afterwards discovered that, although the character was useful otherwise, it was one of only secondary importance, as the long-headed as well as the short-headed group both included races otherwise of the strongest dissimilarity.

In all classifications, the point to be first ascertained is the fundamental plan of construction; but in cases where the fundamental plan has undergone but little modification, we are obliged to make use of what appear trivial characters, and compensate for their triviality by their number. The more numerous the combinations of specialised characters, by which any species or race differs from its congeners, the more confidence we have in their importance. The separation of what is essential from what is incidental or merely superficial in such characters lies at the root of all the problems of this nature that zoologists are called upon to solve; and in proportion as the difficulties involved in this delicate and often perplexing discrimination are successfully met and overcome will the value of the conclusions be increased. These difficulties, so familiar in zoology, are still greater in the case of anthropology. The differences we have to deal with are often very slight; their significance is at present very little understood. We go on expending time and trouble in heaping up elaborate tables of measurements, and minutely recording every point that is capable of description, with little regard to any conclusions that may be drawn from them. It is certainly time now to endeavour, if possible, to discriminate characters which indicate deep-lying affinity from those that are more transient, variable, or adaptive, and to adjust, as far as may be, the proper importance to be attached to each.

It is, however, quite to be expected that, in the infancy of all sciences, a vast amount of labour must be expended in

learning the methods of investigation. In none has this been
more conspicuous than in the subject under consideration.
Many have come to despair, for instance, of any good,
commensurate with the time it occupies, coming of the minute
and laborious work involved in craniometry. This is because
nearly all our present methods are tentative. We have not
yet learnt, or are only beginning to learn, what lines of
investigation are profitable and what are barren. The results,
even as far as we have gone, are, however, quite sufficient, in
my opinion, to justify perseverance. I am, however, not so
sure whether it be yet time to answer the demand, so eager
and so natural, which is being made in many quarters for the
formulation of a definite plan of examination, measurement, and
description to which all future investigation should rigidly ad-
here. All steps to promote agreement upon fundamental points
are to be cordially welcomed, and meetings or congresses con-
vened for such a purpose will be of use by giving opportunities
for the impartial discussion of the relative value of different
methods; but the agreement will finally be brought about by
the general adoption of those measurements and methods which
experience proves to be the most useful, while others will gradually fall into disuse by a kind of process of natural selection.

The changes and improvements which are being made
yearly, almost monthly, in instruments and in methods, show
what we should lose if we were to stop at any given period,
and decree in solemn conclave that this shall be our final
system, this instrument and this method shall be the only one
used throughout the world, that no one shall depart from it.
We scarcely need to ask how long such an agreement would
be binding. The subject is not sufficiently advanced to be
reduced to a state of stagnation such as this would bring it to.

To take an example from what is perhaps the most important of the anatomical characters by which man is distinguished from the lower animals, the superior from the
inferior races of man; the smaller or greater projection
forwards of the lower part of the face in relation to the skull
proper, or that which contains the brain. From the time
when Camper drew his facial angle, to the present day, the
readiest and truest method of estimating this projection has

occupied the attention of anatomists and anthropologists, and we are still far from any general agreement. Every country, every school, has its own system, so different that comparison with one another is well-nigh impossible. This is undoubtedly an evil; but the question is whether we should all agree to adopt one of the confessedly defective systems now in vogue, or whether we should not rather continue to hope for, and endeavour to find, one which may not be subject to the well-known objections urged against all.

We want, especially in this country, more workers—trained and experienced men, who will take up the subject seriously, and devote themselves to it continuously. Of such we may say, without offence to those few who have done occasional excellent work in physical anthropology, but whose chief scientific activity lies in other fields, we have not one. In the last number of the French *Revue d'Anthropologie*, a reference caught my eye to a craniometrical method in use by the "English school" of anthropologists. It was a reference only to a method which I had ventured to suggest, but which, as far as I know, has not been adopted by any one else. A school is just what we have not, and what we want—a body of men, not only willing to learn, but able to discuss, to criticise, to give their approval to, or reduce to its proper level, the results put forth by our few original investigators and writers. The rapidity with which any one of the most slender pretensions who ventures into the field (I speak from painful experience) is raised to be an oracle among his fellows is one of the most alarming proofs of the present barrenness of the land.

Another most urgent need is the collection and preservation of the evidences of the physical structure of the various modifications of man upon the earth. Especially urgent is this now, as we live in an age in which, in a far greater degree than any previous one, the destruction of races, both by annihilation and absorption, is going on. The world has never witnessed such changes in its ethnology as those now taking place, owing to the rapid extension of maritime discovery and maritime commerce, which is especially affecting the island population among which, more than elsewhere, the solution of the most important anthropological problems may be looked for. If we have at present neither the knowledge nor the leisure to examine and describe, we can at least preserve from destruction the materials for our successors to work upon. Photographs, models, anatomical specimens, skeletons or parts of skeletons, with their histories carefully registered, of any of the so-called aboriginal races, now rapidly undergoing extermination or degeneration, will be hereafter of inestimable value. Drawings, descriptions, and measurements are also useful, though in a far less degree, as allowance must always be made for imperfections in the methods as well as the capacity of the artist or observer. Such collections must be made upon a far larger scale than has hitherto been attempted, as, owing to the difficulties already pointed out in the classification of man, it is only by large numbers that the errors arising from individual peculiarities or accidental admixture can be obviated, and the prevailing characteristics of a race or group truly ascertained. It is only in an institution commanding the resources of the nation that such a collection can be formed, and it may therefore be confidently hoped that the Trustees of the British Museum will appropriate some portion of the magnificent new building, which has been provided for the accommodation of their natural history collections, to this hitherto neglected branch of the subject.

I have mentioned two of the needs of anthropology in this country—more workers and better collections: there is still a third—that of a society or institution in which anthropologists can meet and discuss their respective views, with a journal in which the results of their investigations can be laid before the public, and a library in which they can find the books and periodicals necessary for their study. All this ought to be provided by the Anthropological Institute of Great Britain and Ireland, which originated in the amalgamation of the old Ethnological and Anthropological Societies. But, as I intimated some time ago, the Institute does not at the present time flourish as it should; its meetings are not so well attended as they might be; the journal is restricted in its powers of illustration and printing by want of funds: the library is quite insufficient for the needs of the student.

This certainly does not arise from any want of good

management in the Society itself. Its affairs have been presided over and administered by some of the most eminent and able men the country has produced. Huxley, Lubbock, Busk, Evans, Tylor, and Pitt-Rivers have in succession given their energies to its service, and yet the number of its members is falling away, its usefulness is crippled, and its very existence seems precarious. Some decline to join the Institute, others leave it, upon the plea that, being unable from distance or other causes to attend the meetings, they cannot obtain the full return for their subscriptions; others on the ground that the Journal does not contain the exact information which they require.

There surely is to be found a sufficient number of persons who are influenced by different considerations, who feel that anthropological science is worth cultivating, and that those who are laboriously and patiently tracing out the complex problems of man's diversity and man's early history are doing a good work, and ought to be encouraged by having the means afforded them of carrying on their investigations and of placing the results of their researches before the world—who feel, moreover, that there ought to be some central body, representing the subject, which may, on occasion, influence opinion or speak authoritatively on matters often of great practical importance to the nation.

There must be many in this great and wealthy country who feel that they are helping a good cause in joining such a society, even if they are not individually receiving what they consider a full equivalent for their small subscription—many who feel satisfaction in helping the cause of knowledge, in helping to remove the opprobrium that the British Anthropological Society alone of those of the world is lacking in vitality, and in helping to prevent this country from falling behind all the nations in the cultivation of a science in which for the strongest reasons it might be expected to hold the foremost place. It is a far more grateful task to maintain, extend, and if need be improve, an existing organisation than to construct a new one. I feel, therefore, no hesitation in urging upon all who take interest in the promotion of the study of Anthropology to rally round the Institute, and to support the endeavours of the present excellent president to increase its usefulness.

PRESIDENTIAL ADDRESS TO THE SECTION OF ANTHROPOLOGY [1]

IT is not usual for the President of a Section of this Association to think it necessary to give any explanation of the nature of the subjects brought under its cognisance, or to emphasise their importance among other branches of study; but so general is the ignorance, or at all events vagueness of information, among otherwise well-instructed persons, that I will ask your permission to devote the short time accorded to me before the actual work of the Section begins to giving some account of the history and present position of the study of Anthropology in this country, and especially to indicate what this Association has done in the past, and is still doing, to promote it.

It is only ten years since the section in which we are now taking part acquired a definite and assured position in the organisation of the Association. The subject, of course, existed long before that time, and was also recognised by the Association, though with singular vicissitudes of fortune and position. It first appeared officially in 1846, when the "Ethnological sub-Section of Section D" (then called "Zoology and Botany") was constituted. This lasted till 1851, when Geography parted company from Geology, with which it had been previously associated in Section C, and became Section E, under the title of "Geography and Ethnology." In 1866 Section D changed its name to "Biology," with Physiology and Anthropology (the first occurrence of this word in our official proceedings) as separate "Departments"; but the latter does not seem to have regained its definite footing as a

[1] British Association for the Advancement of Science (Oxford Meeting), 9th August 1894.

branch of Biological Science until three years later (1869), when Section E, dropping Ethnology from its title, henceforward became Geography alone. The Department for the first two years (1869 and 1870) was conducted under the title of Ethnology, but in 1871 it resumed the name of Anthropology, given to it in 1866, and it flourished to such an extent, attracting so many papers and such large audiences, that it was finally constituted into a distinct Section, to which the letter H was assigned, and which had its first session at the memorable meeting at Montreal, exactly ten years ago, under the fitting and auspicious presidency of Dr. E. B. Tylor.

The history of the gradual recognition of Anthropology as a distinct subject by this Association is an epitome of the history of its gradual growth, and the gradual recognition of its position among other sciences in the world at large, a process still in operation and still far from complete. Although the word Anthropology had certainly existed, but used in a different sense, it was not till well into the middle of the present century that it, or any other word, had been thought of to designate collectively the scattered fragments of various kinds of knowledge bearing upon the natural history of man, which were beginning to be collected from so many diverse sources. Indeed, as I have once before upon a similar occasion remarked, one of the great difficulties with regard to making Anthropology a special subject of study, and devoting a special organisation to its promotion, is the multifarious nature of the knowledge comprehended under the title. This very ambition, which endeavours to include such an extensive range of subjects, ramifying in all directions, illustrating and receiving light from so many other sciences, appears often to overleap itself, and give a looseness and indefiniteness to the aims of the individual or the institution proposing to cultivate it. The old term Ethnology, or the study of peoples or races, has a limited and definite meaning. It treats of the resemblances and modifications of the different groups of the human species in their relations to each other, but Anthropology, as now understood, has a far wider scope. It treats of mankind as a whole. It investigates its origin and his relations to the rest of the universe. It invokes the aid of the sciences of zoology, comparative anatomy, and physiology, in its attempts to estimate the distinctions and resemblances between man and his nearest allies, and in fixing his place in the scale of living beings. In endeavouring to investigate the origin and antiquity of man, geology must lend its assistance to determine the comparative ages of the strata in which the evidences of his existence are found, and researches into his early history soon trench upon totally different branches of knowledge. In tracing the progress of the race from its most primitive condition, the characteristics of its physical structure and relations with the lower animals are soon left behind, and it is upon evidence of a kind peculiar to the human species, and by which man is so pre-eminently distinguished from all other living beings, that our conclusions mainly rest. The study of the works of our earliest known forefathers—"prehistoric archæology" as it is commonly called—is now almost a science by itself. It investigates the origin of all human culture, endeavours to trace to their common beginning the sources of our arts, customs, and history. The difficulty is, what to include and where to stop; as, though the term prehistoric may roughly indicate an artificial line between the province of the anthropologist and that which more legitimately belongs to the archæologist, the antiquary, and the historian, it is perfectly evident that the studies of the one pass insensibly into those of the others. Knowledge of the origin and development of particular existing customs throws immense light upon their real nature and importance; and conversely, it is often only from a profound acquaintance with the present or comparatively modern manifestations of culture that we are able to interpret the slight indications afforded us by the scanty remains of primitive civilisation.

It is considerations such as these that have caused the gradual introduction of the term Anthropology as a substitute for Ethnology, which I have traced in the history of this Association, and which is seen in other organisations for the cultivation of our science.

The first general association for the study of man in this country was founded in 1843, under the name of the "Ethnological Society" (three years, therefore, before the Ethnological sub-Section of Section D of this Association).   It did excellent work for many years under that title, but partly from personal considerations, and partly from a desire to undertake a wider and somewhat different field of research, another and in some senses a rival society, which adopted the name of "Anthropological," was founded in 1863.   For some years these existed side by side, each representing in its most active supporters different schools of the science.   This arrangement naturally involved a waste of strength, and it was felt that the interests of the subject would be promoted by an amalgamation of the two societies.   Many difficulties, chiefly, as is usual in such cases, of a personal nature, had to be overcome, one of the principal being the selection of a name for the united society. It was generally felt that "Anthropological" would be most appropriate, but the members of the old Ethnological Society could not bring themselves absolutely to sink the fact of their priority of existence, and all that they had done for science for so many years, by merging their society into that of their younger and active rivals; so after much discussion a compromise was effected, and the new organisation which arose from the coalescence of the two societies adopted the rather cumbrous title of *Anthropological Institute of Great Britain and Ireland*.   This was in 1871, and since that period the Society, as it is to all intents and purposes both in structure and function, has pursued a peaceful and useful course of existence, holding meetings at stated periods throughout the session, at which papers are read, and subjects of interest to anthropologists exhibited and discussed.   It has also published a quarterly journal, which has been the principal means in this country of communicating new information upon such subjects. The Institute has for twenty-three years performed this duty in a business-like and unostentatious manner, the only remarkable circumstance connected with its history being the singular want of interest taken by the outside world in its proceedings, considering their intrinsic importance to society, especially in an empire like ours, which more than any other affords a field for the study of man, under almost every aspect of diversity of race, climate, and culture.   At the present time it numbers only 305 ordinary members, whose subscriptions afford barely sufficient means to maintain the library and journal in a state of efficiency. The kindred Geographical and Zoological Societies have respectively 3775 and 2985 fellows, so far greater is the interest taken in the surface of the earth itself, and in the animals which dwell upon it, than in its human inhabitants!

Societies similar in their object to that the history of which I have just sketched have sprung up, and are now in a more or less flourishing condition, in every civilised country of the world. But confining our retrospect to our own country, we may take a glance at what has been done in recent years to promote the organised study of Anthropology otherwise than by means of this Association (to which I shall refer again later) or the Society of which I have just spoken.

One of the most potent means of registering facts, and making them available for future study and reference, is to be found in actual collections of tangible objects. To very considerable branches of anthropological science this method of fixing the evidence upon which our knowledge of the subject is based is particularly applicable. These branches are mainly two, very distinct from each other, and each representing one of the principal sides in which Anthropology presents itself.

I. Collections illustrating the physical structure of man, and its variations in the different races.

II. Collections showing his characteristic customs and methods of living, his arts, arms, and costumes, as developed under different circumstances and also modified by different racial conditions.

It is very rarely that these two are combined in one general arrangement, and they are almost always studied apart, the characteristics of mind, the general education and special training which are required for the successful cultivation of either being rarely combined in a single individual; and yet the complete history of any race of mankind, especially with regard to its relation to other races, must be based upon a knowledge both of its physical and psychical characteristics, and customs, habits, language, and tradition

largely help, when anatomical characters fail to separate and
define.

The anthropological museums of this country, as well as
elsewhere, are all of recent growth, and they are making
progress everywhere with steadily accelerating speed. This
cannot be better illustrated than in the place where we are
at the present time. Many of those who are now in this
room can remember when the materials for the study of
either branch of the subject in Oxford were absolutely non-
existent. I can myself recall the time when the site of the
handsome building which now houses the scientific treasures
of the University was a bare field. All who know the
modern history of Oxford must be aware that it was mainly
owing to the enthusiastic zeal and steady perseverance in the
cause of scientific education of one who is happily still among
us, the veteran Regius Professor of Medicine, Sir Henry
Acland, that that building was erected. The possession of a
well-selected and representative collection illustrating the
anatomical characters of the human species is chiefly owing
to the energetic labours of Professor Rolleston, one of the
brightest and noblest of Oxford's sons, a man of whom I
cannot speak without feelings of the strongest affection and
most profound regret for his untimely loss to the University
and the world.

The collection illustrating the arts and customs of primitive
people the University owes to the ingenuity and munificence
of General Pitt-Rivers, who not only provided the material
on which it is based, but also the original and unique scheme
of arrangement, which adds so greatly to its value as a means
of education, and is so admirably calculated to awaken an
interest in the subject, even in the minds of the most superficial visitor. In speaking thus of the method of displaying
the Pitt-Rivers collection, I must not be supposed to imply
any disparagement of others arranged on different plans.
Provided there is a definite and consistent arrangement of
some sort, it is well that there should be a diversity in the
treatment of different collections, and for such a vast and
exhaustive collection as that under the care of Sir Wollaston
Franks, at the British Museum, the geographical system

which has been adopted is certainly the best. In it every specimen of whatever nature at once finds a place, in which it can at any time be discovered and recognised.

In referring to our great national collection, I cannot refrain from saying that there seemed till lately to be only one element wanting to make it all that could be desired, and that was space, not only for the proper preservation and exhibition of what it already contains, but also for its inevitable future expansion. The provision in this respect was totally inadequate to do justice to the importance of the subject. Happily this consideration will be no longer a bar to the development of the collection. The provident action of the authorities of the museum, aided by the liberality of the Duke of Bedford, and the wisdom of Her Majesty's Government, has secured for many years to come the necessary room for the expansion of the grandest of our national institutions.

More modern even than museums has been the introduction of any systematic teaching of Anthropology into this country. This is certainly most remarkable, considering that there is no nation to which the subject is of such great importance. Its importance to those who have to rule—and there are few of us now who are not called upon to bear our share of the responsibilities of government—can scarcely be overestimated in an empire like this, the population of which, as I have just said, is composed of examples of almost every diversity under which the human body and mind can manifest itself. The physical characteristics of race, so strongly marked in many cases, are probably always associated with equally or more diverse characteristics of temper and intellect. In fact, even when the physical divergences are weakly shown, as in the different races which contribute to make up the home portion of the Empire, the mental and moral characteristics are still most strongly marked. As the wise physician will not only study the particular kind of disease under which his patient is suffering before administering the approved remedies for such disease, but will also take into careful account the peculiar idiosyncrasy and inherited tendencies of the individual, which so greatly modify both the course of the disease and

the action of remedies, so it is absolutely necessary for the
statesman who would govern successfully not to look upon
human nature in the abstract and endeavour to apply
universal rules, but to consider the special moral, intellectual,
and social capabilities, wants, and aspirations of each particular
race with which he has to deal.   A form of government under
which one race would live happily and prosperously may to
another be the cause of unendurable misery.   All these
questions then should be carefully studied by those who have
any share in the government of people belonging to races alien
to themselves.   A knowledge of their special characters and
relations to one another has a more practical object than the
mere satisfaction of scientific curiosity; it is a knowledge
upon which the happiness and prosperity or the reverse of
millions of our fellow-creatures may depend.   The ignorance
often shown upon these subjects, even in so select an assembly
as the House of Commons, would be ludicrous if it were not
liable to lead to disastrous results.

Now let us consider what, amid all the complex, diverse,
and costly machinery of education in this country, is being
done to satisfy the demands for such knowledge.   We may
say at once, as regards all institutions for primary and secondary
education, absolutely nothing.   The inhabitants of the various
regions of our own earth are treated with no more consideration and interest in all such institutions than if they lived
on the moon or the planets.   We must turn straight to the
higher intellectual centres in the hope of finding any anthropological teaching.   Here, at Oxford, if anywhere, we may
expect to find it, and here, first among the British Universities,
have we seen, since the year 1883, among the list of the subjects taught the word "Anthropology," but the teacher, though
one of the most learned of men in the subject the country has
produced, still only bears the modest title of "Reader."   A
professorship of Anthropology does not exist at present in the
British Isles, and even here the subject, though recognised as
a "special," offers little field for distinction in the examinations
for degrees, and has therefore never been taken up in a
thorough manner by students.   Dr. Tylor's lectures must,
however, have done much to have spread an intelligent

interest in some branches of Anthropology, and have proved a
valuable complement to the Pitt-Rivers collection, as have
also the courses which have been given by Mr. Henry Balfour
upon the arts of mankind and their evolution, one of which I
am glad to see is announced among the advantages offered
to the University Extension students at present with us.
Physical Anthropology has also been taken up by Professor
A. Thomson, who, I understand, gives instructive lectures
upon it, open to the members of his class of human anatomy.
At the opposite end almost of the subject must be mentioned
the extension and organisation of the Ashmolean Museum
under the care of Mr. Arthur Evans, which has a bearing
upon some branches of Anthropology, and the foundation of
the Indian Institute under the auspices of Sir Monier Monier-
Williams, which must give an impetus to the study of the
characteristics of the races of our great Empire in the East.
Last, but by no means least in its bearing upon the origin,
divisions, and diffusion of races, is the world-famous linguistic
work of Professor Max Müller and Professor Sayce, both of
whom have presided over this Section at former meetings of
the Association.

Of the sister University I wrote thus in 1884 : " In
Cambridge there are many hopeful signs.   The recently
appointed Professor of Anatomy, Dr. Macalister, is known to
have paid much attention to Anatomical Anthropology, and
has already intimated that he proposes to give instruction in
it during the summer term.   An Ethnological and Archæological
Museum is also in progress of formation, which, if not destined
to rival that of Oxford, already contains many objects of
great value, and a guarantee of its good preservation and
arrangement may be looked for in the appointment of Baron
Anatole von Hügel as its first curator."

Ten years have passed, and it is satisfactory to know that
the teaching of Anthropology has not only been fairly established, but the subject has also found a place in the scheme
of University examination.   The learned Professor of Human
Anatomy continues to take a wide view of his functions,
giving a course during the Easter term on the methods of
Physical Anthropology and also museum demonstrations on

S

craniometry and osteometry, by the aid of a greatly increased
and continually augmenting collection of specimens. Those
students who take anatomy as their subject for the second
part of the Natural Science Tripos have both paper work and
practical examination in Anthropology, each man having a
skull placed in his hands, of which he is expected to make a
complete diagnostic description. For the first part of the
tripos each candidate has one or more questions on the broad
general principles of the subject. Professor Macalister informs
me that he has always at least six men who go through a very
thorough practical course with their own hands. There has
also lately been established a course of lectures on the
Natural History of the Races of Man, delivered during the
Michaelmas and Lent terms by Dr. Hickson, of Downing
College, and Baron von Hügel gives a course of museum
demonstrations on the weapons, ornaments, and other objects
in the Ethnological Museum, which is open to all students,
and of which many take advantage.

In London, owing to the chaotic condition of all forms of
higher instruction, which has been brought so prominently
into notice by the universal demand for a teaching University
(an aspiration which the labours of the late Gresham Commission certainly seem to have brought nearer to realisation
than ever appeared possible before), all systematic anthropological teaching has been entirely neglected. The great
collections to which I have already alluded, that of arts and
customs at the British Museum, and that of osteological
specimens at the Royal College of Surgeons, have by their
steady augmentation done valuable service in preserving a
vast quantity of material for future investigation and instruction, and students have at present all reasonable facilities for
pursuing their own researches in them. Lectures have never
formed any part of the official programme of the British
Museum, but at the College of Surgeons it is otherwise, and
though the contents of the collections are specially indicated
as the subject on which they should be delivered, for the last
ten years at least, Anthropology, notwithstanding the magnificent material at hand for its illustration, has had no place in
the annual syllabus. It is also entirely ignored in the

examination scheme of the University of London, an institution which prides itself as being on a level with modern educational requirements; and the managers of the new Imperial Institute, casting about in all directions for some worthy object to occupy their energies and their spacious buildings, do not appear to have taken into serious consideration the value to the world and the appropriateness to their original design of a great central school of Anthropology, from which might emanate a full and satisfying knowledge of the characteristics of all the various races of which the Empire is composed.

In Scotland the recent Universities Commission has recognised Physical Anthropology as a branch of human anatomy in their scheme for graduation in pure science, the examination on this subject embracing a knowledge of race characters as found in the skull and other parts of the skeleton, in the skin, eyes, hair, features, and the external configuration of the body generally; the methods of anthropometrical measurement, both of the living body and the skeleton; the possible influence of use and of external surroundings in producing modifications in the physical characters of man, and an acquaintance with the "types" of mankind and the structural relations of man to the higher mammals. These regulations came into operation in the University of Edinburgh in 1892, and in accordance with them Professor Sir William Turner delivers a special course of twenty-five lectures on Physical Anthropology, and in addition ten practical demonstrations on osteometry. The museum under his charge has greatly increased of late in number and value of the specimens. But "Human Anatomy, including Anthropology," being only one of a series of nine subjects in any three or more of which a final science examination on a higher standard has to be passed, there is not at present any considerable number of students who take it up, and the other Scotch Universities have not yet thought it necessary to establish distinct courses of Physical Anthropology, although it is becoming more and more a regular part of the anatomical teaching to advanced students.

For the following account of what is being done to further

the knowledge of our subject in the sister isle I am indebted
to Professor D. J. Cunningham. The only place in Ireland
where anthropological work is done is Trinity College. For
many years those in charge of the museum have been collecting
skulls, and they were fortunate in obtaining the greater part
of Sir William Wilde's collection. To these great additions
have been recently made, principally in the form of Irish
crania from different districts. All the anthropological specimens are lodged in one large room, which is also used as an
anthropometric laboratory. Though there has never been
any systematic teaching of Anthropology in Trinity College,
Dr. C. R. Browne (Professor Cunningham's able assistant),
who takes charge of the laboratory, attends for two hours on
three days a week, and gives demonstrations in anthropological
methods to any students who are interested in the subject.
The laboratory was opened in June 1891, the instruments
being provided by a grant from the Royal Irish Academy,
and about 500 individuals have already been measured, the
greater number of them students of the College. This is,
however, only part of the work carried out by the laboratory.
Every year the instruments are taken to some selected district
in Ireland, and a systematic study of the inhabitants is made.
The Aran Islands, and also the islands of Inishbofin and
Inishshark, have been already worked out, and this year
excursions are organised to Kerry, to a district in Wicklow,
and to another in the west of Ireland. The Academy makes
yearly grants to the Committee for carrying on this work,
the results of which have been published in admirable
memoirs by Professor A. C. Haddon and Dr. C. R. Browne.
The Science and Art Museum in Dublin, under the direction
of Dr. V. Ball, contains a small collection, arranged with a
view to general instruction, showing by means of skulls and
casts the physical characteristics of the different races of man,
those of each race being explained by a short printed label,
and its range shown on a map.

Though the development of anthropological science has
thus not been greatly advanced, in this country at least, by
means of endowments, or by aid of the State, or, till very
recently, by our great scholastic institutions, but has been

mainly left to the unorganised efforts of amateurs of the
subject, its progress in recent years has been undeniably
great.   I will give an instance of the strides that have been
made in one of its most important branches.

Physical or Anatomical Anthropology, or the study of the
modifications of the human body under its various aspects, the
modifications dependent upon sex and age, the modifications
dependent upon race, and those dependent upon individual
variability, studied not many years ago in a vague and loose
manner, has gradually submitted to a rigorous and, therefore,
strictly scientific method of treatment.   The generalities
which were formerly used to express the differences that were
recognised between the various subjects compared with each
other have been replaced by terms conveyed in almost mathe-
matical precision.   No one acquainted with the history of
the development of this branch of Anthropology can fail to
recognise how much it was accelerated by the genius of Broca,
and the school which he established in France, although all
cultivated nations are now vying with each other in the
practice of exactitude in anthropological research, and the
time seems rapidly approaching when a common agreement
will be arrived at, by which all the observations which may
be made, under whatever diverse circumstances, and by whatever different individuals, will be available for comparison one
with another.

This branch of our science has received the name of
" Anthropometry."   Although, as the name implies, measurement is one of its principal features, it includes such other
methods of comparison as can be reduced to a definite standard,
or to which definite tests can be applied, such as the colour of
the hair, eyes, and complexion, the form of the ear and nose.
The great desiderata that have been sought for, and gradually
attained, in measuring either the skeleton or the living person
have been two in number : 1. Exact definition of the points
between which the measurements should be taken.   2. Exact
methods and instruments of measurement.   In both these
cases the object looked for has been not only that the
measurements taken by the same observer at different times
and under different circumstances should coincide, but also

that those taken by different observers should be comparable. These requirements seem so simple and natural at first sight that the majority of persons whom I am addressing will wonder that I should allude to them. Only those who are seriously occupied, or perhaps I should rather say, only those who were seriously occupied a few years ago with the endeavour to solve these problems can have any idea of their difficulty. The amount of time and labour that has been spent upon them is enormous, but the result has, I think, been quite commensurate with it.

We have attained at last to methods of measurement and standards of comparison which, in the hands of persons of ordinary intelligence, and with a moderate amount of training, will give data which may be absolutely depended upon. From these we hope to be able to formulate accurate information as to the physical conformation of all the groups into which mankind is divided, and so gradually to arrive at a natural classification of those groups, and a knowledge of their affinities one to another.

But the exact methods of modern Anthropometry are not only important on account of the aid they give in studying the race characteristics of man. As has so often happened when scientific observation has been primarily carried out for its own sake, it ultimately leads to practical applications undreamt of by its earlier cultivators. The application of Anthropometry not to the comparison of races, but to elucidate various social problems—as the laws of growth, of heredity, of comparative capacities of individuals within a community, and the effects of different kinds of education and occupation, as worked out first by Quetelet in Belgium, and subsequently by Francis Galton, Roberts, and others in this country, and its still more concrete application as an aid in administering justice by methods perfected by Bertillon in France, are striking illustrations of the practical utility of labours originally undertaken under the influence of devotion to science pure and simple.

The importance of being able to determine the identity of an individual under whatever circumstances of disguise he may be presented for examination has, of course, long been

apparent to all who have had anything to do with the administration of the criminal law, and rough and ready methods of
recognition, depending mainly upon the more or less acute
faculty of perception and recollection of differences and resemblances, possessed by the persons upon whom the duty of
identification has devolved, have long been in operation. The
general conformation, height, form of features, and colour of
complexion, hair, and eyes, have also been noted. Much
additional assistance has been obtained by the registration of
definite physical characteristics, the results either of natural
conformation, or of injury, such as mutilations, tattoo-marks,
and scars, inflicted by accident or design. The application of
one of the most important scientific discoveries of the age,
photography, was eagerly seized upon as a remedy for the
difficulties hitherto met with in tracing personal identity, and
enormous numbers of photographs were taken of persons the
peculiarities of whose career led them to fall into the hands of
the police, and who were likely to be wanted again on some
future occasion. No doubt much help has been derived from
this source, but also much embarrassment. Even among
photographic portraits of one's own personal friends, taken
under most favourable circumstances, and with no intention of
deception, we cannot often help exclaiming how unlike they
are to the person represented. With portraits of criminals,
the varying expression of the face, changes in the mode of
wearing the hair and beard, differences of costume, the effects
of a long lapse of time, years perhaps passed in degradation
and misery, may make such alterations that recognition
becomes a matter at least of uncertainty. That photographs
are extremely valuable as aids to identification, when their
true position in the process is recognised, cannot be doubted,
but as a primary method they have been found to be quite
inapplicable, owing partly to the causes just indicated, but
mainly to the difficulty, if not impossibility, of classifying
them. The enormous expenditure of time and trouble that
must be consumed in making the comparison between any
suspected person and the various portraits of the stock which
accumulates in prison bureaus may be judged of from the fact
that, in Paris alone, upwards of 100,000 such portraits of

persons interesting to the police have been taken in a period of ten years.

The primary desideratum in a system of identification is a ready means of classifying the data upon which it is based. To accomplish this is the aim of the Bertillon system. Exact measurements are taken between certain well-known and fixed points of the bony framework of the body, which are known not to change under different conditions of life. The length and breadth of the head, the length of the middle finger, the length of the foot, and the length of the forearm, are considered the best, though others are added for greater certainty, as the height, span of arms, length of ear, colour of eyes, etc. All these particulars of every individual examined are recorded upon a card, and by dividing each measurement into three classes, long, medium, and short, and by classifying the various combinations thus obtained, the whole mass of cards, kept arranged in drawers in the central bureau, is divided up into groups, each containing a comparatively small number, and therefore quite easily dealt with. When the card of a new prisoner is brought in, a few minutes suffice to eliminate the necessity of comparison with any but one small batch, which presents the special combination. Then photographs and other means of recognition, as distinctive marks and form of features, are brought into play, and identification becomes a matter of certainty. On the other hand, if the combination of measurements upon a new card does not coincide with any in the classed collection in the bureau, it is known with absolute certainty that the individual being dealt with has never been measured before.

One of the most striking results of the introduction of this system into France has been that, since it has been brought fully into operation, a large proportion of old offenders, knowing that concealment is hopeless, admit their identity at once, and save a world of trouble and expense to the police by ceasing to endeavour to conceal themselves under false names.

Various representations upon this subject have been addressed to the Home Secretary of our own Government during the last few years, and among others one from the Council of this Association, which originated in a resolution of this Section, adopted by the General Committee at the meeting at Edinburgh in 1892, to this effect:

"That the Council be requested to draw the attention of Her Majesty's Government to the Anthropometric Method for the measurement of criminals, which is successfully in operation in France, Austria, and other continental countries, and which has been found effective in the identification of habitual criminals, and consequently the prevention and repression of crime."

In consequence of these representations a Committee was appointed on 21st October 1893 by Mr. Asquith, consisting of Mr. C. E. Troup, of the Home Office, Major Arthur Griffiths, Inspector of Prisons, and Mr. Melville Leslie Macnaghten, Chief Constable in the Metropolitan Police Force, with Mr. H. B. Simpson, of the Home Office, as Secretary, "to inquire (*a*) into the method of registering and identifying habitual criminals now in use in England; (*b*) into the 'Anthropometric' system of classified registration and identification in use in France and other countries; (*c*) into the suggested system of identification by means of a record of finger marks; to report whether the anthropometric system or the finger-mark system can with advantage be adopted in England either in substitution for or to supplement the existing methods; and, if so, what arrangements should be adopted for putting them into practice, and what rules should be made under Section 8 of the Penal Servitude Act, 1891, for the photographing and measuring of prisoners."

The Report of this Committee, with minutes of evidence and appendices, was issued as a Parliamentary Blue-book in March last, and not only contains a lucid and concise description of the methods of identification already in use in this country, but also most striking testimony from impartial but well-qualified persons to the value of a more scientific mode of dealing with the subject. No pains seem to have been spared to obtain, both by personal observation and by the examination of competent witnesses, a thorough knowledge of the advantages of the Bertillon system as practised in France, and the result has been the recommendation of that system, with certain modifications, for adoption in this country, with the addition of the remarkably simple, ingenious, and certain method of personal identification first used in India by Sir William Herschel, but fully elaborated in this country by Mr. Francis Galton—that called the "finger-mark system," about which I shall have a few more words to say presently.

With the concluding words of the Committee's Report I most fully concur: "We may confidently anticipate that, if fairly tried, it will show very satisfactory results within a few years in the metropolis; but the success of its application in the country generally will depend on the voluntary co-operation of the independent county and borough police forces. This, we feel sure, will not be withheld. When the principles of the system are understood and its usefulness appreciated, we believe it will not only save much time and labour to the police in the performance of an important duty, but will give them material assistance in tracing and detecting the antecedents of the guilty, and will afford, so far as its scope extends, an absolute safeguard to the innocent."

It is very satisfactory to be able to add that in the House of Commons on 26th June, in answer to a question from Colonel Howard Vincent, the Home Secretary announced that the recommendations of the Committee have been adopted; and that, in order to facilitate research into the judicial antecedents of international criminals, the registers of measurements would be kept on the same plan as that adopted with such success in France, and also in other continental countries.

I have just mentioned the "finger-mark system," and of all the various developments of Anthropology in recent times none appears to me more interesting than the work done by Mr. Galton upon this subject; for though, as indicated above, he is not quite the first who has looked into the question or shown its practical application in personal identification, he has carried his work upon it far beyond that of any of his predecessors, both in its practical application and into regions of speculation unthought of by any one else. Simple and insignificant as in the eyes of all the world are the little ridges and furrows which mark the skin of the under-surface of our fingers, existing in every man, woman, and child born into the world, they have been practically unnoticed by every one until Mr. Galton has shown, by a detailed and persevering study of their peculiarities, that they are full of significance, and amply repay the pains and time spent upon their study. It is not to be supposed that all the knowledge that may be obtained from a minute examination of them is yet by any means exhausted, but they have already given valuable data for the study of such subjects as variation unaffected by natural or any other known form of selection, and the difficult problems of heredity, in addition to their being one of the most valuable means hitherto discovered of fixing personal identity.

As an example of the importance of some ready method to prove identity, apart from its application to the detection, punishment, and prevention of crime, to which I have already referred, I may recall to your recollection that remarkable trial which agitated the length and breadth of the land rather more than twenty years ago ; a trial which occupied so many months of the precious time of our most eminent judges and counsel, and cost the country, as well as several innocent persons—I am afraid to say how many— thousands of pounds, all upon an issue which might have been settled in two minutes if Roger Tichborne, before starting on his voyage, had but taken the trouble to imprint his thumb upon a piece of blackened paper. It is wonderful to me, on reading again the reports of the trial, to see how comparatively little attention was paid by counsel, judge, or jury, to the extremely different physical characteristics of the two persons claimed to be identical, but which were so strongly marked that they ought to have disposed of the claim, without any hesitation, at the very opening of the case. It was not until the 102nd day of the first trial that the attention of the jury was pointedly called to the fact that it was known that Sir Roger Tichborne had been tattooed on the left arm with a cross, anchor, and a heart, and that the Claimant exhibited no such marks. When this was clearly brought out and proved, the case broke down at once. The second trial for perjury occupied the court 188 days, the Lord Chief-Justice's charge alone lasting eight days. The issues were, however, more complex than in the first trial, as it was not only necessary to prove that the Claimant was not Tichborne, but also to show that he was some one else.   I feel convinced that at the present time the greater confidence that is reposed in the methods of Anthropometry or close observance of physical characters, and in the persistence of such characters through life, would have greatly simplified the whole case; and I would strongly recommend all who have nothing about their lives they think it expedient to conceal to place themselves under the hands of Mr. Galton, or one of his now numerous disciples, and get an accurate and unimpeachable register of all those characteristics which will make loss of identity at any future period a sheer impossibility.

Partly with this object in view, the Association has, for several years past, during each of its meetings, opened, under the superintendence of Dr. Garson, an Anthropometric Laboratory, on the plan of the admirable institution of the same name which has been carried on in the South Kensington Museum since the beginning of the year 1888, under the direction and at the sole cost of Mr. Francis Galton, in which up to the present time more than 7000 complete sets of measurements have been made and recorded. The results obtained at the British Association meetings have been published in the Annual Reports of the Association, and though on a smaller scale than Mr. Galton's, the operations of the laboratory have been most useful in diffusing a knowledge of the value of anthropometric work, and of the methods by which it is carried on.

For many years an "Anthropometric" Committee of the Association, in which the late Dr. W. Farr, Mr. F. Galton, Mr. C. Roberts, Dr. Beddoe, Sir Rawson Rawson, and others, took an active part, was engaged in collecting statistical information relating to the physical characters, including stature, weight, chest-girth, colour of eyes and hair, strength of arms, etc., of the inhabitants of the British Isles; and their reports, illustrated by maps and diagrams, were published in the annual volume issued by the Association. This Committee terminated its labours in 1883, although, as was fully acknowledged in the concluding report, the subject was by no means completely exhausted.

A great and important work which the Association has now in hand, in some sense a continuation of that of the Anthropometric Committee, though with a more extended scope of operation, is the organisation of a complete ethnographical survey of the United Kingdom based upon scientific principles. In this work the Association has the co-operation of the Society of Antiquaries of London, the Folk-lore Society, the Dialect Society, and the Anthropological Institute. Representatives of these different bodies have been formed into a Committee, of which Mr. E. W. Brabrook is now chairman. It is proposed to record in a systematic and uniform character, for certain typical villages and the neighbouring districts (1), the Physical Types of the Inhabitants, (2) their current Traditions and Beliefs, (3) Peculiarities of Dialect, (4) Monumental and other Remains of Ancient Culture, and (5) Historical Evidence as to Continuity of Race. The numerous Corresponding Societies of the Association scattered over various parts of the country have been invited to co-operate, and the greater number of them have cordially responded, and special local committees have been formed in many places to carry out the work.

The result of a preliminary inquiry as to the places in the United Kingdom which appeared especially to deserve ethnographic study, mainly on account of the stationary nature of the population for many generations back, was given in the first Report of the Committee presented at the Nottingham meeting of the Association last year, in which it was shown that in the British Isles there are more than 250 places which, in the opinion of competent authorities, would be suitable for ethnographic survey, and in which, notwithstanding the rapid changes which have taken place during the last fifty years in all parts of the country, much valuable material remains for the Committee to work upon. Without doubt, as interest in the subject is aroused, this number will be greatly increased.

A most important step in securing the essential condition that the information obtained should be of the nature really required for the purpose, and that the records of different observers should be as far as possible of equal value and comparable one with another, has been the compilation of a very elaborate and carefully prepared schedule of questions and directions for distribution among those who have signified their willingness to assist, and as a guarantee that the answers obtained to the questions in the schedules will be utilised to the fullest extent, certain members of the Committee, specially qualified for each branch of the work, have undertaken to examine and digest the reports when received.

It may be remarked in passing that the Anthropological Society of Paris has within the past year formed a Commission of its members to collect, in a systematic manner, the scattered data which, when united and digested, shall form " une anthropologie véritablement nationale de la France," and has issued a circular with schedules of the required observations.   These are, however, at present limited to the physical characters of the population.

Among the many services rendered to the science of Anthropology by the British Association, not the least has been the aid it has afforded in the publication of that most useful little manual entitled *Notes and Queries on Anthropology*, of which the first edition was brought out exactly twenty years ago (1874), under the supervision and partly at the expense of General Pitt-Rivers.   Since that time the subject has made such great advances that a second edition, brought up to the requirements of the present time, was urgently called for.   A Committee of the British Association, appointed to consider the report upon the best means of doing this, recommended that the work should be placed in the hands of the Anthropological Institute of Great Britain and Ireland. This recommendation was approved by the Association, and grants amounting to £70 were made to assist in defraying the cost of publication.   The Council of the Anthropological Institute appointed a Committee of its members to undertake the revision of the different subjects, with Dr. J. G. Garson and Mr. C. H. Read as editors respectively of the two parts into which it is divided.   The work was published at the end of the year 1892, and is invaluable to the traveller or investigator in pointing out the most important subjects of inquiry, and in directing the observations he may have the means of making into a methodical and systematic channel.

Besides those I have already mentioned, the Association has aided many other anthropological investigations by the appointment of Committees to carry them out, and in some cases by the more substantial method of giving grants from its funds, and by defraying the cost of publication of the results in its journal. Among these I may specially mention the series of very valuable Reports upon the Physical Characters, Languages, and Industrial and Social Condition of the North-Western Tribes of the Dominion of Canada, drawn up by Mr. Horatio Hale, Dr. F. Boas, and others, the importance of which has been recognised by the Canadian Government in the form of a grant in aid of the expenses.

Another very interesting investigation into the Habits, Customs, Physical Characteristics, and Religion of the Natives of Northern India, initiated by Mr. H. H. Risley, and carried on under his supervision by the Indian Government, though it has received little more than moral support from the Association, may be mentioned here on account of the illustration it affords of the value of exact anthropometric methods in distinguishing groups of men. Although a practised eye can frequently tell at a glance the tribe or caste of a man brought before him for the first time, the special characters upon which the opinion is based have not hitherto been reduced to any definite and easily comparable method of description. In Mr. Risley's examination, the nose, for instance (which I have always held to be one of the most important of features for classificatory purposes), instead of being vaguely described as broad or narrow, is accurately measured, and the proportion of the greatest width to the length (from above downwards), or the " nasal index," as it is termed (though it must not be confounded with the nasal index as defined by Broca upon the skull), gives a figure by which the main elements of the composition of this feature in any individual may be accurately described. The average or mean nasal indices of a large number of individuals of any race, tribe, or caste offer means of comparison which bring out most interesting results. By this character alone the Dravidian tribes of India are easily separated from the Aryan. " Even more striking," as Mr. Risley remarks, " is the curiously close correspondence between the

gradations of racial type indicated by the nasal index and certain
of the social data ascertained by independent inquiry.   If we
take a series of castes in Bengal, Behar, or the North-Western
Provinces, and arrange them in the order of the average nasal
index, so that the caste with the finest nose shall be at the
top, and that with the coarsest at the bottom of the list, it
will be found that this order substantially corresponds with
the accepted order of social precedence.   The casteless tribes
—Kols, Korwas, Mundas, and the like—who have not yet
entered the Brahmanical system occupy the lowest place in
both series.   Then come the vermin-eating Musubars, and the
leather-dressing Chamárs.   The fisher castes of Bauri, Bind,
and Kewat are a trifle higher in the scale; the pastoral Goala,
the cultivating Kurmi, and a group of cognate castes—from
whose hands a Brahman may take water—follow in due order;
and from them we pass to the trading Khatris, the land-
holding Bábhans, and the upper crust of Hindu society.   Thus,
it is scarcely a paradox to lay down as a law of the caste
organisation in Eastern India that a man's social status varies
in inverse ratio to the width of his nose."   The results
already obtained by this method of observation have been so
important and interesting that it is greatly to be hoped that
the inquiry may be extended throughout the remainder of our
Indian Empire.

But for want of time I might here refer to the valuable
work done in relation to the natives of the Andaman Islands,
a race in many respects of most exceptional interest, first by
Mr. E. H. Man, and more recently by Mr. M. V. Portman, and
for the same reason can scarcely glance at the great progress
that is being made in anthropological research in other
countries than our own.   The numerous workers on this
subject in the United States of America are, with great
assistance from the Government, very properly devoting themselves to exploring, collecting, and publishing, in a systematic
and exhaustive manner, every fact that can still be discovered
relating to the history, language, and characters of the
aboriginal population of their own land.   They have in this
a clear duty set before them, and they are doing it in splendid
style.   I wish we could say that the same has been done

with all the native populations in various parts of the world
which have been, to use a current phrase, " disestablished and
disendowed" by our own countrymen.   We are, however, now,
as I have shown, not altogether unmindful of what is our
duty to posterity in this respect,—a duty, perhaps, more urgent
than that of any other branch of scientific investigation, as it
will not wait.   It must be done, if ever, before the rapid
spread of civilised man all over the world, one of the most
remarkable characteristics of the age in which we live, has
obliterated what still remains of the original customs, arts, and
beliefs of primitive races ; if, indeed, it has not succeeded—as
it too often does—in obliterating the races themselves.

XVIII

ON THE CLASSIFICATION OF THE VARIETIES OF THE HUMAN SPECIES [1]

ON the occasion of the Anniversary Meeting of the Institute last year I endeavoured to sum up in a few words the principal aims and scope of the science of Anthropology as now understood.

I then gave reasons for my belief that the discrimination and description of the characteristics of the various races of men is one of, if not the most practically important of the different branches into which the whole of the great subject is divided. It was also stated that although other characters, such as those derived from language, social customs, traditions, religious beliefs, and from intellectual and moral attributes, were by no means to be neglected, structural or anatomical characters are those upon which in the end most reliance must be placed in discriminating races.

I propose now to give a brief summary of the results attained up to the present time by the study of the racial characters of the human species, and to show what progress has been made towards arriving at a natural classification of the varieties into which the species may be divided.

The most ordinary observation is sufficient to demonstrate the fact that certain groups of men are strongly marked from others by definite characters common to all members of the group, and transmitted regularly to their descendants by the laws of inheritance. The Chinaman and the Negro, the native of Patagonia and the Andaman Islander, are as distinct

---

[1] Address delivered at the Anniversary Meeting of the Anthropological Institute of Great Britain and Ireland, 27th January 1885.

from each other structurally as are many of the so-called
species of any natural group of animals. Indeed, it may be
said with truth that their differences are even greater than
those which mark the groups called genera by many naturalists
of the present day. Nevertheless, the difficulty of parcelling
out all the individuals composing the human species into
certain definite groups, and of saying of each man that he
belongs to one or other of such groups, is insuperable. No
such classification has ever, or indeed can ever, be obtained.
There is not one of the most characteristic, most extreme
forms, like those I have just named, from which transitions
cannot be traced by almost imperceptible gradations to any
of the other equally characteristic, equally extreme, forms.
Indeed, a large proportion of mankind is made up, not of
extreme or typical, but of more or less generalised or intermediate, forms, the relative numbers of which are continually
increasing, as the long-existing isolation of nations and races
breaks down under the ever-extending intercommunication
characteristic of the period in which we live.

The difficulties of framing a natural classification of man,
or one which really represents the relationship of the various
minor groups to each other, are well exemplified by a study
of the numerous attempts which have been made from the
time of Linnæus and Blumenbach downwards. Even in the first
step of establishing certain primary groups of equivalent
rank there has been no accord. The number of such groups
has been most variously estimated by different writers from
two up to sixty or more, although it is important to note that
there has always been a tendency to revert to the four
primitive types sketched out by Linnæus — the European,
Asiatic, African, and American — expanded into five by
Blumenbach by the addition of the Malay,[1] and reduced by
Cuvier to three by the suppression of the last two. After a
perfectly independent study of the subject, extending over
many years, I cannot resist the conclusion, so often arrived at
by various anthropologists, and so often abandoned for some
more complex system, that the primitive man, whatever he

---

[1] The Malay of Blumenbach was a strange conglomeration of the then little
known Australian, Papuan, and true Malay types.

may have been, has in the course of ages divaricated into three extreme types, represented by the Caucasian of Europe, the Mongolian of Asia, and the Ethiopian of Africa, and that all existing individuals of the species can be ranged around these types, or somewhere or other between them. Large numbers are doubtless the descendants of direct crosses in varying proportions between well-established extreme forms; for, notwithstanding opposite views formerly held by some authors on this subject, there is now abundant evidence of the wholesale production of new races in this way. Others may be the descendants of the primitive stock, before the strongly marked existing distinctions had taken place, and therefore present, though from a different cause from the last, equally generalised characters. In these cases it can only be by most carefully examining and balancing all characters, however minute, and finding out in what direction the preponderance lies, that a place can be assigned to them. It cannot be too often insisted on that the various groups of mankind, owing to their probable unity of origin, the great variability of individuals, and the possibility of all degrees of intermixture of races at remote or recent periods of the history of the species, have so much in common that it is extremely difficult to find distinctive characters capable of strict definition, by which they may be differentiated. It is more by the preponderance of certain characters in a large number of members of a group, than by the exclusive or even constant possession of these characters, in each of its members, that the group as a whole must be characterised.

Bearing these principles in mind, we may endeavour to formulate, as far as they have as yet been worked out, the distinctive features of the typical members of the three great divisions, and then show into what subordinate groups each of them seems to be divided.

To begin with the Ethiopian, Negroid or Melanian, or "black" type. It is characterised by a dark, often nearly black, complexion; black hair, of a kind called "frizzly" or, incorrectly, "woolly," *i.e.*, each hair is closely rolled up upon itself, a condition always associated with a more or less flattened or elliptical transverse section; a moderate or scanty development of beard; an almost invariably dolichocephalic skull; small and moderately retreating malar bones (mesopic face); a very broad and flat nose, platyrhine in the skeleton; moderate or low orbits; prominent eyes; thick, everted lips; prognathous jaws; large teeth (macrodont); a narrow pelvis (index in the male 90 to 100); a long forearm (humero-radial index 80), and certain other proportions of the body and limbs which are being gradually worked out and reduced to numerical expression as material for so doing accumulates.

The most characteristic examples of the second great type, the Mongolian or Xanthous, or "yellow," have a yellow or brownish complexion; black, coarse, straight hair, without any tendency to curl, and nearly round in section, on all other parts of the surface except the scalp, scanty and late in appearing; a skull of variable form, mostly mesocephalic (though extremes both of dolichocephaly and brachycephaly are found in certain groups of this type); a broad and flat face, with prominent, anteriorly - projecting malar bones (platyopic face); nose small, mesorhine or leptorhine; orbits high and round, with very little development of glabella or supraciliary ridges; eyes sunken, and with the aperture between the lids narrow; in the most typical members of the group with a vertical fold of skin over the inner canthus, and with the outer angle slightly elevated; jaws mesognathous; teeth of moderate size (mesodont). The proportions of the limbs and form of the pelvis have yet to be worked out, the results at present obtained showing great diversity among different individuals of what appear to be well-marked races of the group, but this is perhaps due to the insufficient number of individuals as yet examined with accuracy.

The last type, which, for want of a better name, I still call by that which has the priority, Caucasian or "white," has usually a light-complexioned skin (although in some, in so far aberrant cases, it is as dark as in the Negroes); hair fair or black, soft, straight, or wavy, in section intermediate between the flattened and cylindrical form; beard fully developed; form of cranium various, mostly mesocephalic; malar bones retreating; face narrow and projecting in the middle line (pro-opic); orbits moderate; nose narrow and prominent (leptorhine); jaws orthognathous; teeth small (microdont); pelvis broad (pelvic index of male 80); forearm short relatively to humerus (humero-radial index 74).

In endeavouring further to divide up into minor groups the numerous and variously modified individuals which cluster around one or other of these great types—a process quite necessary for many practical or descriptive purposes—the distinctions afforded by the study of physical characters are often so slight that it becomes necessary to take other considerations into account, among which geographical distribution and language hold an important place.

I. The Ethiopian or Negroid races may be primarily divided as follows :—

A. African or typical Negroes—inhabitants of all the central portion of the African continent, from the Atlantic on the west to the Indian Ocean on the east, greatly mixed all along their northern frontier with Hamitic and Semitic Melanochroi, a mixture which, taking place in various proportions and under varied conditions, has given rise to many of the numerous races and tribes inhabiting the Soudan.

A branch of the African Negroes are the Bantu—distinguished chiefly, if not entirely, by the structure of their language. Physically indistinguishable from the other negroes where they come in contact in the equatorial regions of Africa, the Southern Bantu, or Kaffirs, as they are generally called, show a marked modification of type, being lighter in colour, having a larger cranial capacity, less marked prognathism, and smaller teeth. Some of these changes may possibly be due to crossing with the next race.

B. The Hottentots and Bushmen form a very distinct modification of the Negro race. They formerly inhabited a much larger district than at present; but, encroached upon by the Bantu from the north and the Dutch and English from the south, they are now greatly diminished, and indeed threatened with extinction. The Hottentots especially are much mixed with other races, and under the influence of a civilisation which has done little to improve their moral condition they have lost most of their distinctive peculiarities. When purebred they are of moderate stature, have a yellowish-brown

complexion, with very frizzly hair, which, being less abundant
than that of the ordinary negro, has the appearance of growing
in separate tufts. The forehead and chin are narrow, and the
cheekbones wide, giving a lozenge shape to the whole face.
The nose is very flat, and the lips prominent. In their
anatomical peculiarities, and in almost everything except size,
the Bushmen agree with the Hottentots; they have, however,
some special characters, for, while they are the most platyrhine
of races, the prognathism so characteristic of the Negro type is
nearly absent. This, however, may be the retention of an
infantile character so often found in races of diminutive stature,
as it is in all the smaller species of a natural group of animals.
The cranium of a Bushman, taken altogether, is one of the best
marked of any race, and could not be mistaken for that of any
other. Their relation to the Hottentots, however, appears to
be that of a stunted and outcast branch, living the lives of the
most degraded of savages among the rocky caves and mountains
of the land of which the comparatively civilised and pastoral
Hottentots inhabited the plains.

Perhaps the Negrillos of Hamy, certain diminutive round-
headed people of Central and Western Equatorial Africa, may
represent a distinct section of the Negro race, but their numbers
are few, and they are very much mixed with the true Negroes
in the districts in which they are found. They form the only
exceptions to the general dolichocephaly of the African branch
of the Negroid race.[1]

C. *Oceanic Negroes* or *Melanesians.*—These include the
Papuans of New Guinea and the majority of the inhabitants
of the islands of the Western Pacific, and form also a substratum of the population, greatly mixed with other races, of
regions extending far beyond the present centre of their area
of distribution.

They are represented, in what may be called a hypertypical
form, by the extremely dolichocephalic Kai Colos, or moun-
taineers of the interior of the Fiji Islands, although the
coast population of the same group has lost their distinctive

---

[1] Further information upon the Negrillos and Bushmen, and later views as
to their relations to each other and to the true Negroes, will be found in the
article which follows " On the Pygmy Races of Men."

characters by crossing. In many parts of New Guinea and the great chain of islands extending eastwards and southwards, ending with New Caledonia, they are found in a more or less pure condition, especially in the interior and more inaccessible portions of the islands, almost each of which shows special modifications of the type recognisable in details of structure. Taken altogether, their chief physical distinction from the African Negroes lies in the fact that the glabella and supra-orbital ridges are generally well developed in the males, whereas in Africans this region is usually smooth and flat. The nose, also, especially in the northern part of their geographical range, New Guinea, and the neighbouring islands, is narrower (often mesorhine) and prominent. The cranium is generally higher and narrower. It is, however, possible to find African and Melanesian skulls quite alike in essential characters.

The now extinct inhabitants of Tasmania were probably pure, but aberrant, members of the Melanesian group, which have undergone a modification from the original type, not by mixture with other races, but in consequence of long isolation, during which special characters have gradually developed. Lying completely out of the track of all civilisation and commerce, even of the most primitive kind, they were little liable to be subject to the influence of any other race, and there is, in fact, nothing among their characters which could be accounted for in this way, as they were intensely, even exaggeratedly, Negroid in the form of nose, projection of mouth, and size of teeth, typically so in character of hair, and aberrant chiefly in width of skull in the parietal region. A cross with any of the Polynesian or Malay races sufficiently strong to produce this would, in all probability, have also left some traces on other parts of their organisation.

On the other hand, in many parts of the Melanesian region there are distinct evidences of large admixture with Negrito, Malay, and Polynesian elements in varying proportions, producing numerous physical modifications. In many of the inhabitants of the great island of New Guinea itself and of those lying around it this mixture can be traced. In the people of Micronesia in the north, and New Zealand in the south, though the Melanesian element is present, it is completely overlaid by the Polynesian, but there are probably few, if any, of the islands of the Pacific in which it does not form some factor in the composite character of the natives.

The inhabitants of the continent of Australia have long been a puzzle to ethnologists. Of Negroid complexion, features, and skeletal characters, yet without the characteristic frizzly hair, their position has been one of great difficulty to determine. They have, in fact, been a stumbling-block in the way of every system proposed. The solution, supported by many considerations too lengthy to enter into here, appears to lie in the supposition that they are not a distinct race at all—that is, not a homogeneous group formed by the gradual modification of one of the primitive stocks, but rather a cross between two already formed branches of these stocks. According to this view, Australia was originally peopled with frizzly-haired Melanesians, such as those who still do, or did before the European invasion, dwell in the smaller islands which surround the north, east, and southern portions of the continent, but that a strong infusion of some other race, probably a low form of Caucasian Melanochroi, such as that which still inhabits the interior of the southern parts of India, has spread throughout the land from the north-west, and produced a modification of the physical characters, especially of the hair. This influence did not extend across Bass's Straits into Tasmania, where, as just said, the Melanesian element remained in its purity. It is more strongly marked in the northern and central parts of Australia than on many portions of the southern and western coasts, where the lowness of type and more curly hair, sometimes closely approaching to frizzly, show a stronger retention of the Melanesian element. If the evidence should prove sufficiently strong to establish this view of the origin of the Australian natives, it will no longer be correct to speak of a primitive Australian, or even Australoid, race or type, or look for traces of the former existence of such a race anywhere out of their own land. Absolute proof of the origin of any race is, however, very difficult, if not impossible to obtain, and I know nothing to exclude the possibility of the Australians being mainly the direct descendants of a very primitive human type, from which the frizzly-haired Negroes may be an offset.   This character of hair must be a specialisation, for it seems very unlikely that it was the attribute of the common ancestors of the whole human race.

D. The fourth branch of the Negroid race consists of the diminutive round-headed people called Negritos, still found in a pure or unmixed state in the Andaman Islands, and forming a substratum of the population, though now greatly mixed with invading races, especially Malays, in the Philippines, and many of the islands of the Indo-Malayan Archipelago, and perhaps of some parts of the southern portion of the mainland of Asia.   They also contribute to the varied population of the great island of Papua or New Guinea, where they appear to merge into the taller, longer-headed and longer-nosed Melanesians proper.   They show, in a very marked manner, some of the most striking anatomical peculiarities of the Negro race, the frizzly hair, the proportions of the limbs, especially the humero-radial index, and the form of the pelvis; but they differ in many cranial and facial characters, both from the African Negroes on the one hand, and the typical Oceanic Negroes, or Melanesians, on the other, and form a very distinct and well-characterised group.

II.—The principal groups that can be arranged round the Mongolian type are—

A. The Eskimo, apparently a branch of the typical North Asiatic Mongols, who, in their wanderings northwards and eastwards across the American continent, isolated almost as perfectly as an island population would be, hemmed in on one side by the eternal Polar ice, and on the other by hostile tribes of American Indians, with which they rarely, if ever, mingled, have gradually developed characters most of which are strongly-expressed modifications of those seen in their allies, who still remain on the western side of Behring's Straits.   Every special characteristic which distinguishes a Japanese from the average of mankind is seen in the Eskimo in an exaggerated degree, so that there can be no doubt about their being derived from the same stock.   It has also been shown that these special characteristics gradually increase from west to east, and are seen in their greatest perfection in the inhabitants of Greenland; at all events, in those where no crossing with the Danes has taken place. Such scanty remains as have yet been discovered of the early inhabitants of Europe present no structural affinities to the Eskimo, although it is not unlikely that similar external conditions may have led them to adopt similar modes of life. In fact, the Eskimo are an intensely specialised race, perhaps the most specialised of any in existence, and therefore probably of comparatively late origin. In this case they were not as a race contemporaries with the men whose rude flint tools found in our drifts excite so much interest and speculation as to the makers, who have been sometimes, though with little evidence to justify such an assumption, reputed to be related to the present inhabitants of the northernmost parts of America.

B. The typical Mongolian races constitute the present population of Northern and Central Asia. They are not very distinctly, but still conveniently for descriptive purposes, divided into two groups, the Northern and the Southern.

(*a*) The former, or Mongolo-Altaic group, are united by the affinities of their language. These people, from the cradle of their race in the great central plateau of Asia, have at various times poured out their hordes upon the lands lying to the west, and have penetrated almost to the heart of Europe. The Finns, the Magyars, and the Turks, are each the descendants of one of these waves of incursion, but they have for so many generations intermingled with the peoples through whom they have passed in their migrations, or have found in the countries in which they have ultimately settled, that their original physical characters have been completely modified. Even the Lapps, that diminutive tribe of nomads inhabiting the most northern parts of Europe, supposed to be of Mongolian descent, show so little of the special attributes of that branch that it is difficult to assign them a place in it in a classification based upon physical characters. The Japanese are said by their language to be allied rather to the Northern than to the following branch of the Mongolian stock.

(*b*) The southern Mongolian group, divided from the former chiefly by language and habits of life, includes the greater part of the population of China, Thibet, Burmah, and Siam.

C. The next great division of Mongoloid people is the Malay

sub-typical it is true, but to which an easy transition can be traced from the most characteristic members of the type.

D. The brown Polynesians, Malayo-Polynesians, Mahoris, Sawaioris, or Kanakas, as they have been variously called, seen in their greatest purity in the Samoan, Tongan, and Eastern Polynesian Islands, are still more modified, and possess less of the characteristic Mongolian features; but still it is difficult to place them anywhere else in the system. The large infusion of the Melanesian element throughout the Pacific must never be forgotten in accounting for the characters of the people now inhabiting the islands, an element in many respects so diametrically opposite to the Mongolian that it would materially alter the characters, especially of the hair and beard, which has been with many authors a stumbling-block to the affiliation of the Polynesian with the Mongolian stock. The mixture is physically a fine one, and in some proportions produces a combination, as seen, for instance, in the Maories of New Zealand, which in all definable characters approaches quite as near, or nearer, to the Caucasian type than to either of the stocks from which it may be presumably derived. This resemblance has led some writers to infer a real extension of the Caucasian element at some very early period into the Pacific Islands, and to look upon their inhabitants as the product of a mingling of all three great types of men. Though this is a very plausible theory, it rests on little actual proof, as the combination of Mongolo-Malayan and Melanesian characters in different degrees, together with the local variations certain to arise in communities so isolated from each other and exposed to such varied conditions as the inhabitants of the Pacific Islands, would probably account for all the modifications observed among them.

E. The native populations (before the changes wrought by the European conquest) of the great continent of America, excluding the Eskimo, present, considering the vast extent of the country they inhabit and the great differences of climate, and other surrounding conditions, a remarkable similarity of essential characters, with much diversity of detail.

The construction of the numerous American languages, of which as many as twelve hundred have been distinguished, is said to point to unity of origin, as, though widely different in many respects, they are all, or nearly all, constructed on the same general grammatical principle—that called *polysynthesis*—which differs from that of the languages of any of the Old World nations. The mental characteristics of all the American tribes have much that is in common ; and the very different stages of culture to which they had attained at the time of the conquest, as that of the Incas and Aztecs, and the hunting or fishing tribes of the north and south, which have been quoted as evidence of diversities of race, were not greater than those between different nations of Europe, as Gauls and Germans on the one hand, and Greeks and Romans on the other, in the time of Julius Cæsar. Yet all these were Aryans, and in treating the Americans as one race it is not intended that they are more closely allied than the different Aryan people of Europe and Asia. The best argument that can be used for the unity of the American race—using the word in a broad sense—is the great difficulty of forming any natural divisions founded upon physical characters. The important character of the hair does not differ throughout the whole continent. It is always straight and lank, long and abundant on the scalp, but sparse elsewhere. The colour of the skin is practically uniform, notwithstanding the enormous differences of climate under which many members of the group exist. In the features and cranium certain special modifications prevail in different districts, but the same forms appear at widely-separated parts of the continent. I have examined skulls from Vancouver's Island, from Peru, from Patagonia, and from Jamaica, which were almost undistinguishable from one another.

Naturalists who have admitted but three primary types of the human species have always found a difficulty with the Americans, hesitating between placing them with the Mongolian or so-called "yellow" races, or elevating them to the rank of a primary group. Cuvier does not seem to have been able to settle this point to his own satisfaction, and leaves it an open question. Although the large majority of Americans have in the special form of the nasal bones, leading to the characteristic high bridge of the nose of the living face, in the well-developed superciliary ridge and re-treating forehead, characters which distinguish them from the typical Asiatic Mongol, in many other respects they resemble them so much that, although admitting the difficulties of the case, I am inclined to include them as aberrant members of the Mongolian type. It is, however, quite open to any one adopting the Negro, Mongolian, and Caucasian as primary divisions to place the Americans apart as a fourth.

Now that the high antiquity of man in America, perhaps as high as that which he has in Europe, has been discovered, the puzzling problem, from which part of the Old World the people of America have sprung has lost its significance. It is quite as likely that the people of Asia may have been derived from America as the reverse. However this may be, the population of America had been, before the time of Columbus, practically isolated from the rest of the world, except at the extreme north. Such visits as those of the early Norsemen to the coasts of Greenland, Labrador, and Nova Scotia, or the possible accidental stranding of a canoe containing survivors of a voyage across the Pacific or the Atlantic, can have had no appreciable effect upon the characteristics of the people. It is difficult, therefore, to look upon the anomalous and special characters of the American people as the effects of crossing, as was suggested in the case of the Australians, a consideration which gives more weight to the view of treating them as a distinct primary division.

III. The Caucasian, or white division, according to my view, includes the two groups called by Professor Huxley Xanthochroi and Melanochroi, which, though differing in colour of eyes and hair, agree so closely in all other anatomical characters, as far, at all events, as has at present been demonstrated, that it seems preferable to consider them as modifications of one great type than as primary divisions of the species. Whatever their origin, they are now intimately blended, though in different proportions, throughout the whole of the region of the earth they inhabit; and it is to the rapid extension of both branches of this race that the great changes now taking place in the ethnology of the world are mainly due.

A. The Xanthochroi, or blonde type, with fair hair, eyes, and complexion, chiefly inhabit Northern Europe—Scandinavia,

Scotland, and North Germany—but, much mixed with the next group, they extend as far as Northern Africa and Afghanistan. Their mixture with Mongoloid people has given rise to the Lapps, Finns, and some of the tribes of Northern Siberia.

B. Melanochroi, with black hair and eyes, and skin of almost all shades from white to black. They comprise the great majority of the inhabitants of Southern Europe, Northern Africa, and South-west Asia, and consist mainly of the Aryan, Semitic, and Hamitic families. The Dravidians of India, the Veddahs of Ceylon, and probably the Ainos of Japan, and the Maoutze of China, also belong to this race, which may have contributed something to the mixed character of some tribes of Indo-China and the Polynesian Islands, and, as before said, given at least the characters of the hair to the otherwise Negroid inhabitants of Australia. In Southern India they are probably mixed with a Negrito element, and in Africa, where their habitat becomes conterminous with that of the Negroes, numerous cross races have sprung up between them all along the frontier line. The ancient Egyptians were nearly pure Melanochroi, though often showing in their features traces of their frequent intermarriages with their Ethiopian neighbours to the south. The Copts and fellahs of modern Egypt are their little-changed descendants.

In offering this scheme of classification of the human species I have not thought it necessary to compare it in detail with the numerous systems suggested by previous anthropologists. These will all be found in the general treatises on the subject. As I have remarked before, in its broad outlines it scarcely differs from that proposed by Cuvier nearly sixty years ago, and that the enormous increase of our knowledge since that time should have caused such little change is the best testimony to its being a truthful representation of the facts. Still, however, it can only be looked upon as an approximation. Whatever care be bestowed upon the arrangement of already acquired details, whatever judgment be shown in their due subordination one to another, the acquisition of new knowledge may at any time call for a complete or partial rearrangement of our system.

Happily such knowledge is being abundantly brought in by workers in many lands, and, among others, by members of our own Institute, whose contributions, published in our *Journal*, form no mean addition to the general advancement of the science.

This leads me to speak of some of our own more immediate affairs.   During the past year two members of our Council have been removed by death.   Dr. Allen Thomson was for many years an eminent and successful teacher of human anatomy in the University of Glasgow.   His researches into the history of the early stages of development of the embryo gained him a world-wide reputation, and he was beloved by all who knew him personally for the singular modesty and gentleness of his nature.   He had been a Vice-President of the Institute, and a contributor to its proceedings.   Mr. Alfred Tylor, the brother of our distinguished former President, Dr. E. B. Tylor, though greatly interested in many branches of Anthropology, and a frequent attendant at our meetings, was better known as a geologist.   He died at his residence at Carshalton, on the last day of 1884, in the sixty-first year of his age.

At the conclusion of my address last year I announced that a critical time was coming for the Institute, as circumstances had rendered a change of domicile a necessity.   The rooms in St. Martin's Place, in which the Institute had met since its foundation, were required for Government purposes, and we were obliged to move elsewhere.   I think it will be generally admitted that the accommodation we have succeeded in obtaining is in every way superior to that which we left behind, and the annual cost will be but very trifling in excess of that we were paying before.   The expenses of moving and of new fittings have, however, made a heavy inroad in our slender income, and notwithstanding the special assistance of some of our members to meet it, it was necessary to sell out a portion of our capital stock.   We ought to replace this, if possible; and, what is still more important, we ought to have the means of spending more money upon our publications, especially in illustrations, and more upon our library, for the increase of which we are chiefly dependent upon donations.

Binding our serial publications is an item for which provision should especially be made. We should also look forward to the time when the inevitable extension of our collection of books will require additional accommodation. For all these necessities, we need, as I have often said before, additional members. At the present time, as you will have gathered from the Report of the Council, we are stationary in this respect ; but our change of abode ought to be a starting point for acquiring a wider circle of interest in our work.

Under the guidance of our able and painstaking Director, Mr. Rudler, and presided over by the gentleman who, I trust, you will elect in my place, the Institute cannot but flourish. Mr. Francis Galton is a man of most versatile genius. He is well known as an explorer of regions where man may be studied under conditions most opposite to those which obtain in our island. In one of his early adventurous expeditions he visited Khartoum, a place then as unknown to English ears as it is now unhappily familiar. His subsequent journey in the opposite extremity of the African continent led to the publication of very useful observations, and also to the work called *The Art of Travel.* His ingenious researches on the subject of characteristics transmitted by inheritance, and his methods of testing physical capabilities, have frequently been brought before the notice of the Institute. His anthropometric laboratory, organised last year at the International Health Exhibition, brought before thousands the interest and importance of the subject. I have much satisfaction in resigning into his hands the office with which you have honoured me for two consecutive years.

THE PYGMY RACES OF MEN [1]

IT is well known that the nations of antiquity entertained a
widespread belief in the existence of a race or races of human
beings of exceedingly diminutive stature, who dwelt in some
of the remote and unexplored regions of the earth. These
were called *Pygmies*, a word said to be derived from πυγμή,
which means a fist, and also a measure of length, the distance
from the elbow to the knuckles of an ordinary-sized man,
or rather more than 13 inches.

In the opening of the third book of the *Iliad* the Trojan
hosts are described as coming on with noise and shouting,
"like the cranes which flee from the coming of winter and
sudden rain, and fly with clamour towards the streams of
ocean, bearing slaughter and fate to the Pygmy men, and in
early morn offer cruel battle," or, as Pope has it,

> So when inclement winters vex the plain,
> With piercing frosts, or thick descending rain,
> To warmer seas the cranes embodied fly,
> With noise and order through the midway sky,
> To Pygmy nations wounds and death they bring,
> And all the war descends upon the wing."

The combats between the pygmies and the cranes are often
alluded to by later classical writers, and are not unfrequently
depicted upon Greek vases. In one of these in the Hope
Collection at Deepdene, in which the figures are represented
with great spirit, the pygmies are dwarfish-looking men with
large heads, negro features, and close woolly or frizzly hair.
They are armed with lances. Notices of a less poetical and

[1] Lecture at the Royal Institution of Great Britain, 13th April 1888.

apparently more scientific character of the occurrence of races
of very small human beings are met with in Aristotle,
Herodotus, Ctesias, Pliny, Pomponius Mela, and others.
Aristotle places his pygmies in Africa, near the sources of the
Nile, while Ctesias describes a race of dwarfs in the interior
of India. The account in Herodotus is so circumstantial, and
has such an air of truthfulness about it, especially in connection with recent discoveries, that it is worth quoting in full.[1]

" I did hear, indeed, what I will now relate, from certain
natives of Cyrênê. Once upon a time, they said, they were
on a visit to the oracular shrine of Ammon, when it chanced
that, in the course of conversation with Etearchus, the
Ammonian king, the talk fell upon the Nile, how that its
sources were unknown to all men. Etearchus upon this
mentioned that some Nasamonians had once come to his
court, and when asked if they could give any information
concerning the uninhabited parts of Libya, had told the
following tale. (The Nasamonians are a Libyan race vho
occupy the Syrtes, and a tract of no great size towards the
east.) They said there had grown up among them some
wild young men, the sons of certain chiefs, who, when they
came to man's estate, indulged in all manner of extravagances,
and among other things drew lots for five of their number to
go and explore the desert parts of Libya, and try if they
could not penetrate further than any had done previously.
The young men, therefore, dispatched on this errand by their
comrades with a plentiful supply of water and provisions,
travelled at first through the inhabited region, passing which
they came to the wild beast tract, whence they finally entered
upon the desert, which they proceeded to cross in a direction
from east to west. After journeying for many days over a
wide extent of sand, they came at last to a plain where they
observed trees growing; approaching them, and seeing fruit
on them, they proceeded to gather it. While they were
thus engaged, there came upon them some dwarfish men,
under the middle height, who seized them and carried them
off. The Nasamonians could not understand a word of their
language, nor had they any acquaintance with the language of

_____
[1] Herodotus, Book ii. 32, Rawlinson's translation, p. 47.

the Nasamonians. They were led across extensive marshes, and finally came to a town, where all the men were of the height of their conductors, and black-complexioned. A great river flowed by the town, running from west to east, and containing crocodiles."

It is satisfactory to know that the narrative concludes by saying that these pioneers of African exploration,—forerunners of Bruce and Park, of Barth, Livingstone, Speke, Grant, Schweinfurth, Stanley, and the rest,—"got safe back to their country."

Extension of knowledge of the natural products of the earth, and a more critical spirit on the part of authors, led to attempts to account for this belief, and the discovery of races of monkeys—of the doings of which, it must be said, more or less fabulous stories were often reported by travellers— generally sufficed the commentators and naturalists of the last century to explain the origin of the stories of the pygmies. To this view the great authority of Buffon was extended.

Still more recently-acquired information as to the actual condition of the human population of the globe has, however, led to a revision of the ideas upon the subject, and to more careful and critical researches into the ancient documents. M. de Quatrefages, the eminent and veteran Professor of Anthropology at the Muséum d'Histoire Naturelle of Paris, especially, has carefully examined and collated all the evidence bearing upon the question, and devoted much ingenuity of argument to prove that the two localities in which the ancient authors appear to place their pygmies, the interior of Africa near the sources of the Nile, and the southernmost parts of Asia, and the characters they assign to them, indicate an actual knowledge of the existence of the two groups of small people which still inhabit these regions, the history of which will form the subject of this lecture. The evidence which has convinced M. de Quatrefages, and which, I have no doubt, will suffice for those who take pleasure in discovering an underlying truth in all such legends and myths, or in the more grateful task of rehabilitating the veracity of the fathers of literature and history, will be found collected in a very

readable form in a little book published last year in the
" Bibliothèque scientifique contemporaine," called *Les Pygmées*,
to which I refer my hearers for fuller information upon the
subject of this discourse, and especially for numerous references
to the literature of the subject, which, as the book is accessible
to all who wish to pursue it further, I need not give here.

It is still, however, to my mind, an open question whether
these old stories may not be classed with innumerable others,
the offspring of the fertile invention of the human brain, the
potency of which as an origin of myths has, I think, sometimes been too much underrated. I shall, therefore, now
take leave of them, and confine myself to giving you, as far as
the brief space of time at my disposal admits, an account of
our actual knowledge of the smallest races of men either
existing, or, as far as we know, ever having existed on earth,
and which may, therefore, taking the word in its current
though not literal sense, be called the "pygmies" of the
species.

Among the various characters by which the different races
of men are distinguished from one another, *size* is undoubtedly
one of considerable importance. Not but what in each race
there is much individual variation, some persons being taller
and some shorter; yet these variations are, especially in the
purer or less mixed races, restricted within certain limits, and
there is a general average, both for men and women, which
can be ascertained when a sufficient number of accurate
measurements have been recorded. That the prevailing size
of a race is a really deeply-seated, inherited characteristic,
and depends but little on outward conditions, as abundance of
food, climate, etc., is proved by well-known facts. The tallest
and the shortest races in Europe are respectively the Norwegians and the Lapps, living in almost the same region. In
Africa, also, the diminutive Bushmen and the tallest race of
the country, the Kaffirs, are close neighbours. The natives
of the Andaman Islands and those of many islands of the
equatorial region of the Pacific, in which the conditions are
similar, or if anything more favourable to the former, are at
opposite ends of the scale of height. Those not accustomed
to the difficulties both of making and recording such measure-

ments will scarcely be prepared, however, to learn how meagre, unsatisfactory, and unreliable our knowledge of the stature of most of the races of mankind is at present, although unquestionably it has been considerably increased within recent years. We must, however, make use of such material as we possess, and trust to the future correction of errors when better opportunities occur.

It is convenient to divide men, according to their height, into three groups—tall, medium, and short; in Topinard's system,[1] the first being those the average height (of the men) of which is above 1·700 metres (5 feet 7 inches), the last those below 1·600 metres (5 feet 3 inches), and the middle division those between the two. In the short division are included certain of the Mongolian or yellow races of Asia, as the Samoyedes, the Ostiaks, the Japanese, the Siamese, and the Annamites; also the Veddahs of Ceylon and certain of the wild hill-tribes of Southern India. These all range between 1·525 and 1·600 metres—say between 5 feet and 5 feet 3 inches.

It is of none of these people that I am going to speak to-day. My pygmies are all on a still smaller scale, the average height of the men being in all cases below 5 feet—in some cases, as we shall see, considerably below.

Besides their diminutive size, I may note at the outset that they all have in a strongly-marked degree the character of the hair distinguished as frizzly—i.e. growing in very fine, close curls, and flattened or elliptical in section, and therefore, whatever other structural differences they present, they all belong to the same primary branch of the human species as the African negro and the Melanesian of the Western Pacific.

I will first direct your attention to a group of islands in the Indian Ocean—the Andamans—where we shall find a race in many respects of the greatest possible interest to the anthropologist.

These islands are situated in the Bay of Bengal between the 10th and 14th parallels of north latitude, and near the meridian 93° east of Greenwich, and consist of the Great and Little Andamans. The former is about 140 miles long, and

---

[1] *Eléments d'Anthropologie Général*, p. 463. Paris, 1885.

has a breadth nowhere exceeding 20 miles.   It is divided by narrow channels into three, called respectively North, Middle, and South Andaman, and there are also various smaller islands belonging to the group.   Little Andaman is a detached island lying about 28 miles to the south of the main group, about 27 miles in length and 10 to 18 in breadth.

Although these islands have been inhabited for a very great length of time by people whose state of culture and customs have undergone little or no change, as proved by the examination of the contents of the old kitchen-middens, or refuse heaps, found in many places in them, and although they lie so near the track of civilisation and commerce, the islands and their inhabitants were practically unknown to the world until so recently as the year 1858.   It is true that their existence is mentioned by Arabic writers of the ninth century, and again by Marco Polo, and that in 1788 an attempt was made to establish a penal colony upon them by the East India Company, which was abandoned a few years after; but the bad reputation the inhabitants had acquired for ferocious and inhospitable treatment of strangers brought by accident to their shores, caused them to be carefully avoided, and no permanent settlement or relations of anything like a friendly nature, or likely to afford any useful information as to the character of the islands or the inhabitants, were established.   It is fair to mention that this hostility to foreigners, which for long was one of the chief characteristics by which the Andamanese were known to the outer world, found much justification in the cruel experiences they suffered from the mal-practices, especially kidnapping for slavery, of the Chinese and Malay traders who visited the islands in search of *bêche de mer* and edible birds'-nests.   It is also to this characteristic that the inhabitants owe so much of their interest to us from a scientific point of view, for we have here the rare case of a population, confined to a very limited space, and isolated for hundreds, perhaps thousands, of years from all contact with external influence, their physical characters unmixed by crossing, and their culture, their beliefs, their language, entirely their own.

In 1857, when the Sepoy mutiny called the attention of

the Indian Government to the necessity of a habitation for their numerous convict prisoners, the Andaman Islands were again thought of for the purpose, A commission, consisting of Dr. F. J. Mouat, Dr. G. Playfair, and Lieutenant J. A. Heathcote, was sent to the islands to report upon their capabilities for such a purpose; and, acting upon its recommendations, early in the following year the islands were taken possession of in the name of the East India Company by Captain (now General) H. Man, and the British flag hoisted at Port Blair, near the southern end of Great Andaman, which thenceforth became the nucleus of the settlement of invaders, now numbering about 15,000 persons, of whom more than three-fourths are convict prisoners, the rest soldiers, police, and the usual accompaniments of a military station.

The effect of this inroad upon the unsophisticated native population, who, though spread over the whole area of the islands, were far less numerous, may easily be imagined. It is simply deterioration of character, moral and physical decay, and finally extinction. The newly-introduced habits of life, vices, and diseases are spreading at a fearful rate and with deadly effect. In this sad history there are, however, two redeeming features which distinguish our occupation of the Andamans from that of Tasmania, where a similar tragedy was played out during the present century. In the first place, the British governors and residents appear to have used every effort to obtain for the natives the most careful and considerate treatment, and to alleviate as much as possible the evils which they have unintentionally been the means of inflicting upon them. Secondly, most careful records have been preserved of the physical characters, the social customs, the arts, manufactures, traditions, and language of the people while still in their primitive condition. For this most important work—a work which, if not done, would have left a blank in the history of the world which could never have been replaced—we are indebted almost entirely to the scientific enthusiasm of one individual, Mr. Edward Horace Man, who most fortunately happened to be in a position (as assistant superintendent of the islands, and specially in charge of the natives) which enabled him to obtain the required information with facilities which probably no one else could have had, and whose observations " On the Aboriginal Inhabitants of the Andaman Islands," published by the Anthropological Institute of Great Britain and Ireland, are most valuable, not only for the information they contain, but as correcting the numerous erroneous and misleading statements circulated regarding these people by previous and less well-informed or less critical authors.

The Arab writer of the ninth century previously alluded to states that " their complexion is frightful, their hair frizzled, their countenance and eyes frightful, their feet very large, and almost a cubit in length, and they go quite naked," while Marco Polo (about 1285) says that " the people are no better than wild beasts, and I assure you all the men of this island of Angamanain have heads like dogs, and teeth and eyes likewise ; in fact, in the face they are just like big mastiff dogs." These specimens of mediæval anthropology are almost rivalled by the descriptions of the customs and moral character of the same people published as recently as 1862, based chiefly on information obtained from one of the runaway sepoy convicts, and which represent them as among the lowest and most degraded of human beings.

The natives of the Andamans are divided into nine distinct tribes, each inhabiting its own district. Eight of these live upon the Great Andaman Islands, and one upon the hitherto almost unexplored Little Andaman. Although each of these tribes possesses a distinct dialect, these are all traceable to the same source, and are all in the same stage of development. The observations that have been made hitherto relate mostly to the tribe inhabiting the south island, but it does not appear that there is any great variation either in physical characters or manners, customs, and culture among them.

With regard to the important character of size, we have more abundant and more accurate information than of most other races. Mr. Man gives the measurements of forty-eight men and forty-one women, making the average of the former 4 feet $10\frac{3}{4}$ inches, that of the latter 4 feet $7\frac{1}{4}$ inches, a difference therefore of $3\frac{1}{2}$ inches between the sexes. The tallest man was 5 feet $4\frac{1}{4}$ inches, the shortest 4 feet 6 inches. The tallest woman 4 feet $11\frac{1}{2}$ inches, the shortest 4 feet 4 inches. Measurements made upon the living subject are always liable to errors, but it is possible that in so large a series these will compensate each other, and that therefore the averages may be relied upon. My own observations, based upon the measurements of the bones alone of as many as twenty-nine skeletons, give smaller averages, viz. 4 feet $8\frac{1}{2}$ inches for the men, and 4 feet $6\frac{1}{2}$ inches for the women ; but these, it must be recollected, are calculated from the length of the femur, upon a ratio which, though usually correct for Europeans, may not hold good in the case of other races. The hair is fine, and very closely curled—woolly, as it is generally called, or rather, frizzly—and elliptical in section, as in the negroes. The colour of the skin is very dark, although not absolutely black. The head is of roundish (brachycephalic) form, the cephalic index of the skull being about 82. The other cranial characters are fully described in the papers referred to below. The teeth are large, but the jaws are only slightly prognathous. The features possess little of the negro type—at all events, little of the most marked and coarser peculiarities of that type. The projecting jaws, the prominent thick lips, the broad and flattened nose of the genuine negro are so softened down in the Andamanese as scarcely to be recognised ; and yet in the relative proportions of the limb-bones, especially in the shortness of the humerus compared with the forearm, and in the form of the pelvis, negro affinities are most strongly indicated.[1]

In speaking of the culture of the Andamanese, of course I only refer to their condition before the introduction of European civilisation into the islands. They live in small villages or encampments, in dwellings of simple and rude construction, built only of branches and leaves of trees. They are entirely ignorant of agriculture, and keep no poultry or domestic animals. They make rude pots of clay, sun-dried, or partially

---

[1] See "On the Osteology and Affinities of the Natives of the Andaman Islands" (*Journal Anthropological Institute*, vol. ix. p. 108, 1879) ; and " Additional Observations on the Osteology of the Natives of the Andaman Islands " (*ibid.* vol. xiv. p. 115, 1884).

baked in the fire, but these are hand-made, as they are ignorant of the use of the potter's wheel. Their clothing is of the scantiest description, and what little they have serves chiefly for decorative or ornamental purposes, and not for keeping the body warm. They make no use of the skins of animals. They have fairly well-made dug-out canoes and outriggers, but fit only for navigating the numerous creeks and straits between the islands, and not for voyages in the open sea. They are expert swimmers and divers. Though constantly using fire, they are quite ignorant of the art of producing it, and have to expend much care and labour in keeping up a constant supply of burning or smouldering wood. They are ignorant of all metals, but for domestic purposes make great use of shells, especially a species of *Cyrene* found abundantly on the shores of the islands, also quartz chips and flakes, and bamboo knives. They have stone anvils and hammers, and they make good string from vegetable fibres, as well as baskets, fishing nets, sleeping mats, etc. Their principal weapons are the bow and arrow, in the use of which they are very skilful. They have harpoons for killing turtle and fish, but no kind of shield or breastplate for defence when fighting. The natural fertility of the island supplies them with abundance and variety of food all the year round. This consists of pigs (*Sus andamanensis*), which are numerous on the islands, paradoxures, dugongs, and occasionally porpoises, iguanas, turtles, turtles' eggs, many kinds of fish, prawns, mollusks, larvæ of large wood-boring and burrowing beetles, honey, and numerous roots (as yams), fruits, and seeds. The food, the purveying of which affords occupation and amusement for the greater part of the male population, is invariably cooked before eating, and generally taken when extremely hot. They were ignorant of all stimulants or intoxicating drinks—in fact, water was their only beverage; and tobacco, or any substitute for it, was quite unknown till introduced by Europeans.

As with all other human beings existing at present in the world, however low in the scale of civilisation, the social life of the Andamanese is enveloped in a complex maze of un-written law or custom, the intricacies of which are most

difficult for any stranger to unravel.   The relations they may
or may not marry, the food they are obliged or forbidden to
partake of at particular epochs of life or seasons of the year,
the words and names they may or may not pronounce,—all
these, as well as their traditions, superstitions, and beliefs,
their occupations, games, and amusements, of which they seem
to have had no lack, would take far too long to describe here ;
but, before leaving these interesting people, I may quote an
observation of Mr. Man's, which, unless he has seen them with
too *couleur-de-rose* eyesight, throws a very favourable light
upon the primitive, unsophisticated life of these poor little
savages, now so ruthlessly broken into and destroyed by the
exigencies of our ever-extending empire.

"It has been asserted," Mr. Man says, "that the 'com-
munal marriage' system prevails among them, and that
'marriage is nothing more than taking a female slave' ; but,
so far from the contract being regarded as a merely temporary
arrangement, to be set aside at the will of either party, no
incompatibility of temper or other cause is allowed to dissolve
the union ; and while bigamy, polygamy, polyandry, and
divorce are unknown, conjugal fidelity till death is not the
exception but the rule, and matrimonial differences, which,
however, occur but rarely, are easily settled with or without
the intervention of friends."   In fact, Mr. Man goes on to
say, "One of the most striking features of their social relations
is the marked equality and affection which subsists between
husband and wife," and "the consideration and respect with
which women are treated might with advantage be emulated
by certain classes in our own land."

It should also be mentioned that cannibalism and infanti-
cide, two such common incidents of savage life, were never
practised by them.

We must now pass to the important scientific question,
Who are the natives of the Andaman Islands, and where,
among the other races of the human species, shall we look for
their nearest relations ?

It is due mainly to the assiduous researches into all the
documentary evidence relating to the inhabitants of Southern
Asia and the Indian Archipelago, conducted through many

years by M. de Quatrefages, in some cases with the assistance of his colleague, M. Hamy, that the facts I am about to put before you have been prominently brought to light, and their significance demonstrated.

It is well known that the greater part of the large island of New Guinea, and of the chain of islands extending eastwards and southwards from it, including the Solomon Islands, the New Hebrides, and New Caledonia, and also the Fijis, are still inhabited mainly by people of dark colour, frizzly hair, and many characteristics allying them to the negroes of Africa. These constitute the race to which the terms Melanesian, negroes, or Oceanic, are commonly applied in this country. They are the "Papouas" of Quatrefages. Their area at one time was more extensive than it is now, and has been greatly encroached upon by the brown, straight-haired Polynesian race with Malay affinities, now inhabiting many of the more important islands of the Pacific, and the mingling of which with the more aboriginal Melanesians in various proportions has been a cause, among others, of the diverse aspect of the population on many of the islands in this extensive region. These Papouas, or Melanesians, however, differ greatly from the Andamanese in many easily-defined characters, which are, especially, their larger stature, their long, narrow, and high skulls, and their coarser and more negrolike features. Although undoubtedly allied, we cannot look to them as the nearest relations of our little Andamanese.

When the Spaniards commenced the colonisation of the Philippines, they met with, in the mountainous region in the interior of the Island of Luzon, besides the prevailing native population, consisting of Tagals of Malay origin, very small people, of black complexion, with the frizzly hair of the African negroes. So struck were they with the resemblance that they called them "Negritos del Monte" (little negroes of the mountain). Their local name was Aigtas, or Inagtas, said to signify "black," and from which the word Aëta, generally now applied to them, is derived. These people have lately been studied by two French travellers, M. Marche and Dr. Montano; the result of their measurements gives 4 feet $8\frac{3}{4}$ inches as the average height of the men, and 4 feet

$6\frac{1}{2}$ inches the average for the women. In many of their
moral characteristics they resemble the Andamanese. The
Aëtas are faithful to their marriage vows, and have but one
wife. The affection of parents for children is very strong,
and the latter have for their father and mother much love
and respect. The marriage ceremony, according to M. Montano,
is very remarkable. The affianced pair climb two flexible
trees placed near to each other. One of the elders of the
tribe bends them towards each other. When their heads
touch, the marriage is legally accomplished. A great *fête*,
with much dancing, concludes the ceremony.

It was afterwards found that the same race existed in
other parts of the archipelago, Panay, Mindanao, etc., and
that they entirely peopled some little islands—among others,
Bougas Island, or " Isla de los Negros."

As the islands of these eastern seas have become better
known, further discoveries of the existence of a small Negroid
population have been made in Formosa, in the interior of
Borneo, Sandalwood Island (Sumba), Xulla, Bourou, Ceram,
Flores, Solor, Lomblem, Pantar, Ombay, the eastern peninsula
of Celebes, etc. In fact, Sumatra and Java are the only large
islands of this great area which contain no traces of them
except some doubtful cross-breeds, and some remains of an
industry which appears not to have passed beyond the Age of
Stone.

The Sunda Islands form the southern limit of the Negrito
area; Formosa, the last to the north, where the race has
preserved all its characters. But beyond this, as in Loo Choo,
and even in the south-east portion of Japan, it reveals its
former existence by the traces it has left in the present
population. That it has contributed considerably to form
the population of New Guinea is unquestionable. In many
parts of that great island, small, round-headed tribes live
more or less distinct from the larger and longer-headed people
who make up the bulk of the population.

But it is not only in the islands that the Negrito race
dwell. Traces of them are found also on the mainland of
Asia, but everywhere under the same conditions; in scattered
tribes, occupying the more inaccessible mountainous regions

of countries otherwise mainly inhabited by other races, and generally in a condition more or less of degradation and barbarism, resulting from the oppression with which they have been treated by their invading conquerors; often, moreover, so much mixed that their original characters are scarcely recognisable. The Semangs of the interior of Malacca in the Malay peninsula, the Sakays from Perak, the Moys of Annam, all show traces of Negrito blood. In India proper, especially among the lowest and least civilised tribes, not only of the central and southern districts, but almost to the foot of the Himalayas, in the Punjab, and even to the west side of the Indus, according to Quatrefages, frizzly hair, negro features, and small stature are so common that a strong argument can be based on them for the belief in a Negrito race forming the basis of the whole pre-Aryan, or Dravidian as it is generally called, population of the peninsula. The crossing that has taken place with other races has doubtless greatly altered the physical characters of this people, and the evidences of such alteration manifest themselves in many ways; sometimes the curliness of the hair is lost by the admixture with straight-haired races, while the black complexion and small stature remain; sometimes the stature is increased, but the colour, which seems to be one of the most persistent of characteristics, remains.

The localities in which the Negrito people are found in their greatest purity, either in almost inaccessible islands, as on the Andamans, or elsewhere in the mountainous ranges of the interior only, and their social condition and traditions, wherever they exist,—all point to the fact that they were the earliest inhabitants; and that the Mongolian and Malay races on the east, and the Aryans on the west, which are now so rapidly exterminating and replacing them, are later comers into the land, exactly as, in the greater part of the Pacific Ocean, territory formerly occupied by the aboriginal dark, frizzly-haired Negroid Melanesians has been gradually and slowly invaded by the brown Polynesians, who in their turn, but by a much more rapid process, are being replaced by Europeans.

We now see what constitutes the great interest of the

Andamanese natives to the student of the ethnological history
of the Eastern world.   Their long isolation has made them
a remarkably homogeneous race, stamping them all with a
common resemblance not seen in the mixed races generally
met with in continental areas.   For although, as with most
savages, marriages within the family (using the term in a very
wide sense) are most strictly forbidden, all such alliances have
necessarily been confined to natives of the islands.   They are
the least modified representatives of the people who were, so
far as we know, the primitive inhabitants of a large portion
of the earth's surface, but who are now verging on extinction.
It is, however, not necessary to suppose that the Andaman
Islanders give us the exact characters and features of all the
other branches of the race.   Differences in detail doubtless
existed—differences which are almost always sure to arise
whenever races become isolated from each other for long
periods of time.

   In many cases the characters of the ancient inhabitants of
a land have been revealed to us by the preservation of their
actual remains.   Unfortunately we have as yet no such
evidence to tell us of the former condition of man in Southern
Asia.   We may, however, look upon the Andamanese, the
Aëtas, and the Semangs, as living fossils, and by their aid
conjecture the condition of the whole population of the land
in ancient times.   It is possible, also, to follow Quatrefages, and
to see in them the origin of the stories of the Oriental pygmies
related by Ctesias and by Pliny.

   We now pass to the continent of Africa, in the interior of
which the pygmies of Homer, Herodotus, and Aristotle have
generally been placed.   Africa, as is well known, is the
home of another great branch of the black, frizzly-haired,
or Ethiopian division of the human species, which does, or
did till lately, occupy the southern two-thirds of this great
continent, the northern third being inhabited by Hamite and
Semite branches of the great white or Caucasian primary
division of the human species, or by races resulting from the
mixture of these with the Negroes.   But besides the true
Negro there has long been known to exist in the southern
part of the continent a curiously modified type, consisting of

the Hottentots and the Bushmen—Bosjesmen (men of the woods) of the Dutch colonists—the latter of whom, on account of their small size, come within the scope of the present subject. They lead the lives of the most degraded of savages, dwelling among the rocky and more inaccessible mountains of the interior, making habitations of the natural caves, subsisting entirely by the chase, being most expert in the use of the bow and arrow, and treated as enemies and outcasts by the surrounding and more civilised tribes, whose flocks and herds they show little respect for when other game is not within reach. The physical characters of these people are well known, as many specimens have been brought to Europe alive for the purpose of exhibition. The hair shows the extreme of the frizzly type; being shorter and less abundant than that of the ordinary Negro, it has the appearance of growing in separate tufts, which coil up together into round balls compared to "peppercorns." The yellow complexion differs from that of the Negro, and, combined with the wide cheek-bones and form of the eyes, so much recalls that of certain of the pure yellow races that some anthropologists are inclined to trace true Mongolian affinities and admixture, although the extreme crispness of the hair makes such a supposition almost impossible. The width of the cheek-bones and the narrowness of the forehead and the chin give a lozenge-shape to the front view of the face. The forehead is prominent and straight; the nose extremely flat and broad, more so than in any other race, and the lips prominent and thick, although the jaws are less prognathous than in the true Negro races. The cranium has many special characters by which it can be easily distinguished from that of any other. It has generally a very feminine, almost infantile, appearance, though the capacity of the cranial cavity is not the smallest, exceeding that of the Andamanese. In general form the cranium is rather oblong than oval, having straight sides, a flat top, and especially a vertical forehead, which rises straight from the root of the nose. It is moderately dolichocephalic or rather mesaticephalic, the average index of ten specimens being 75·4. The height is in all considerably less than the breadth, the average index

being 71·1.   The glabella and supra-orbital ridge are little
developed, except in the oldest males.   The malar bones
project much forwards, and the space between the orbits is
very wide and flat.   The nasal bones are extremely small
and depressed, and the aperture wide ; the average nasal
index being 60·8, so they are the most platyrhine of races.

With regard to the stature, we have not yet sufficient
materials for giving a reliable average.   Quatrefages, following
Barrow, gives 4 feet 6 inches for the men, and 4 feet for the
women, and speaks of one individual of the latter sex, who
was the mother of several children, measuring only 3 feet
9 inches in height, but later observations (still, however,
insufficient in number) give a rather larger stature ; thus
Topinard places the average at 1·404 metres, or 4 feet $7\frac{1}{2}$
inches ; and Fritsch, who measured six male Bushmen in
South Africa, found their mean height to be 1·444 metres,
or nearly 4 feet 9 inches.   It is probable that, taking
them all together, they differ but little in size from the
Andamanese, although in colour, in form of head, in features,
and in the proportions of the body, they are widely removed
from them.

There is every reason to believe that these Bushmen represent the earliest race of which we have, or are ever likely
to have, any knowledge, inhabiting the southern portion
of the African continent, but that long before the advent of
Europeans upon the scene, they had been invaded from the
north by Negro tribes, who, being superior in size, strength,
and civilisation, had taken possession of the greater part of
their territories, and, mingling freely with the aborigines, had
produced the mixed race called Hottentots, who retained the
culture and settled pastoral habits of the Negroes, with many
of the physical features of the Bushmen.   These, in their turn,
encroached upon by the pure-bred Bantu Negroes from the
north, and by the Dutch and English from the south, are now
greatly diminished, and indeed threatened with the same fate
that will surely soon befall the scanty remnant of the early
inhabitants who still retain their primitive type.

At present the habitat of the Bushman race is confined to
certain districts in the south-west of Africa, from the confines

of the Cape Colony, as far north as the shores of Lake Ngami.
Farther to the north the great equatorial region of Africa is
occupied by various Negro tribes, using the term in its broadest
sense, but belonging to the divisions which, on account of
peculiarities of language, have been grouped together as
Bantu.   They all present the common physical characteristics
typical of the Negro race, only two of which need be specially
mentioned here—medium or large stature, and dolichocephalic
skull (average cranial index about 73·5).

It is at various scattered places in the midst of these that
the only other small people of which I shall have to speak,
the veritable pygmies of Homer, Herodotus, and Aristotle,
according to Quatrefages, are still to be met with.[1]

The first notice of the occurrence of these in modern times
is contained in " The strange adventures of Andrew Battell of
Leigh in Essex, sent by the Portugals prisoner to Angola, who
lived there and in the adjoining regions near eighteen years "
(1589 to 1607), published in *Purchas his Pilgrimes* (1625),
lib. vii. chap. iii. p. 983 :—

" To the north-east of *Mani-Kesock* are a kind of little
people, called *Matimbas ;* which are no bigger than Boyes of
twelve yeares old, but very thicke, and live only upon flesh,
which they kill in the woods with their bows and darts.
They pay tribute to *Mani - Kesock*, and bring all their
Elephants' teeth and tayles to him.   They will not enter
into any of the *Maramba's* houses, nor will suffer any to
come where they dwell.   And if by chance any *Maramba*
or people of *Longo* pass where they dwell, they will forsake
that place and go to another.   The women carry Bows and
Arrows as well as the men.   And one of these will walk in
the woods alone and kill the Pongos with their poysoned
Arrows."

Battell's narrative, it should be said, is generally admitted
to have an air of veracity about it not always conspicuous
in the stories of travellers of his time.   In addition to the

---

[1] The scattered information upon this subject was first collected together by
Hamy in his " Essai de co-ordination des Matériaux récemment recueillis sur
l'ethnologie des Négrilles ou Pygmées de l'Afrique équatoriale," *Bull. Soc.
d'Anthropologie de Paris*, tome ii. (ser. iii.), 1879, p. 79.

observations on the human inhabitants, it contains excellent descriptions of animals, as the pongo or gorilla, and the zebra, now well known, but in his day new to Europeans.

Dapper, in a work called *Description de la Basse Ethiopie,* published in Amsterdam in 1686, speaks of a race of dwarfs inhabiting the same region, which he calls *Mimos* or *Bakke-Bakke,* but nothing further was heard of these people until quite recent times.   A German scientific expedition to Loango, the results of which were published in the *Zeitschrift für Ethnologie,* 1874, and in Hartmann's work, *Die Negritier,* obtained, at Chinchoxo, photographs and descriptions of a dwarf tribe called "Baboukos," whose heads were proportionally large and of roundish form (cephalix index of skull, 78 to 81).   One individual, supposed to be about forty years of age, measured 1·365 metres, rather under 4 feet 6 inches.

Dr. Touchard, in a "Notice sur le Gabon," published in the *Revue Maritime et Coloniale* for 1861, describes the recent destruction of a population established in the interior of this country, and to which he gives the name of "Akoa."   They seem to have been exterminated by the M'Pongos in their expansion towards the west.   Some of them, however, remained as slaves at the time of the visit of Admiral Fleuriot de Langle, who in 1868 photographed one (measuring about 4 feet 6 inches high) and brought home some skulls, which were examined by Hamy, and all proved very small and sub-brachycephalic.

Another tribe, the M'Boulous, inhabiting the coast north of the Gaboon river, have been described by M. Marche as probably the primitive race of the country.   They live in little villages, keeping entirely to themselves, though surrounded by the larger negro tribes, M'Pongos and Bakalais, who are encroaching upon them so closely that their numbers are rapidly diminishing.   In 1860 they were not more than 3000; in 1879 they were much less numerous.   They are of an earthy-brown colour, and rarely exceed 1·600 metres in height (5 feet 3 inches).   In the rich collections of skulls made by Mr. R. B. Walker and by M. Du Chaillu, from the coast of this region, are many which are remarkable for their small size and round form.   Of many other notices of tribes of negroes

of diminutive size, living near the west coast of Equatorial
Africa, I need only mention that of Du Chaillu, who gives
an interesting account of his visit to an Obongo village in
Ashango-land, between the Gaboon and the Congo; although
unfortunately, owing to the extreme shyness and suspicion of
the inhabitants, he was allowed little opportunity for anthropological observations. He succeeded, however, in measuring
one man and six women; the height of the former was 4 feet
6 inches, the average of the latter 4 feet 8 inches.[1]

Farther into the interior, towards the centre of the
region contained in the great bend of the Congo or Livingstone River, Stanley heard of a numerous and independent
population of dwarfs, called " Watwas," who, like the Batimbas
of Battell, are great hunters of elephants, and use poisoned
arrows.   One of these he met with at Ikondu was 4 feet $6\frac{1}{2}$
inches high, and of a chocolate brown colour.[2]   More recently
Dr. Wolff describes, under the name of " Batouas " (perhaps the
same as Stanley's Watwas), a people of lighter colour than
other negroes, and never exceeding 1·40 metres (4 feet 7
inches) high, but whose average is not more than 1·30 (4 feet
3 inches), who occupy isolated villages scattered through the
territory of the Bahoubas, with whom they never mix.[3]

Penetrating into the heart of Africa from the north-east,
in 1870, Dr. Schweinfurth first made us acquainted with a
diminutive race of people who have since attained a considerable
anthropological notoriety.   They seem to go by two names in
their own country, *Akka* and *Tikki-tikki*, the latter reminding
us curiously of Dapper's Bakke-bakke, and the former, more
singularly still, having been read by the learned Egyptologist,
Mariette, by the side of the figure of a dwarf in one of the
monuments of the early Egyptian empire.

It was at the court of Mounza, king of the Monbuttu, that
Schweinfurth first met with the Akkas.   They appear to live
under the protection of that monarch, who had a regiment of
them attached to his service, but their real country was
farther to the south and west, about 3° N. lat. and 25° E.

[1] *A Journey to Ashango-land*, 1867, p. 315.
[2] *Through the Dark Continent*, vol. ii.
[3] *La Gazette Géographique*, 1887, p. 153, quoted by Quatrefages.

long.   From the accounts the traveller received, they occupy a considerable territory, and are divided into nine distinct tribes, each having its own king or chief.   Like all the other pygmy African tribes, they live chiefly by the chase, being great hunters of the elephant, which they attack with bows and arrows.

In exchange for one of his dogs, Schweinfurth obtained from Mounza one of these little men, whom he intended to bring to Europe, but who died on the homeward journey at Berber.   Unfortunately all the measurements and observations which were made in the Monbuttu country by Schweinfurth perished in the fire which destroyed so much of the valuable material he had collected.   His descriptions of their physical characters are therefore chiefly recollections.   Other travellers —Long, Marno, and Vossion—though not penetrating as far as the Akka country, have given observations upon individuals of the race they have met with in their travels.   The Italian Miani, following the footsteps of Schweinfurth into the Monbuttu country, also obtained by barter two Akka boys, with the view of bringing them to Europe.   He himself fell a victim to the fatigues of the journey and climate, but left his collections, including the young Akkas, to the Italian Geographical Society.   Probably no two individuals of a savage race have been so much honoured by the attentions of the scientific world.   First at Cairo, and afterwards in Italy, Tebo (or Thibaut) and Chairallah, as they were named, were described, measured, and photographed, and have been the subjects of a library of memoirs, their biographers including the names of Owen, Panceri, Cornalia, Mantegazza, Giglioli, Zannetti, Broca, Hamy, and de Quatrefages,   On their arrival in Italy, they were presented to the king and queen, introduced into the most fashionable society, and finally settled down as members of the household of Count Miniscalchi Erizzo, at Verona, where they received a European education, and performed the duties of pages.

In reply to an inquiry addressed to my friend, Dr. Giglioli, of Florence, I hear that Thibaut died of consumption on 28th January 1883, being then about twenty-two years of age, and was buried in the cemetery at Verona.   Unfortunately no

scientific examination of the body was allowed, but whether
Chairallah still lives or not I have not been able to learn. As
Giglioli has not heard of his death, he presumes that he is
still living in Count Miniscalchi's palace.

One other specimen of this race has been the subject of
careful observation by European anthropologists—a girl named
Saida, brought home by Romolo Gessi (Gordon's lieutenant),
and who is still, or was lately, living at Trieste as servant to
M. de Gessi.

The various scattered observations hitherto made are obviously insufficient to deduce a mean height for the race,
but the nearest estimate that Quatrefages could obtain is
about 4 feet 7 inches for the men, and 4 feet 3 inches for
the women, decidedly inferior, therefore, to the Andamanese.
With regard to their other characters, their hair is of the
most frizzly kind, their complexion lighter than that of most
negroes, but the prognathism, width of nose, and eversion
of lips characteristic of the Ethiopian branch of the human
family are carried to an extreme degree, especially if Schweinfurth's sketches can be trusted. The only essential point of
difference from the ordinary negro, except the size, is the
tendency to shortening and breadth of the skull, although it
by no means assumes the "almost spherical" shape attributed
to it by Schweinfurth.

Some further information about the Akkas will be found
in the work, just published, of the intrepid and accomplished
traveller, in whose welfare we are now so much interested,
Dr. Emin Pasha, Gordon's last surviving officer in the Soudan,
who, in the course of his explorations, spent some little time
lately in the country of the Monbuttu. Here he not only
met with living Akkas, one of whom he apparently still
retains as a domestic in his service, and of whose dimensions
he has sent me a most detailed account, but he also, by
watching the spots where two of them had been interred,
succeeded in obtaining their skeletons, which, with numerous
other objects of great scientific interest, safely arrived at the
British Museum in September of last year. I need hardly
say that actual bones, clean, imperishable, easy to be measured
and compared, not once only, but any number of times, furnish

the most acceptable evidence that an anthropologist can possess of many of the most important physical characters of a race. There we have facts which can always be appealed to in support of statements and inferences based on them. Height, proportions of limbs, form of head, characters of the face even, are all more rigorously determined from the bones than they can be on the living person. Therefore, the value of these remains, imperfect as they unfortunately are, and of course insufficient in number for the purpose of establishing average characters, is very great indeed.

As I have entered fully into the question of their peculiarities elsewhere,[1] I can only give now a few of the most important and most generally to be understood results of their examination. The first point of interest is their size. The two skeletons are both those of full-grown people, one a man, the other a woman. There is no reason to suppose that they were specially selected as exceptionally small; they were clearly the only ones which Emin had an opportunity of procuring; yet they fully bear out, more than bear out, all that has been said of the diminutive size of the race. Comparing the dimensions of the bones, one by one, with those of the numerous Andamanese that have passed through my hands, I find both of these Akkas smaller, not than the average, but smaller than the smallest,—smaller also than any Bushman whose skeleton I am acquainted with, or whose dimensions have been published with scientific accuracy. In fact, they are both, for they are nearly of a size, the smallest normal human skeletons which I have seen, or of which I can find any record. I say normal, because they are thoroughly well grown and proportioned, without a trace of the deformity almost always associated with individual dwarfishness in a taller race. One only, that of the female, is sufficiently perfect for articulation. After due allowance for some missing vertebræ, and for the intervertebral spaces, the skeleton measures from the crown of the head to the ground exactly 4 feet, or 1·218 metres. About three-quarters

---

[1] In a paper read before the Anthropological Institute of Great Britain and Ireland, 14th February 1888, and published in the August number of the Journal.

of an inch more for the thickness of the skin of the head and
soles of the feet would complete the height when alive.	The
other (male) skeleton was, judging by the length of the femur,
about a quarter of an inch shorter.

The full-grown woman of whom Emin gives detailed
dimensions is stated to be only 1·164 metres, or barely 3
feet 10 inches.[1]	These heights are all unquestionably less
than anything that has been yet obtained based upon such
indisputable data.	One very interesting and almost unexpected
result of a careful examination of these skeletons is that they
conform in the relative proportions of the head, trunk, and
limb, not to dwarfs, but to full-sized people of other races,
and they are therefore strikingly unlike the stumpy, long-
bodied, short-limbed, large-headed pygmies so graphically
represented fighting with their lances against the cranes on
ancient Greek vases.

The other characters of these skeletons are Negroid to an
intense degree, and quite accord with what has been stated of
their external appearance.	The form of the skull, too, has
that sub-brachycephaly which has been shown by Hamy to
characterise all the small Negro populations of Central Africa.
It is quite unlike that of the Andamanese, quite unlike that
of the Bushmen.	They are obviously Negroes of a special
type, to which Hamy has given the appropriate term of
Negrillo.	They seem to have much the same relation to the
larger long-headed African Negroes that the small round-
headed Negritos of the Indian Ocean have to their larger
long-headed Melanesian neighbours.

At all events, the fact now seems clearly demonstrated
that at various spots across the great African continent,
within a few degrees north and south of the equator, extending from the Atlantic coast to near the shores of the Albert
Nyanza (30° E. long.), and perhaps, from some indications which
time will not allow me to enter into now (but which will be
found in the writings of Hamy and Quatrefages), even farther

---

[1] In his letters Emin speaks of an Akka man as "3 feet 6 inches" high,
though this does not profess to be an observation of scientific accuracy as does
the above.	He says of this man that his whole body was covered by thick, stiff
hair, almost like felt, as was the case with all the Akkas he had yet examined.

to the east, south of the Galla land, are still surviving, in scattered districts, communities of these small negroes, all much resembling each other in size, appearance, and habits, and dwelling mostly apart from their larger neighbours, by whom they are everywhere surrounded. Our information about them is still very scanty, and to obtain more would be a worthy object of ambition for the anthropological traveller. In many parts, especially at the west, they are obviously holding their own with difficulty, if not actually disappearing, and there is much about their condition of civilisation, and the situations in which they are found, to induce us to look upon them, as in the case of the Bushmen in the south and the Negritos in the east, as the remains of a population which occupied the land before the incoming of the present dominant races. If the account of the Nasamonians related by Herodotus be accepted as historical, the river they came to, "flowing from west to east," must have been the Niger, and the northward range of the dwarfish people far more extensive twenty-three centuries ago than it is at the present time.[1]

This view opens a still larger question, and takes us back to the neighbourhood of the south of India as the centre from which the whole of the great Negro race spread, east over the African continent, and west over the islands of the Pacific, and to our little Andamanese fellow - subjects as probably the least modified descendants of the primitive members of the great branch of the human species characterised by their black skins and frizzly hair.

---

[1] Since this was written indications of the existence of a dwarf tribe in Mount Atlas have been collected and published by Mr. R. G. Haliburton, and Sergi and Kollman have adduced evidence tending to show the former extension of small races into several regions of Southern and Central Europe.

XX

FASHION IN DEFORMITY

AS ILLUSTRATED IN THE CUSTOMS OF BARBAROUS AND CIVILISED RACES [1]

What I here present you with is an Enditement framed against most of the Nations under the Sun ; whereby they are arraigned at the tribunal of Nature, as guilty of High Treason, in Abasing, Counterfeiting, Defacing, and Clipping her Coin instampt with her Image and Superscription on the Body of Man.

J. BULWER, *Anthropometamorphosis*, 1650.

THE propensity to *deform*, or alter from the natural form, some part of the body is one which is common to human nature in every aspect in which we are acquainted with it, the most primitive and barbarous, and the most civilised and refined.

The alterations or deformities which it is proposed to consider in this essay are those which are performed, not by isolated individuals, or with definite motives, but by considerable numbers of members of a community, simply in imitation of one another—in fact, according to *fashion*, "that most inexorable tyrant, to which the greater part of mankind are willing slaves."

Fashion is now often associated with change, but in less civilised conditions of society fashions of all sorts are more permanent than with us ; and in all communities such fashions as those here treated of are, for obvious reasons, far less likely to be subject to the fluctuations of caprice than

---

[1] The substance of this essay was delivered as a lecture at the Royal Institution of Great Britain, on the evening of Friday, 7th May 1880, and subsequently published in the *Proceedings* of that body. It was republished as a separate work with additional illustrations (Macmillan & Co., 1881).

those affecting the dress only, which, even in Shakespeare's time, changed so often that "the fashion wears out more apparel than the man." Alterations once made in the form of the body cannot be discarded or modified in the lifetime of the individual, and therefore, as fashion is intrinsically imitative, such alterations have the strongest possible tendency to be reproduced generation after generation.

The origins of these fashions are mostly lost in obscurity, all attempts to solve them being little more than guesses. Some of them have become associated with religious or superstitious observances, and so have been spread and perpetuated; some have been vaguely thought to be hygienic in motive; most have some relation to conventional standards of improved personal appearance; but whatever their origin, the desire to conform to common usage, and not to appear singular, is the prevailing motive which leads to their continuance. They are perpetuated by imitation, which, as Herbert Spencer says, may result from two widely divergent motives. It may be prompted by reverence for one imitated, or it may be prompted by the desire to assert equality with him.

Before treating of the subject in its application to the human body, it will be well to glance, in passing, at the fact that a precisely similar propensity has impelled man, at various ages of the world's history, and under various conditions of society, to interfere in the same manner with the natural conformation of many of the animals which have come under his influence through domestication.

The Hottentots, objecting to symmetry of growth in the horns of their cattle, twist them while young and pliant, so that ultimately they are made to assume various fantastic and unnatural directions. Sheep with multiple horns are produced in some parts of Africa, by splitting with a knife the budding horn of the young animal. Hotspur's exclamation: "What horse? a roan, a *crop-ear*, is it not?" points to a custom not yet extinct in England. Docking horses' tails—that is, cutting off about half the length, not of the hair only, but of the actual flesh and bone, and *nicking*, or dividing the tendons of the under side, so that the paralysed stump is always carried in an unnatural or "cocked" position—was

common enough a generation ago, as seen in all equestrian
pictures of the period, and is still occasionally practised.
In spite of all warnings of common sense and experience, we
continue, solely because it is the fashion, to torture and deform
our horses' mouths and necks with tight bearing-reins, which,
though only temporarily keeping the head in a constrained
and unnatural, and therefore inelegant position, produce many
permanent injuries.[1]   Dogs may still be seen with the natural
form of their ears and tails "improved" by mutilation.

Besides these and many other modifications of the form
given by nature, practised upon the individual animal, selective
breeding through many generations has succeeded in producing
inherited structural changes, sometimes of very remarkable
character.   These have generally originated in some accidental,
perhaps slight, peculiarity, which has been taken advantage of,
perpetuated, and increased.   In this way the race of bull-dogs,
with their shortened upper jaws, bandy legs, and twisted tails,
have been developed.   The now fashionable "dachshund" is
another instance.   In this category may also be placed polled
and humped cattle, tailless cats of the Isle of Man and
Singapore, lop-eared rabbits; tailless, crested, or other strange
forms of fowls, pouter, tumbler, feather-legged, and other
varieties of pigeons, and the ugly double-tailed and prominent-
eyed goldfish which delight the Chinese.   Thus the power
which, when judiciously exercised, has led to the vast improvement seen in many domestic species over their wild progenitors,
has also ministered to strange vagaries and caprices, in the
production and perpetuation of monstrous forms.

To return to man, the most convenient classification of our
subject will be one which is based upon the part of the body
affected, and I will begin with the treatment of the hair and
other appendages of the skin as the more superficial and comparatively trivial in its effects.

Here we are at once introduced to the domain of fashion
in her most potent sway.   The facility with which hair lends
itself to various methods of treatment has been a temptation
too great to resist in all known conditions of civilisation.
Innumerable variations of custom exist in different parts of

[1] See *Bits and Bearing-Reins*, by Edward Fordham Flower: London, 1879.

the world, and marked changes in at least all more or less
civilised communities have characterised successive epochs of
history.   Not only the length and method of arrangement,
but even the colour of the hair, is changed in obedience to
caprices of fashion.   In many of the islands of the Western
Pacific the naturally jet black hair of the natives is converted
into a tawny brown by the application of lime, obtained by
burning the coral found so abundantly on their shores; and
not many years since similar means were employed for producing the same result among the ladies of Western Europe
—a fact which considerably diminishes the value of an idea
entertained by many ethnologists, that community of custom
is evidence of community of origin or of race.

Notwithstanding the painful and laborious nature of the
process, when conducted with no better implements than flint
knives, or pieces of splintered bone or shell, the custom of
keeping the head closely shaved prevails extensively among
savage nations.   This, doubtless, tends to cleanliness, and
perhaps comfort, in hot countries; but the fact that it is in
many tribes practised only by the women and children shows
that these considerations are not those primarily engaged in
its perpetuation.   In some cases, as among the Fijians, while
the heads of the women are commonly cropped or closely
shaved, the men cultivate, at great expense of time and
attention, a luxuriant and elaborately arranged mass of hair,
exactly reversing the conditions met with in the most highly
civilised nations.

In some regions of Africa it is considered necessary to
female beauty carefully to eradicate the eyebrows, special
pincers for the purpose forming part of the appliances of the
toilette; while the various methods of shaving and cutting
the beard among men of all nations are too well known to
require more than a passing notice.   The treatment of finger
nails, both as to colour and form, has also been subject to
fashion; but the practical inconveniences attending the in-
ordinate length to which these are permitted to grow in some
parts of the east of Asia appear to have restricted the custom
to a few localities.   (See Fig. 13, p. 319.)

It may be objected to the introduction of this illustration

here, that such nails should not be considered deformities, but rather as natural growth, and that to clip and mutilate them as we do is the departure from nature's intention. But this is not so. It is only by constant artificial care and protection that such an extraordinary and inconvenient length can be obtained. When the hands are subjected to the normal amount of use, the nails break or wear away at their free ends in a ratio equal to their growth, as with the claws or hoofs of animals in a wild state.

The exceedingly widespread custom of tattooing [1] the skin may also be alluded to here, as the result of the same propensity as that which produces the more serious deformations presently to be spoken of. The rudest form of the art was practised by the now extinct Tasmanians and some tribes of Australians, whose naked bodies showed linear or oval raised scars, arranged in a definite manner on the shoulders and breast. These were produced by gashes inflicted with sharp stones, into which wood-ashes were rubbed,

Fig. 13.—Hand of Chinese ascetic. From Tylor's *Anthropology*.

so as to allow of healing only under unfavourable conditions, leaving permanent large and elevated cicatrices, conspicuous from being of a lighter colour than the rest of the skin. From this it is a considerable step in decorative art to the elaborate and often beautiful patterns, wreaths, scrolls, spirals, zigzags' etc., sometimes confined to the face, and sometimes covering the whole body from head to foot, seen in the natives of many of the Polynesian Islands. These are permanently im-

---

[1] A word used by the natives of Tahiti, spelt *tattowing* by Cook, who gives a minute account of the method in which it is performed in that island. *First Voyage*, vol. ii. p. 191.

pressed upon the skin by the introduction of colouring matter, generally some kind of lamp-black, by means of an instrument made of a piece of shell cut into a number of fine points, or a bundle of sharp needles. When the custom of the land demands that the surface to be treated thus is a large one, the process is not only very tedious, but entails an amount of suffering painful to think of. When completed it answers part at least of the purpose of dress with us, as an untattooed

FIG. 14.—Australian native, with bone nose-ornament.

skin exhibited to society is looked upon much as an unclothed one would be in more civilised communities. The natural colour of the skin seems to have influenced the method and extent of tattooing, as in the black races it is limited to such scars as those spoken of above; which, variously arranged in lines or dots, become tribal distinctions among African negroes. In Europe tattooing on the same principle as that of the Polynesians is confined almost exclusively to sailors, among whom it is kept up obviously by imitation or fashion.

The nose, the lips, and the ears have in almost all races offered great temptations to be used as foundations for the display of ornament, some process of boring, cutting, or alteration of form being necessary to render them fit for the purpose. When Captain Cook, exactly one hundred years ago, was describing the naked savages of the east coast of Australia,[1] he says: "Their principal ornament is the bone which they thrust through the cartilage which divides the nostrils from each other. What perversion of taste could make them think this a decoration, or what could prompt them, before they had worn it or seen it worn, to suffer the pain and inconvenience that must of necessity attend it, is perhaps beyond the power of human sagacity to determine. As this bone is as thick as a man's finger, and between five and six inches long, it reaches quite across the face, and so effectually stops up both the nostrils that they are forced to keep their mouths wide open for breath, and snuffle so when they attempt to speak that they are scarcely intelligible even to each other. Our seamen, with some humour, called it their spritsail-yard; and indeed it had so ludicrous an appearance, that till we were used to it we found it difficult to refrain from laughter."

Fig. 15.—Tortoise-shell lip ornament of the Moskito Indians. From Dampier.

Eight years later, on his visit to the north - west coast (of America, Captain Cook found precisely the same custom prevailing among the natives of Prince William's Sound, whose mode of life was in most other respects quite dissimilar to that of the Australians, and who belong ethnologically to a totally different branch of the human race.

In 1681 Dampier[2] thus describes a custom which he found existing among the natives of the Corn Islands, off the Moskito Coast, in Central America: "They have a fashion to cut holes in the Lips of the Boys when they are young, close to their Chin, which they keep open with little Pegs till they are 14 or 15 years old; then they

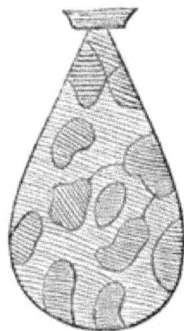

[1] *First Voyage*, vol. ii. p. 633.
[2] *Voyage Round the World*, ed. 1717, vol. i. p. 32.

Y

wear Beards in them, made of Turtle or Tortoise-shell, in the
form you see in the Margin. (See Fig. 15, p. 321.) The
little notch at the upper end they put in through the Lip,
where it remains between the Teeth and the Lip; the under
part hangs down over their Chin. This they commonly wear
all day, and when they sleep they take it out. They have
likewise holes bored in their Ears, both Men and Women,
when young, and by continual stretching them with great

FIG. 16.—Botocudo Indian.   From Bigg-Wither's
Pioneering in South Brazil (1878).

Pegs, they grow to be as big as a mill'd five Shilling Piece.
Herein they wear pieces of Wood, cut very round and smooth,
so that their Ear seems to be all wood, with a little Skin
about it."

It is very remarkable that an almost exactly similar
custom still prevails among a tribe of Indians inhabiting the
southern part of Brazil—the Botocudos, so called from a Portu-
guese word (*botoque*), meaning a plug or stopper. Among these
people the lip-ornament consists of a conical piece of hard
and polished wood, frequently weighs a quarter of a pound,

and drags down, elongates, and everts the lower lip, so as to
expose the gums and teeth, in a manner which to our taste is
hideous, but with them is considered an essential adjunct to
an attractive and correct appearance.

In the extreme north of America, the Eskimo "pierce the
lower lip under one or both corners of the mouth, and insert
in each aperture a double-headed sleeve-button or dumb-bell-
shaped labret, of bone, ivory, shell, stone, glass, or wood.
The incision when first made is about the size of a quill, but
as the aspirant for improved beauty grows older, the size of
the orifice is enlarged until it reaches the width of half to
three-quarters of an inch."[1]   These operations appear to be
practised only on the men, and are supposed to possess some
significance other than that of mere ornament.   The first
piercing of the lip, which is accompanied by some solemnity
as a religious feast, is performed on approaching manhood.

But the people who, among the various American tribes,
have carried these strange customs to the greatest excess
are the Thlinkeets, who inhabit the south-eastern shores of
Alaska.[2]   "Here it is the women who, in piercing the nose
and ears, and filling the apertures with bones, shells, sticks,
pieces of copper, nails, or attaching thereto heavy pendants,
which drag down the organs and pull the features out of
place, appear to have taxed their inventive powers to the
utmost, and, with a success unsurpassed by any nation in the
world, to produce a model of hideous beauty.   This success is
achieved in their wooden lip-ornament, the crowning glory
of the Thlinkeet matron, described by a multitude of eye-
witnesses.   In all female free-born Thlinkeet children a slit
is made in the under lip, parallel with the mouth, and about
half an inch below it.   A copper wire, or a piece of shell or
wood, is introduced into this, by which the wound is kept
open and the aperture extended.   By gradually introducing
larger objects the required dimensions of the opening are
produced.   On attaining the age of maturity, a block of wood

---

[1] H. H. Bancroft, *Native Races of the Pacific States of North America*,
vol. i. 1875.

[2] See Bancroft, *op. cit.* vol. i., for numerous citations from original observers
regarding these customs.

is inserted, usually oval or elliptical in shape, concave on the sides, and grooved like the wheel of a pulley on the edge in order to keep it in place. The dimensions of the block are from two to six inches in length, from one to four inches in width, and about half an inch thick round the edge, and it is highly polished. Old age has little terror in the eyes of a Thlinkeet belle; for larger lip-blocks are introduced as years advance, and each enlargement adds to the lady's social status, if not to her facial charms. When the block is withdrawn, the lip drops down upon the chin like a piece of leather, displaying the teeth, and presenting altogether a ghastly spectacle. The privilege of wearing this ornament is not extended to female slaves."

In this method of adornment the North Americans are, however, rivalled, if not eclipsed, by the negroes of the heart of Africa.

"The Bongo women (says Schweinfurth [1]) delight in distinguishing themselves by an adornment which to our notion is nothing less than a hideous mutilation. As soon as a woman is married, the operation commences of extending her lower lip. This, at first only slightly bored, is widened by inserting into the orifice plugs of wood, gradually increasing in size, until at length the entire feature is enlarged to five or six times its original proportions. The plugs are cylindrical in form, not less than an inch thick, and are exactly like the pegs of bone or wood worn by the women of Musgoo. By this means the lower lip is extended horizontally till it projects far beyond the upper, which is also bored and fitted with a copper plate or nail, and now and then by a little ring, and sometimes by a bit of straw, about as thick as a lucifer-match. Nor do they leave the nose intact; similar bits of straw are inserted into the edges of the nostrils, and I have seen as many as three of these on each side. A very favourite ornament for the cartilage between the nostrils is a copper ring, just like those that are placed in the noses of buffaloes and other beasts of burden for the purpose of rendering them more tractable. The greatest coquettes among the ladies wear a clasp, or clamp, at the corners of the

[1] *Heart of Africa*, vol. i. p. 297.

mouth, as though they wanted to contract the orifice, and literally to put a curb upon its capabilities. These subsidiary ornaments are not, however, found at all universally among the women, and it is rare to see them all at once upon a single individual; the plug in the lower lip of the married women is alone a *sine quâ non*, serving, as it does, for an artificial distinction of race."

The slightest fold or projection of the skin furnishes an excuse for boring a hole, and inserting a plug or a ring.

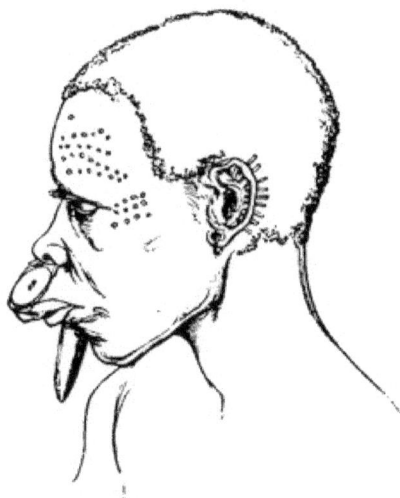

Fig. 17.—Loobah woman. From Schweinfurth's *Heart of Africa*.

There are women in the country whose bodies are pierced in some way or other in little short of a hundred different places, and the men are often not far behind in the profusion with which this kind of adornment is carried out.

"The whole group of the Mittoo exhibits peculiarities by which it may be distinguished from its neighbours. The external adornment of the body, the costume, the ornaments, the mutilations which individuals undergo — in short, the general fashions—have all a distinctive character of their own. The most remarkable is the revolting, because unnatural, manner in which the women pierce and distort their lips; they seem to vie with each other in their mutilation;

and their vanity in this respect, I believe, surpasses anything that may be found throughout Africa. Not satisfied with piercing the lower lip, they drag out the upper lip as well for the sake of symmetry.[1] . . . Circular plates, nearly as large as a crown piece, made variously of quartz, of ivory, or of horn, are inserted into the lips that have been stretched by the growth of years, and then often bent in a position that is all but horizontal; and when the women want to drink they have to elevate the upper lip with their fingers, and to pour the draught into their mouth.

"Similar in shape is the decoration which is worn by the women of Maganya; but though it is round, it is a ring and not a flat plate; it is called 'pelele,' and has no object but to expand the upper lip. Some of the Mittoo women, especially the Loobah, not content with the circle or the ring, force a cone of polished quartz through the lips as though they had borrowed the idea from the rhinoceros. This fashion of using quartz belemnites of more than two inches long is in some instances adopted by the men."

The traveller who has been the eye-witness of such customs may well add, "Even amongst these uncultured children of nature, human pride crops up amongst the fetters of fashion, which, indeed, are fetters in the worst sense of the word; for fashion in the distant wilds of Africa tortures and harasses poor humanity as much as in the great prison of civilisation."

It seems, indeed, a strange phenomenon that in such different races, so far removed in locality, customs so singular —to our ideas so revolting and unnatural, and certainly so painful and inconvenient—should either have been perpetuated for an enormous lapse of time, if the supposition of a common origin be entertained, or else have developed themselves independently.

These are, however, only extreme or exaggerated cases of the almost universal custom of making a permanent aperture through the lobe of the ear for the purpose of inserting some adventitious object by way of adornment, or even for utility, as in the man of the island of Mangea, figured in Cook's voyages,

[1] The mutilation of both lips was also observed by Rohlfs among the women of Kadje, in Segseg, between Lake Tsad and the Benwe.

who carries a large knife through a hole in the lobe of the right
ear.   The New Zealanders of both sexes, when first visited
by Europeans, all had holes bored through their ears, and
enlarged by stretching, and which in their domestic economy
answered the purpose of our pockets.   Feathers, bones, sticks,
talc chisels and bodkins, the nails and teeth of their deceased
relations, the teeth of dogs, and in fact anything which they
could get that they thought curious or valuable, were thrust
through or suspended to them.   The iron nails given them
by the English sailors were at once conveyed to these
miscellaneous receptacles.[1]   The Zulus lately exhibited in
London carried their cigars in the same manner.   Mr.
Wilfred Powell informs me that he met with a man on
one of the islands near New Guinea, the holes in whose ears
had been extended to such an extent that the lobes had been
converted into great pendent rings of skin, through which he
could easily pass his arms !

Among ourselves the custom of wearing earrings still
survives, even in the highest grades of society, although it
has been almost entirely abandoned by one-half of the
community, and in the other the perforation is reduced to
the smallest size compatible with the purpose of carrying the
ornament suspended from it.   Nose-rings are not now the
fashion in Europe, but the extent to which they are admired
in the East may be judged of by the frequency with which
they are worn by the ayahs or female servants who so often
accompany English families returning from India.

The teeth, although allowed by the greater part of the
world to retain their natural beauty and usefulness of form,
still offer a field for artificial alterations according to fashion,
which has been made use of principally in two distinct
regions of the world and by two distinct races.   It is, of
course, only the front teeth, and mainly the upper incisors,
that are available for this purpose.   Among various tribes of
negroes of Equatorial Africa different fashions of modifying
the natural form of these teeth prevail, specimens of which
may be found in any large collection of crania of these
people.   One of the simplest consists of chipping and filing

---

[1] *Cook's First Voyage*, vol. iii. p. 456.

away a large triangular piece from the lower and inner edge of each of the central incisors, so that a gap is produced in the middle of the row in front (Fig. 18, 1). Another fashion is to shape all the incisors into sharp points by chipping off the corners, giving a very formidable crocodilian appearance to the jaws (2); and another is to file out either a single or a double notch in the cutting edge of each tooth, producing a serrated border to the whole series (3).

The Malays, however, excel the Africans both in the universality and in the fantastic variety of their supposed

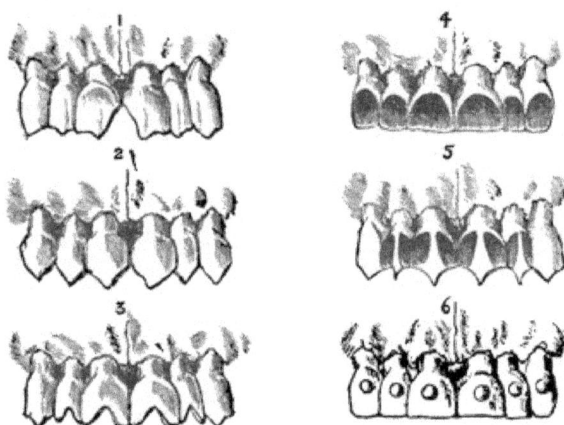

FIG. 18.—Upper front teeth altered according to fashion.
1, 2, 3, African ; 4, 5, 6, Malay.

improvements upon nature. While the natural whiteness of the surface of these organs is always admired by us, and by most people, the Malays take the greatest pains to stain their teeth black, which they consider greatly adds to their beauty. White teeth are looked upon with perfect disgust by the Dayaks of the neighbourhood of Sarawak. In addition to staining the teeth, filing the surface in some way or other is almost always resorted to. The nearly universal custom in Java is to remove the enamel from the front surface of the incisors, and often the canine teeth, hollowing out the surface, sometimes so deeply as to penetrate the pulp cavity (Fig. 18, 4). The cutting edges are also worn down to a level line with pumice-stone. Another and less common, though more elaborate fashion, is to point the teeth, and file out notches from the anterior surface of each side of the upper part of the crown, so as to leave a lozenge-shaped piece of enamel untouched; as this receives the black stain less strongly than the parts from which the surface is removed, an ornamental pattern is produced (5). In Borneo a still more elaborate process is adopted, the front surface of each of the teeth is drilled near its centre with a small round hole, and into this a plug of brass with a round or star-shaped knob is fixed (6). This is always kept bright and polished by the action of the lip over it, and is supposed to give a highly attractive appearance when the teeth are displayed. A skull with the teeth treated in this way may be seen in the Barnard Davis Collection, now in the museum of the Royal College of Surgeons.

The Javan practice appears also to prevail in fashionable circles in the neighbouring parts of the mainland of Asia. The Siamese envoy who visited this country in 1880 had his upper incisor teeth treated like No. 4, Fig. 18, and one of his suite had them pointed.

Perhaps the strange custom, so frequently adopted by the natives of Australia, and of many islands of the Pacific, of knocking out one or more of the front teeth might be mentioned here, but it is usually associated with some other idea than ornament or even mere fashion. In the first-named country it constitutes part of the rites by which the youth are initiated into manhood, and in the Sandwich Islands it is performed as a propitiatory sacrifice to the spirits of the dead.

The projection forwards of the front upper teeth, which we think unbecoming, is admired by some races, and among the negro women of Senegal it is increased by artificial means employed in childhood.[1]

All these modifications of form of comparatively external and flexible parts are, however, trivial in their effects upon the body to those to be spoken of next, which induce permanent structural alterations both upon the bony framework and upon the important organs within.

[1] Hamy, *Revue d'Anthropologie*, January 1879, p. 22.

Whatever might be the case with regard to the hair, the ears, the nose, and lips, or even the teeth, it might have been thought that the actual shape of the head, as determined by the solid skull, would not have been considered a subject to be modified according to the fashion of the time and place. Such, however, is far from being the case. The custom of artificially changing the form of the head is one of the most ancient and widespread with which we are acquainted. It is far from being confined, as many suppose, to an obscure tribe of Indians on the north-west coast of America, but is found under various modifications at widely different parts of the earth's surface, and among people who can have had no intercourse with one another. It appears, in fact, to have originated independently in many quarters, from some natural impulse common to the human race. When it once became an established custom in any tribe, it was almost inevitable that it should continue, until put an end to by the destruction either of the tribe itself, or of its peculiar institutions, through the intervention of some superior force; for a standard of excellence in form, which could not be changed in those who possessed it, was naturally followed by all who did not wish their children to run the risk of the social degradation which would follow the neglect of such a custom. " Failure properly to mould the cranium of her offspring gives to the Chinook [1] matron the reputation of a lazy and undutiful mother, and subjects the neglected children to the ridicule of their young companions, so despotic is fashion." [2]   A traveller, who mentions that he occasionally saw Chinooks with heads of the ordinary shape, sickness or some other cause having prevented the usual distortion in infancy, adds that such individuals could never attain to any influence or rise to any dignity in their tribe, and were not unfrequently sold as slaves [3]

It is related in the narrative of Commodore Wilkes' United States Exploring Expedition,[4] that " at Niculuita Mr.

---

[1] A tribe of Indians inhabiting the neighbourhood of the Columbia River, North America.

[2] Bancroft, *op. cit.* vol. i. p. 238.

[3] T. K. Townsend, *Journey to the Columbia River*, p. 175.

[4] Vol. iv. p. 388.

Drayton obtained the drawing of a child's head, of the Walla-walla tribe (Fig. 19), that had just been released from its bandages, in order to secure its flattened shape. Both the parents showed great delight at the success they had met with in effecting this distortion."

Endeavours have been made to trace the origin of this and many analogous customs to a desire to intensify or exaggerate any prevailing natural peculiarity of conformation. Thus races in which the forehead is naturally low are supposed to have admired, and then to have artificially imitated, those individuals in which the peculiarity was most pronounced. But this assumption does not rest upon any strong basis of fact. The motives assigned by the native Peruvians for their interference with the natural form of their children's heads, as reported by the early Spanish historians, were very various. Some said that it contributed to health, and enabled them to bear greater burdens; others that it increased the ferocity of the countenance in war.[1] These were all probably excuses for a blind adherence to custom or the imperious demands of fashion.

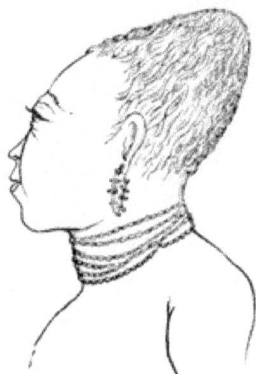

Fig. 19.—Flat-headed Indian child.

Many of the less severe alterations of the form to which the head is subjected are undesigned, resulting only from the mode in which the child is carried or dressed during infancy. Thus habitually carrying the child on one arm appears to produce an obliquity in the form of the skull which is retained to a greater or less degree all through life. The practice followed by nomadic people of carrying their infants fastened to stiff pillows or boards commonly causes a flattening of the occiput; and the custom of dressing the child's head with tightly-fitting bandages, still common in many parts of the Continent, and even used in England within the memory of living people, produces an elongated and laterally constricted

[1] Morton's *Crania Americana*, p. 116.

form.[1]   In France this is well known, and so common is it in the neighbourhood of Toulouse, that a special form of head produced in this manner is known as the "*déformation Toulousaine.*"

Of the ancient notices of the custom of purposely altering the form of the head, the most explicit is that of *Hippocrates,* who in his treatise *De Aëris, Aquis et Locis,* written about 400 B.C., says,[2] speaking of the people near the boundary of Europe and Asia, near the *Palus Mæotis* (Sea of Azoff): " I will pass over the smaller differences among the nations, but will now treat of such as are great either from nature or custom ; and first, concerning the *Macrocephali.*   There is no other race of men which have heads in the least resembling theirs.   At first, usage was the principal cause of the length of their head, but now nature co-operates with usage.   They think those the most noble who have the longest heads.   It is thus with regard to the usage : immediately after the child is born, and while its head is still tender, they fashion it with their hands, and constrain it to assume a lengthened shape by applying bandages and other suitable contrivances, whereby the spherical form of the head is destroyed, and it is made to increase in length.   Thus, at first, usage operated, so that this constitution was the result of force ; but in the course of time it was formed naturally, so that usage had nothing to do with it."

Here, Hippocrates appears to have satisfied himself upon a point which is still discussed with great interest, and still not cleared up—the possibility of transmission by inheritance of artificially-produced deformity.   Some facts seem to show that such an occurrence may take place occasionally, but there is an immense body of evidence against its being habitual.

Herodotus also alludes to the same custom as do, at later dates, Strabo, Pliny, Pomponius Mela, and others, though assigning different localities to the nations or tribes to which

[1] A gentleman of advanced age lately [1879] showed me a circular depression round the upper part of his head, which he believed had been produced in this manner, as the custom was still prevailing at the time of his birth in the district of Norfolk, of which he was a native.

[2] Sydenham Society's edition, by Dr. Adam, vol. i. p. 270.

they refer, and also indicating variations of form in their peculiar cranial characteristics.

Recent archæological discoveries fully bear out these statements. Heads deformed in various fashions, but chiefly of the constricted, elongated shape, have been found in great numbers in ancient tombs, in the very region indicated by Herodotus. They have been found near Tiflis, where as many as 150 were discovered at one time, and at other places in the Caucasus, generally in rock tombs; also in the Crimea, and at different localities along the course of the Danube; in Hungary, Silesia, in the south of Germany, Switzerland, and even in France and Belgium. The people who have left such undoubted evidence of the practice of deforming their heads have been supposed by various authors to have been Avars, Huns, Tartars, or other Mongolian invaders of Europe; but later French authors who have discussed this subject are inclined to assign them to an Aryan race, who, under the name of Cimmerians, spread westward over the part of Europe in which their remains are now found, in the seventh or eighth century before our era. Whether the French habit, scarcely yet extinct, of tightly bandaging the heads of infants, is derived from these people, or is of independent origin, it is impossible to say.

There is no unequivocal proof that the custom of designedly altering the form of the head ever existed in this country, but the singular shape of a skull found in 1853 in a Saxon grave at West Harnham, in Wilts, figured and described in Davis's and Thurnam's *Crania Britannica*, and now in the museum of the College of Surgeons, is apparently due to such a cause.

In Africa and Australia no analogous customs have been shown to exist, but in many parts of Asia and Polynesia, deformations, though usually only confined to flattening of the occiput, are common. Though often undesigned, they are done purposely, I am informed by Mr. H. B. Low, by the Dayaks, in the neighbourhood of Sarawak. Sometimes, in the islands of the Pacific, the head of the new-born infant is merely pressed by the hands into the desired form, in which case it generally soon recovers that which nature intended for it.   In one island alone, Mallicollo, in the New Hebrides, the practice of permanently depressing the forehead is almost universal, and skulls are even found constricted and elongated exactly after the manner of the Aymaras of ancient Peru.   The extraordinary flatness of the forehead, by which the inhabitants of this island differ from those of all around, was noticed by Captain Cook and the two Forsters, who accompanied him as naturalists, but they were not able to ascertain whether it was a natural conformation or due to art.   It is only within the last few years that crania have been sent to England which abundantly confirm the old description of the great navigator, and also prove the artificial character of the deformity.

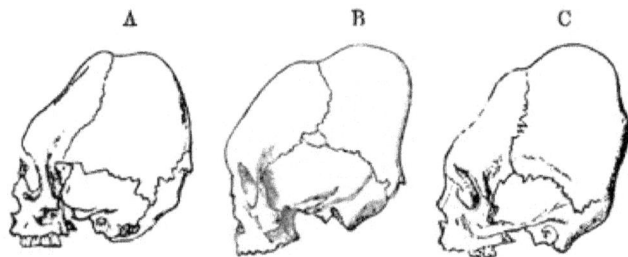

FIG. 20.—Skulls artificially deformed according to similar fashions. A, from an ancient tomb at Tiflis; B, from Titicaca, Peru. C, from the island of Mallicollo, New Hebrides. (From specimens in the Museum of the Royal College of Surgeons.)

Though the Chinese usually allow the head to assume its natural form, confining their attentions to the feet, a certain class of mendicant devotees appear to have succeeded to a remarkable extent in getting their skulls elongated into a conical form, if the figure in Picart's *Histoire des Religions*, vol. iv. plate 131, is to be trusted.

America is, however, or rather has been, the headquarters of all these fantastic practices, and especially along the western coast, and mainly in two regions, near the mouth of the Columbia River in the north and in Peru in the south. The practice also existed among the Indians of the southern parts of what are now the United States, and among the Caribs of the West India Islands.   In ancient Peru, before the time of the Spanish conquest, it was almost universal. In an edict of the ecclesiastical authorities of Lima, issued

in 1585, three distinct forms of deformation are mentioned.
Notwithstanding the severe penalties imposed by this edict
upon parents persisting in the practice, the custom was so
difficult to eradicate that another injunction against it was
published by the Government as late as 1752.

In the West Indies, and the greater part of North America,
the custom has become extinct with the people who used it;
but the Chinook Indians, of the neighbourhood of the
Columbia River, and the natives of Vancouver Island,

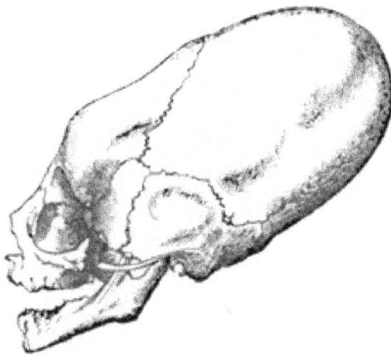

Fig. 21.—Deformed skull of an
infant who had died during the process of flattening; from the Columbia
River. (Mus. Roy. Coll. Surgeons.)

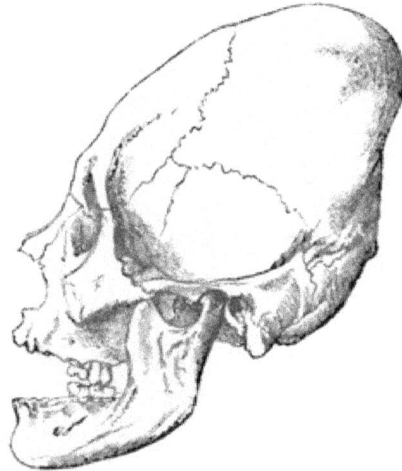

Fig. 22.—Artificially flattened skull
of ancient Peruvian. (Mus. Roy.
Coll. Surgeons.)

continue it to the present day; and this is the last strong-
hold of this strange fashion, though under the influence of
European example and discouragement it is rapidly dying
out. Here the various methods of deforming the head, and
their effects, have been studied and described by numerous
travellers. The process commences immediately after the
birth of the child, and is continued for a period of from
eight to twelve months, by which time the head has permanently assumed the required form, although during subsequent growth it may partly regain its proper shape. " It
might be supposed," observes Mr. Kane, who had large

opportunities of watching the process, "that the operation would be attended with great suffering; but I never heard the infants crying or moaning, although I have seen their eyes seemingly starting out of the sockets from the great pressure; but, on the contrary, when the thongs were loosened and the pads removed I have noticed them cry until they were replaced. From the apparent dulness of the children whilst under the pressure, I should imagine that a

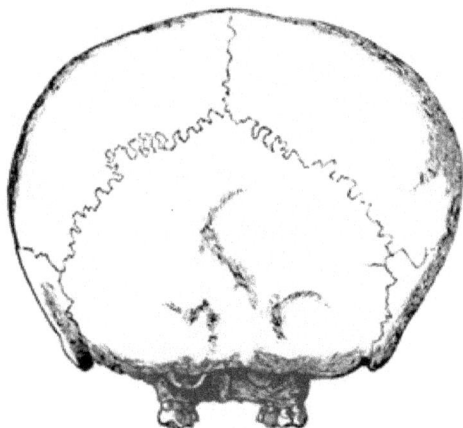

FIG. 23.—Posterior view of cranium, deformed according to the fashion of flattening, with compensatory lateral widening. (Mus. Roy. Coll. Surgeons.)

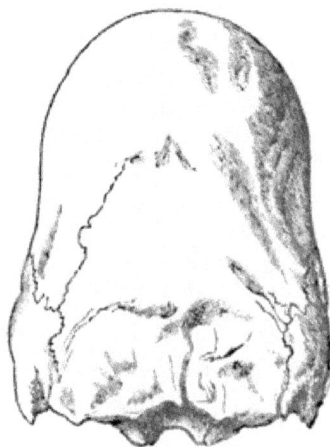

FIG. 24.—Posterior view of cranium deformed according to the fashion of circular constriction and elongation. (Mus. Roy. Coll. Surgeons.)

state of torpor or insensibility is induced, and that the return to consciousness occasioned by its removal must be naturally followed by a sense of pain."

Nearly if not all the different fashions in cranial deformity, observed in various parts of the world, are found associated within a very small compass in British Columbia and Washington Territory, each small tribe having often a particular method of its own. Many attempts have been made to classify these various deformities; but as they mostly pass insensibly into one another, and vary according as the intention has been carried out with a greater or less degree of perseverance and skill, it is not easy to do so. Besides the simple occipital and the simple frontal compressions, all the others may be grouped into two principal divisions. First (Figs. 21, 22, and 23), that in which the skull is flattened between boards or pads made (among the Indians of the Columbia River) of deer-skin stuffed with frayed cedar bark or moss, applied to the forehead and back of the head; and as there is no lateral pressure, it bulges out sideways, as seen in Fig. 23, to compensate for the shortening in the opposite direction. This form is very often unsymmetrical, as the flattening boards, applied to a nearly spherical surface, naturally incline a little to one side or the other: and when this once commences, unless great care is used, it must increase until the very curious oblique flattening so common in these skulls is produced. This is the ordinary form of deformity among the Chinook Indians of the Columbia River, commonly called " Flat-heads." It is also most frequent among the Quichuas of Peru.

The methods by which this particular kind of deformity was produced varied in detail in different tribes. One of the most effective is thus described by Mr. Townsend : " The Wallamet Indians place the infant, soon after birth, upon a board, to the edges of which are attached little loops of hempen cord or leather ; and other similar cords are passed across and back, in a zigzag manner, through these loops, enclosing the child and binding it firmly down. To the upper edge of this board, in which is a depression to receive the back part of the head, another smaller one is attached by hinges of leather, and made to lie obliquely upon the forehead, the force of the pressure being regulated by several strings attached to its edge, which are passed through holes in the board upon which the infant is lying, and secured there."

The second form of deformity (Figs. 20, 24, and 25) is produced by constricting bandages of deer's hide, or other similar material, encircling the head behind the ears, usually passing below the occiput behind, and across the forehead, and again across the vertex, behind the coronal suture, producing a circular depression. The result is an elongation of the head, but with no lateral bulging and with no

z

deviation from bilateral symmetry. This was the form
adopted, with trifling modifications, by the *Macrocephali* of
Herodotus, by the Aymara Indians of Peru, and by certain
tribes, as the Koskeemos, of Vancouver Island. The "*déformation Toulousaine*" is a variation of the same form. Another
modification is thus described in Wilson's *Prehistoric Man*:
"The Newatees, a warlike tribe on the north end of Vancouver's
Island, give a conical shape to the head by means of a thong
of deer's skin, padded with the inner bark of the cedar-tree

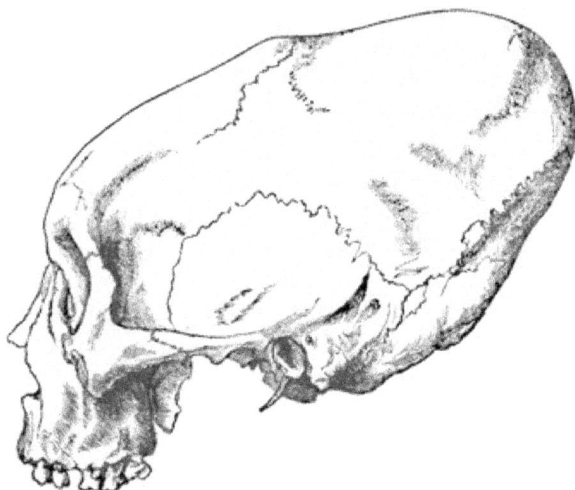

FIG. 25.—Cranium of Koskeemo Indian, Vancouver Island, deformed by circular
constriction and elongation. (Mus. Roy. Coll. Surgeons.)

frayed until it assumes the consistency of very soft tow. This
forms a cord about the thickness of a man's thumb, which
is wound round the infant's head, compressing it gradually
into a uniformly tapering cone. The effect of this singular
form of head is still further increased by the fashion of
gathering the hair into a knot on the crown of the head."
A "sugar-loaf" form of skull has also been found in an
ancient grave in France, at Voiteur in the Department of
Jura.

The brain, of course, has to accommodate itself to the
altered shape of the bony case which contains it; and the

question naturally arises, whether the important functions belonging to this organ are in any way impaired or affected by its change of form. All observations upon the living Indians who have been subjected to it concur in showing that if any modification in mental power is produced, it must be of a very inconsiderable kind, as no marked difference has been detected between them and the people of neighbouring tribes which have not adopted the fashion. Men whose heads have been deformed to an extraordinary extent, as Concomly, a Chinook chief, whose skull is preserved in the museum at Haslar Hospital, have often risen by their own abilities to considerable local eminence; and the fact that the relative social position of the chiefs, in whose families the heads are always deformed, and the slaves on whom it is never permitted, is constantly maintained, proves that the former evince no decided inferiority in intelligence or energy.

Of the Newatees, mentioned above, Wilson says, " The process seems neither to affect the intellect nor the courage of the people, who are remarkable for cunning, as well as fierce daring, and are the terror of surrounding tribes."

Of the Mallicollese it is expressly stated by George Forster that " they are the most intelligent people we have ever met with in the South Seas; they understood our signs and gestures as if they had been long acquainted with them, and in a few minutes taught us a great number of words. . . . Thus what they wanted in personal attraction they made up in acuteness of understanding." Cook gives some remarkable instances of the honesty of the " ape-like nation," as he calls them.

Although the American Indians,—living a healthy life in their native wilds, and under physical conditions which cause all bodily injuries to occasion far less constitutional or local disturbance than is the case with people living under the artificial conditions and the accumulated predisposition to disease which civilisation entails,—thus appear to suffer little, if at all, from this unnatural treatment, it seems to be otherwise with the French, on whom its effects have been watched by medical observers more closely than it can have been on the savages in America. " Dr. Foville proves, by positive and

numerous facts, that the most constant and the most frequent
effects of this deformation, though only carried to a small
degree, are headaches, deafnesses, cerebral congestions, menin-
gitis, cerebritis, and epilepsy; that idiocy or madness often
terminates this series of evils; and that the asylums for
lunatics and imbeciles receive a large number of their inmates
from among these unhappy people."[1]   For this reason the
French physicians have exerted all their influence, and with
great success, to introduce a more rational system in the
districts where the practice of compressing the heads of
infants prevailed."[2]

We may now pass from the head to the extremities, but
there will be little to say about the hands, for the artificial
deformities practised upon those members are confined to
chopping off one or more of the fingers, generally of the left
hand, and usually not so much in obedience merely to fashion,
as part of an initiatory ceremony, or an expiation or oblation
to some superior, or to some departed person.   Such practices
are common among the American Indians, some tribes of
Africans, the Australians, and Polynesians, especially those
greatest of all slaves of ceremonial, the Fijians, where the
amputation of fingers is demanded to appease an angry
chieftain, or voluntarily performed as a token of affection on
the occasion of the death of a relative.

But *per contra*, the feet have suffered more, and altogether
with more serious results to general health and comfort, from
simple conformity to pernicious customs, than any other part
of the body.   And on this subject, instead of relating the
unaccountable caprices of the savage, we have to speak only
of people who have already advanced to a tolerably high
grade of civilisation, and to include all those who are at the
present time foremost in the ranks of intellectual culture.

The most extreme instance of modification of the size and

---

[1] Gosse, "Essai sur les Déformations artificielles du Crane," *Annales d'Hygiene*,
2 ser. tom. iv. p. 8.

[2] Ample references to the literature of artificially produced deformities of the
cranium are given by Prof. Rolleston, in Greenwell's *British Barrows* (1877),
p. 596.   To these may be added, Lenhossek, *Des Déformations artificielles du
Cranes*, etc., Budapest, 1878, and Topinard, "Des Déformations ethniques du
Crane," in the *Revue d'Anthropologie*, July 1879, p. 496.

form of the foot in obedience to fashion, is the well-known
case of the Chinese women, not entirely confined to the
highest classes, but in some districts pervading all grades of
society alike.   The deformity is produced by applying tight
bandages round the feet of the girls when about five years
old.   The bandages are specially manufactured, Miss Norwood [1]
tells us, and are about two inches wide and two yards long
for the first year, five yards long for subsequent years.   The
end of the strip is laid on the inside of the foot at the instep,
then carried over the toes, under the foot, and round the heel,

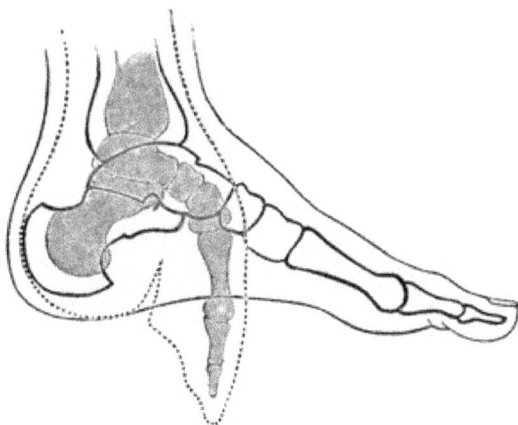

FIG. 26.—Section of natural foot with the bones, and a corresponding section
of a Chinese deformed foot.   The outline of the latter is dotted, and the
bones shaded.

the toes being thus drawn towards and across the sole, while
a bulge is produced in the instep, and a deep indentation in
the sole.   Successive layers of bandage are wound round the
foot until the strip is all used, and the end is then sewn
tightly down.   After a month the foot is put in hot water
to soak some time ; then the bandage is carefully unwound.
Notwithstanding the powdered alum and other appliances
that are used to prevent it, the surface of the foot is generally
found to be ulcerated, and much of the skin and sometimes
part of the flesh of the sole, and even one or two of the toes,

[1] American missionary at Swatow ; *Times*, 2nd September 1880.

may come off with the bandages, in which case the woman afterwards feels repaid by the smallness and more delicate appearance of her feet. Each time the bandage is taken off, the foot is kneaded to make the joints more flexible, and is then bound up again as quickly as possible with a fresh bandage, which is drawn up more tightly. During the first year the pain is so intense that the sufferer can do nothing but lie and cry and moan. For about two years the foot aches continually, and is subject to a constant pain like the pricking of sharp needles. With continued rigorous binding it

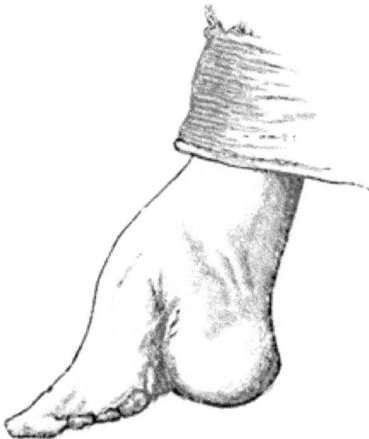

FIG. 27.—Chinese woman's foot, from the inside.
A photograph by Dr. R. A. Jamieson.[1]

FIG. 28.—Sole of
Chinese woman's foot.

ultimately loses its sensibility, the muscles, nerves, and vessels are all wasted, the bones are altered in their relative position to one another, and the whole limb is reduced permanently to a stunted or atrophied condition.

The alterations produced in the form of the foot are —1. Bending the four outer toes under the sole of the foot, so that the first or great toe alone retains its normal position, and a narrow point is produced in front; 2. Com-

---

[1] Dr. Jamieson says, "The fashionable length for a Chinese lady's foot is between 3½ and 4 inches, but comparatively few parents succeed in arresting growth so completely. The above, taken from a woman in the middle station of life, measures almost exactly 5 inches."

pressing the roots of the toes and the heel downwards and towards one another so as greatly to shorten the foot, and produce a deep transverse fold in the middle of the sole (Fig. 28). The whole has now the appearance of the hoof of some animal rather than a human foot, and affords a very inefficient organ of support, as the peculiar tottering gait of those possessing it clearly shows. When once formed, the "golden lily," as the Chinese lady calls her delicate little foot, can never recover its original shape.

But strange as this custom seems to us, it is only a slight step in excess of what the majority of people in Europe subject themselves and their children to. From personal observation of a large number of feet of persons of all ages and of all classes of society in our own country, I do not hesitate to say that there are very few, if any, to be met with that do not, in some degree, bear evidence of having been subjected to a compressing influence more or less injurious. Let any one take the trouble to inquire into what a foot ought to be. For external form look at any of the antique models,—the nude Hercules Farnese or the sandalled Apollo Belvidere ; watch the beautiful freedom of motion in the wide-spreading toes of an infant ; consider the wonderful mechanical contrivances for combining strength with mobility, firmness with flexibility ; the numerous bones, articulations, ligaments ; the great toe, with seven special muscles to give it that versatility of motion which was intended that it should possess ;—and then see what a miserable, stiffened, distorted thing is this same foot when it has been submitted for a number of years to the " improving " process to which our civilisation condemns it. The toes all squeezed and flattened against each other ; the great toe no longer in its normal position, but turned outwards, pressing so upon the others that one or more of them frequently has to find room for itself either above or under its fellows ; the joints all rigid, the muscles atrophied and powerless ; the finely-formed arch broken down ; everything which is beautiful and excellent in the human foot destroyed,—to say nothing of the more serious evils which so generally follow—corns, bunions, ingrowing nails, and all their attendant miseries.

Now, the cause of this will be perfectly obvious to any one who compares the form of the natural foot with the last upon which the shoemaker makes the covering for that foot. This, in the words of the late Mr. Dowie, " is shaped in front like a wedge, the thick part or instep rising in a ridge from the centre or middle toe, instead of the great toe, as in the foot, slanting off to both sides from the middle, terminating at each side and in front like a wedge; that for the inside or great toe being similar to that for the outside or little toe, as if the

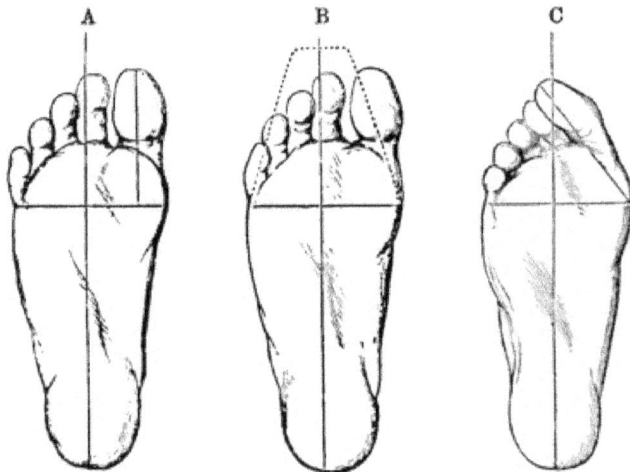

Fig. 29.—A. Natural form of the sole of the foot, the great toe parallel to the axis of the whole foot. B. The same, with outline of ordinary fashionable boot. C. The modification of the form of the foot, necessarily produced by wearing such a boot.

human foot had the great toe in the middle and a little toe at each side, like the foot of a goose!" The great error in all boots and shoes made upon the system now in vogue in all parts of the civilised world lies in this method of construction upon a principle of bilateral symmetry. A straight line drawn along the sole from the middle of the toe to the heel will divide a fashionable boot into two equal and similar parts, a small allowance being made at the middle part, or "waist," for the difference between right and left foot. Whether the toe is made broad or narrow, it is always equally inclined at the sides towards the middle line; whereas in the foot there is no such symmetry. The first or inner toe is much larger than either of the others, and its direction is perfectly parallel with the long axis of the foot. The second toe may be a little longer than the first, as generally represented in Grecian art, but it is more frequently shorter;[1] the others rapidly decrease in size (Fig. 29, A). The modification which must have taken place in the form of the foot and direction of the toes before a boot of the ordinary form can be worn with any approach to ease is shown at Fig. 29, C. (p. 344). Often it will happen

A   B   C

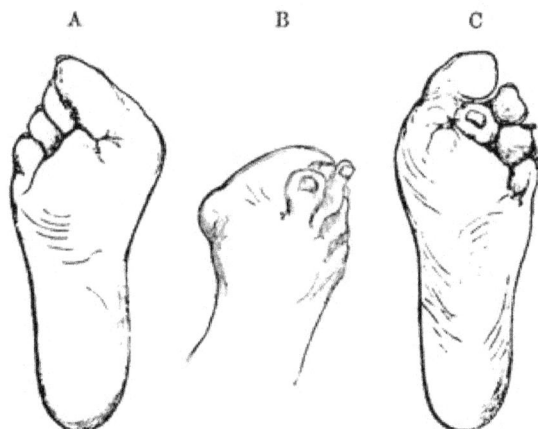

Fig. 30.—English feet deformed by wearing improperly-shaped shoes.
From nature.

that the deformity has not advanced to so great an extent, but every one who has had the opportunity of examining many feet, especially among the poorer classes, must have met with

[1] It seems to be a very common idea with artists and sculptors, as well as anatomists, that the second toe ought to be longer than the first in a well-proportioned human foot, and so it is conventionally represented in art. The idea is derived from the Greek canon, which in its turn was copied from the Egyptian, and probably originally derived from the negro. It certainly does not represent what is most usual in our race and time. Among hundreds of bare and therefore undeformed feet of children I lately examined in Perthshire, I was not able to find one in which the second toe was the longest. As in all apes—in fact, in all other animals—the first toe is considerably shorter than the second, a long great toe is a specially human attribute, and instead of being despised by artists, it should be looked upon as a mark of elevation in the scale of organised beings.

many far worse.   The two figured (Fig. 30), one (C) from a
labouring man, the other (A and B) from a working woman,
both patients at a London hospital, are very ordinary examples
of the European artificial deformity of the foot, and afford good
subjects for comparison with the Chinese foot (Fig. 28).   It
not unfrequently happens that the dislocation of the great toe
is carried so far that it becomes placed almost at a right angle
to the long axis of the foot, lying across the roots of the other
toes.

In walking and especially running, the action of the foot
is as follows : The heel is first lifted from the ground, and
the weight of the body gradually transferred through the
middle to the anterior end of the foot, and the final push
or impulse given with the great toe.   It is necessary then
that these parts should all be in a straight line with one
another.   Any deflection, especially of the great toe, from
its proper direction, or any weakening of its bones, ligaments,
or muscles, must be detrimental to the proper use of the foot
in progression.   Against this it will perhaps be urged that
there are many fairly good walkers and runners among us
whose great toes have been considerably changed from the
normal position in consequence of wearing pointed boots
while young.   This may be perfectly true, but it is also well
known that several persons, as the late Miss Biffin, and an
artist familiar to all frequenters of the Antwerp picture
gallery, have acquired considerable facility in the use of
the brush, though possessing neither hands nor arms, the one
painting only from the shoulder, and the other with the feet.
The compensating power of nature is very wonderful, and when
one part is absent or crippled, other means are found of doing
its work, but always at a disadvantage as compared with those
best fitted for the duty.

The loss of elasticity and motion in the joints of the foot,
as well as the wrong direction acquired by the great toe, are
in most persons seriously detrimental to free and easy progression, and can only be compensated for by a great expenditure of muscular power in other parts of the body,
applied in a disadvantageous manner.   The labouring men of
this country, who from their childhood wear heavy, stiff, and

badly-shaped boots, and in whom, consequently, the play of
the ankle, feet, and toes is lost, have generally small and
shapeless legs and wasted calves, and walk as if on stilts, with
a swinging motion from the hips.   Our infantry soldiers also
suffer much in the same manner, the regulation boots in use
in the service being exceedingly ill-adapted for the development of the feet.   Much injury to the general health—the
necessary consequence of any impediment to freedom of bodily
exercise—must also be attributed to this cause.   Since some
of the leading shoemakers have ventured to deviate a little
from the conventional shape, those persons who can afford to
be specially fitted are better off as a rule than the majority of
poorer people, who, although caring less for appearance, and
being more dependent for their livelihood upon the physical
welfare of their bodies, are obliged to wear ready-made shoes
of the form that an inexorable custom has prescribed.

The changes that a foot has to undergo in order to adapt
itself to the ordinary shape of a shoe could probably not be
effected unless commenced at an early period, when it is young
and capable of being gradually moulded into the required
form.

The English mother or nurse who thrusts the tender feet
of a young child into stiff, unyielding, pointed shoes or boots,
often regardless of the essential difference in form of right and
left, at a time when freedom is especially needed for their
proper growth and development, is the exact counterpart of
the Chinook Indian woman, applying her bandages and boards
to the opposite end of her baby's body, only with considerably
less excuse; for a distorted head apparently less affects health
and comfort than cramped and misshapen feet, and was also
esteemed of more vital importance to preferment in Chinook
society.   Any one who recollects the boots of the late Lord
Palmerston will be reminded that a wide expanse of shoe
leather is in this country, even during the prevalence of
an opposite fashion, quite compatible with the attainment
of the highest political and social eminence.

No sensible person can really suppose that there is anything in itself ugly, or even unsightly, in the form of a
perfect human foot; and yet all attempts to construct shoes

upon its model are constantly met with the objection that something extremely inelegant must be the result. It will perhaps be a form to which the eye is not quite accustomed : but there is no more trite observation than the arbitrary nature of fashion in her dealings with our outward appearance, and we all know how anything which has received her sanction is for the time considered elegant and tasteful, though a few years later it may come to be looked upon as positively ridiculous. That our eye would soon get used to admire a different shape may be easily proved by any one who will for a short time wear shoes constructed upon a more correct principle, when the prevailing pointed shoes, suggestive of cramped and atrophied toes, become positively painful to look upon.

A glance at a series of pictures of costume at various periods of English history will show how fashion has changed at different times with respect to the coverings of the feet. The fact that the excessively pointed, elongated toes of the time of Richard II., for instance, were superseded by the broad, round toed, almost elephantine, but most comfortable shoes seen in the portraits of Henry VIII. and his contemporaries, shows that there is nothing in the former essential to the gratification of the æsthetic instincts of mankind. Each form was doubtless equally admired in the time of its prevalence.

It is not only leather boots and shoes that are to blame for producing alterations in the form of the feet; even the stocking, comparatively soft and pliable as it is, when made with pointed toes and similar form for both sides, must take its share. The continual, steady, though gentle pressure, keeps the toes squeezed together, and especially hinders the recovery of its proper form and mobility, when attempts at curing a misshapen foot are being made by wearing shoes of rational construction. Socks adapted to the different form of the two feet, or "rights and lefts," are occasionally to be met with at hosiers, and it would add greatly to comfort if they were more generally adopted. For some cases it is well to have them made with distinct toes like gloves. With such socks and properly constructed shoes, a much distorted foot, even of a middle-aged person, will recover its power and freedom of motion to a considerable extent.

Only one thing is needed to aggravate the evil effect of a pointed toe, and that is the absurdly high and narrow heel so often seen now on ladies' boots, which throws the whole foot, and in fact the whole body, into an unnatural position in walking, produces diseases well known to all surgeons in large practice, and makes the nearest approach yet effected by any European nation to the Chinese custom, which we generally speak of with surprise and reprobation. And yet this fashion appears just now on the increase among people who boast of the highest civilisation to which the world has yet attained.

The practice of turning out the toes, so much insisted on by dancing masters, when it becomes habitual, is a deformity. Although in standing in an easy position the whole limb may be rotated outwards from the hip, so as to give a broader basis of support, in walking or running the hip, knee, ankle, and joints of the foot are

Fig. 31.—Modern Parisian shoe, copied from a recent advertisement. The nearest European representative of the Chinese deformity depicted in Fig. 27, p. 312.

simple hinges, and it is essential for the proper co-ordination of their actions that they should all work in the same plane, which can only be the case when the toes are pointed directly forward, and the feet nearly parallel to one another. Any deviation from this position must interfere with the true action of the foot when raising and propelling the body, as explained at p. 346. Turning out the toes is, moreover, a common cause of weak ankles, as it throws the weight of the body chiefly on the inside, instead of distributing it equally over all parts of the joint.

I must speak lastly of one of the most remarkable of all the artificial deformities produced by adherence to a conventional standard, in defiance of the dictates of nature and reason.

Of all parts of the body, the elastic and mobile walls of the chest would seem most to need preservation from external constriction, if they are to perform efficiently the important purposes for which their peculiar structure is specially designed. The skull is a solid case, with tolerably uniform walls, the capacity of which remains the same, whatever alteration is made in its shape. Pressure on one part is compensated for by dilatation elsewhere ; the body is not so, it may be compared

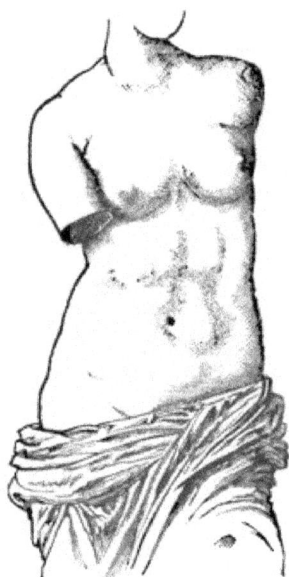

Fig. 32.
Torso of the statue of Venus of Milo.

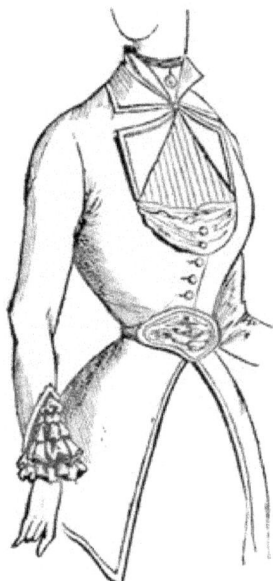

Fig. 33.
Paris fashion, May 1880.

to a cylinder with a fixed length, determined by the vertebral column, and closed above and below by a framework of bone. Circular compression then must actually diminish the area which has to be occupied by some of the most important vital organs. Moreover, the framework of the chest is a most admirable and complex arrangement of numerous pieces of solid bone and elastic cartilage, jointed together in such a manner as to allow of expansion and contraction for the purposes of respiration—expansion and contraction which, if a

function so essential to the preservation of life and health is
to be performed in an efficient manner, should be perfectly
free and capable of variation under different circumstances.
So, indeed, it has been allowed to be in all parts of the world
and in all ages, with one exception.   It was reserved for
mediæval civilised Europe to have invented the system of
squeezing together, rendering immobile, and actually deforming,
the most important part of the human frame ; and the custom
has been handed down to, and flourishes in, our day, notwith-

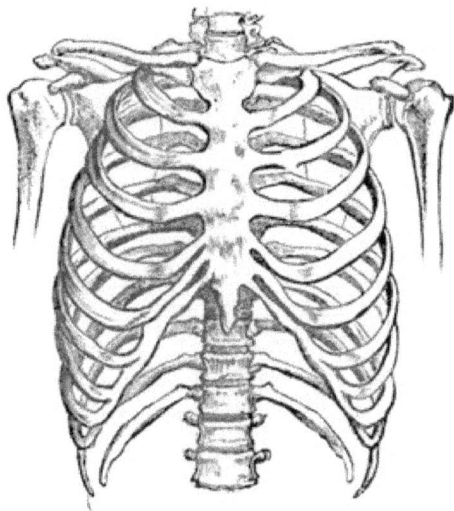

Fig. 34.—Normal form of the skeleton of the chest.

standing all our professed admiration for the models of classical
antiquity, and our awakened attention to the laws of health.

It is only necessary to compare these two Figures (Figs. 32
and 33),—one acknowledged by all the artistic and anatomical
world to be a perfect example of the natural female form,—
to be convinced of the gravity of the structural changes that
must have taken place in such a form before it could be
reduced so far as to occupy the space shown in the second
figure, an exact copy of one of the models now held up for
imitation in the fashionable world.   The actual changes that
have taken place in the bony framework of the chest are seen

by comparing Figures 34 and 35, the one showing the normal
form, the other the result of long-continued tight-lacing.
The alterations in the shape and position of the organs
within need not be dwelt upon here; they and the evil effects
arising from them are abundantly discussed in medical works.
When it is considered that the organs which are affected are
those by which the important functions of respiration, circulation, and digestion are carried on, as well as those essential

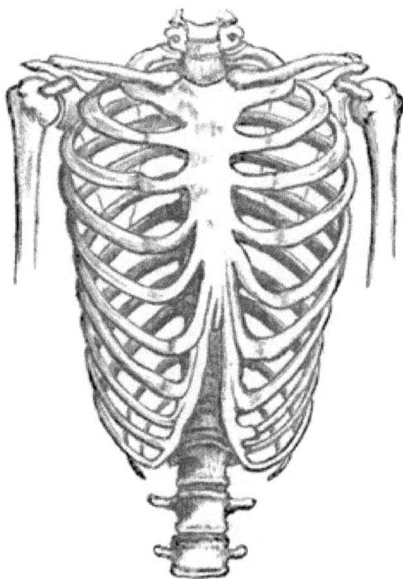

FIG. 35.—Skeleton of the chest of a woman, twenty-three years of age,
deformed by tight-lacing, from Rüdinger's *Anatomie des Menschen.* By no means
an extreme case.

to the proper development and healthy growth of future
generations, it is no wonder that people suffer who have
reduced themselves to live under such conditions.[1]

The true form of the human body is familiar to us, as just
said, from classic models; it is familiar from the works of our

---

[1] See, among many others, the section headed "Improprieties of Dress," in Dr.
Gaillard Thomas's *Practical Treatise on the Diseases of Women* (5th edit. 1881,
p. 45), for convincing proofs (not mere general declamation) of the ill effects
arising from tight-lacing.

greatest modern artists which adorn the Academy walls.   It is, however, quite possible, or even probable, that some of us may think the present fashionable shape the more beautiful of the two.   In such case it would be well to pause to consider whether we are sure that our judgment is sound on the subject. Let us remember that to the Australian the nose-peg is an admired ornament; that to the Thlinkeet, the Botocudo, and the Bongo negro, the lip dragged down by the heavy plug, and the ears distended by huge discs of wood, are things of beauty ; that the Malay prefers teeth that are black to those of the most pearly whiteness; that the native American despises the form of a head not flattened down like a pancake, or elongated like a sugar-loaf,—and then let us carefully ask ourselves whether we are sure that in leaving nature as a standard of the beautiful, and adopting a purely conventional one, we are not falling into an error exactly similar to that of all these people whose tastes we are so ready to condemn.

The fact is, that in admiring such distorted forms as the constricted waist and symmetrically pointed foot, we are opposing our judgment to that of the Maker of our bodies ; we are neglecting the criterion afforded by nature ; we are departing from the highest standard of classical antiquity : we are simply putting ourselves on a level in point of taste with those Australians, Botocudos, and Negroes.   We are taking fashion, and nothing better, higher, or truer, for our guide ; and after the various examples which have now been brought forward, may we not well ask, with Shakespeare,

"Seest thou not, what a deformed thief this fashion is?"

BIOGRAPHICAL SKETCHES

BIOGRAPHICAL NOTICE OF PROFESSOR ROLLESTON [1]

GEORGE ROLLESTON'S death, which took place in Oxford on 16th June 1881, may well be called premature, as he was in the prime of life, and but a few months before seemed to all, except a few closely observant intimate associates, still in the plentitude of his powers, and capable of much good work in time to come.

The son of a Yorkshire clergyman, he was born at Maltby, on 30th July 1829, and had, therefore, not completed his fifty-second year. His early aptitude for classical studies, carried on under the instruction of his father, must have been most remarkable if, as has been stated in one of his biographies, he was able at the age of ten to read any passage of Homer at sight. He was not educated at one of the great public schools, but entered at Pembroke College, Oxford, took a First Class in classics in 1850, and was elected a Fellow of his college in 1851. He then studied medicine at St. Bartholomew's Hospital, joined the staff of the British Civil Hospital at Smyrna during the latter part of the Crimean war, was appointed assistant-physician to the Children's Hospital in London, 1857, but took up his residence again at Oxford in the same year on receiving the appointment of Lee's Reader in Anatomy in Christchurch. In 1860 he was elected to the newly-founded Linacre Professorship of Anatomy and Physiology, which he held to the time of his death. He was elected a Fellow of the Royal Society in 1862, and a Fellow of Merton College, Oxford, in 1872. He was a

[1] From the *Proceedings of the Royal Society*, vol. xxxiii. 1882, with slight alterations.

member of the Council of the University, and its representative
in the General Medical Council, and also an active member of
the Oxford Local Board.

In 1861 he married Grace, daughter of Dr. John Davy,
F.R.S., and niece of Sir Humphrey Davy, and he leaves a
family of seven children.

The duties of the Linacre professorship involved the teaching of a wide range of subjects included under the terms of
physiology and anatomy, human and comparative, to which
he added the hitherto neglected but important subject of
anthropology, as well as the care of a great and ever-growing
museum.   In the present condition of scientific knowledge it
requires a man of very versatile intellect and extensive powers
of reading to maintain anything like an adequate acquaintance
with the current literature of any one of these subjects, much
more to undertake original observations on his own account.
Even a man of Rolleston's powers felt the impossibility of any
one person doing justice to the chair as thus constituted, and
strongly urged the necessity of dividing it into three professorships,—one of physiology, one of comparative anatomy, and one
of human anatomy and anthropology.   The work which he
did, however, contrive to find time to publish, and by which
he will be chiefly known to posterity, is remarkable for its
thoroughness.   He never committed himself to writing without having completely mastered everything that had been
previously written upon the subject, and his memoirs bristle
with quotations from, and references to, authors of all ages
and all nations.   The abundance with which these were
supplied by his wonderful memory, and the readiness with
which, both in speaking and writing, his thoughts clothed
themselves with appropriate words, sometimes made it difficult
for ordinary minds to follow the train of his argument through
long and voluminous sentences, often made up of parenthesis
within parenthesis.

The work which was most especially the outcome of his
professorial duties is the *Forms of Animal Life*, published at
the Clarendon Press in 1870.   Though written chiefly with
a view to the needs of the University students, it is capable
of application to more general purposes, and is one of the

earliest and most complete examples of instruction by the study of a series of types, now becoming so general. As he says in the preface, "The distinctive character of the book consists in its attempting so to combine the concrete facts of zootomy with the outlines of systematic classification, as to enable the student to put them for himself into their natural relations of foundation and superstructure. The foundations may be wide, and the superstructure may have its outlines not only filled up, but even considerably altered by subsequent and more extensive labours; but the mutual relations of the one as foundation and the other as superstructure which this book particularly aims at illustrating, must always remain the same."

Besides this work, Professor Rolleston's principal contributions to comparative anatomy and zoology are the following: "On the Affinities of the Brain of the Orang-Utan," *Nat. Hist. Review*, 1861; "On the Aquiferous and Oviductal System in the Lamellibranchiate Molluscs" (with Mr. C. Robertson), *Phil. Trans.*, 1862; "On the Placental Structures of the Tenrec (*Centetes ecaudatus*), and those of certain other Mammals, with Remarks on the Value of the Placental System of Classification," *Trans. Zool. Soc.* 1866: "On the Domestic Cats of Ancient and Modern Times," *Journal of Anatomy*, 1868; "On the Homologies of certain Muscles connected with the Shoulder-joint," *Trans. Linn. Soc.* 1870; "On the Development of the Enamel in the Teeth of Mammals," *Quart. Journ. Micros. Soc.* 1872; and "On the Domestic Pig in Prehistoric Times," *Trans. Linn. Soc.* 1877.

Latterly he did much admirable work in anthropology, for which he was excellently qualified, being one of the few men who possessed the culture of the antiquarian, historian, and philologist on the one hand, and of the anatomist and zoologist on the other, and could make these different branches of knowledge converge upon the complex problem of man's early history. The chief results of his work of this nature are contained in his contributions to Greenwell's *British Barrows* (1877), a book containing a fund of solid information relating to the early inhabitants of this island. His last publication, and one which is, on the whole, the most characteristic, as exhibiting his vast range of knowledge on many different subjects, was a lecture delivered in 1879 at the Royal Geographical Society, on "The Modifications of the External Aspects of Organic Nature produced by Man's Interference."

That Dr. Rolleston has not left more original scientific work behind him is easily accounted for by the circumstances under which he lived at Oxford. The multifarious nature of the subjects with which the chair was overweighted; the perpetual discussions in which, during the whole term of his office, he was engaged, consequent upon the transitional condition of education, both at Oxford and elsewhere; the immense amount of business thrust upon him, or voluntarily undertaken by him, of the kind which always accumulates round the few men who are at the same time capable and unselfish, such as questions pertaining to the local and especially to the sanitary affairs of the town in which he lived, or questions connected with the reform of the medical profession, arising both within and outside the Medical Council, which latter business constantly brought him to meetings in London; his own wide grasp of interest in social subjects, his deep feeling of the responsibilities of citizenship, and a sense of the duties of social hospitality, which made his house always open to scientific visitors to Oxford,—all these rendered impossible to him that intense concentration which is requisite for carrying out any continuous line of research.

He was often blamed for undertaking so much and such divers kinds of labour, so distracting to his scientific pursuits; but being by constitution a man who could never see a wrong without feeling a burning desire to set it right, who could never "pass by on the other side" when he felt that it was in his power to help, nothing but actual physical impossibility would restrain him. For several years past, when feeling that his health and strength did not respond to the strain he put upon them, he resorted to every hygienic measure suggested but one, and that the one he most required—rest; but this he never would or could take. During the last term he spent at Oxford, before his medical friends positively forced him (though unfortunately too late) to give up his occupations and seek change in a more genial climate, he was working at the highest pressure, rising every morning at six o'clock, to get two uninterrupted hours in which to write the revised edition of the *Forms of Animal Life*, before the regular business of the day commenced.

It is impossible for those who had no personal knowledge of Rolleston to realise what manner of man he was, and how great his loss will be to those who remain behind him. No one can ever have passed an hour in his company, or heard him speak at a public meeting, without feeling that he was a man of most unusual power, of lofty sentiments, generous impulses, marvellous energy, and wonderful command of language. In brilliant repartee, aptness of quotation, and ever-ready illustration from poetry, history, and the literature of many nations and many subjects, besides those with which he was specially occupied, he had few equals. " In God's war slackness is infamy " might well have been his motto, for with Rolleston there was no slackness in any cause which he believed to be God's war. He was impetuous, even vehement, in his advocacy of what appeared to him true and right, and unsparing in denunciation of all that was mean, base, and false. To those points in the faith of his fathers which he believed to be essential he held reverently and courageously, but on many questions, both social and political, he was a reformer of the most advanced type. Often original in his views, always outspoken in giving expression to them, he occasionally met with the fate of those who do not swim with the stream, and was misunderstood; but this was more than compensated for by the affection, admiration, and enthusiasm with which he was regarded by those who were able to appreciate the nobility of his character. The sentiment which guided his life, and which he never lost an opportunity of impressing upon his younger friends, was that contained in the beautiful lines of Wordsworth—

> We live by Admiration, Hope, and Love,
> And even as these are well and wisely fixed,
> In dignity of being we ascend.

When Huxley was spoken of in relation to himself, he at once expressed the chivalrous and loyal feeling he bore towards worthy leaders in science, by quoting the words in which Lancelot speaks of King Arthur—

> In me there dwells
> No greatness, save it be some far-off touch
> Of greatness to know well I am not great :
> There is the man.

The loss of the example afforded by such a nature, and of his elevating influence on younger and weaker men, is to our mind a still greater loss, both within and without the University in which he taught, than the loss of what scientific work he might yet have performed.

Dr. Rolleston's personal appearance corresponded with his character. Of commanding height, broad-shouldered, with a head of unusual size, indicating a volume of brain commensurate with his intellectual power, clear penetrating blue eyes, and strongly-marked and expressive features, in which refinement and vigour were singularly blended, in him we saw just such a man as was described by the public orator at the late Oxford Commemoration, in words with which we may conclude this notice : " Virum excultissimi ingenii, integritatis incorruptissimæ, veritatis amicum, et propugnatorem impavidum."

BIOGRAPHICAL NOTICE OF SIR RICHARD OWEN [1]

RICHARD OWEN was born at Lancaster on 20th July 1804. His father, whose name was also Richard, was engaged in business connected with the West Indies. His mother's name was Catherine Parrin. He was educated at the Grammar School at Lancaster (where one of his schoolfellows was W. Whewell, afterwards Master of Trinity), was apprenticed to a surgeon of the name of Harrison in that town, and studied surgery at the County Hospital. No evidence can now be found for the statement which has appeared in many biographical notices that when a boy he went to sea as a midshipman, nor is there any that at a later period he had an intention to enter the medical service of the Navy, or applied for and obtained an appointment, as has also been stated.

In 1824 he matriculated at the University of Edinburgh, and had the good fortune to attend the anatomical course of Dr. Barclay, then approaching the close of a successful career as an extra-academical lecturer, whose teaching was of a very superior order to that of the third Monro, who, by virtue of hereditary influences, happened at that time to be the University Professor of Anatomy. In his work *On the Nature of Limbs*, Owen refers to "the extensive knowledge of comparative anatomy possessed by my revered preceptor in anatomy, Dr. Barclay," and always spoke of him with affectionate regard.

He did not remain in Edinburgh to take his degree, but removed to St. Bartholomew's Hospital in London, and passed the examination for the membership of the Royal College of Surgeons on 18th August 1826.

[1] From the *Proceedings of the Royal Society*, vol. lv. 1894, with slight alterations.

His first published scientific works were in the direction of surgical pathology, being on encysted calculus, and on the effects of ligature of the internal iliac artery for the cure of aneurism.

At St. Bartholomew's Hospital he soon attracted the attention of the celebrated Abernethy, through whose influence he obtained the appointment of Assistant Conservator to the Hunterian Museum of the Royal College of Surgeons. This was in 1827, and it caused him to abandon the prospect of private practice, to which he had begun to devote himself while living in Serle Street, Lincoln's Inn Fields, for the more congenial pursuit of comparative anatomy. The Conservator of the Museum at that time was William Clift, John Hunter's last and most devoted pupil and assistant, under whose faithful guardianship the collection had been most carefully preserved during the long interval between the death of its founder and its transference to the custody of the College of Surgeons. From him Owen early imbibed an enthusiastic reverence for the great master, which was continually augmented with the closer study of his collection and works, which now became the principal duty of his life. In 1830 and 1831 he visited Paris, where he attended the lectures of Cuvier and Geoffroy St. Hilaire, and worked in the dissecting rooms and public galleries of the Jardin des Plantes. In 1835 he married Clift's only daughter, Caroline, and in 1842 was associated with him as joint Conservator of the Museum. On Clift's retirement soon after, he became sole Conservator, with Mr. J. T. Quekett as assistant.

He was appointed Hunterian Professor of Comparative Anatomy and Physiology in 1835, an office which he held until his retirement from the College in 1856, and from which he took the title of " Professor Owen," by which he was far more widely known than by the knightly addition of his later years.

Until the year 1852, when the Queen gave him the charming cottage called Sheen Lodge, in Richmond Park, where he resided to the end of his life, he occupied small apartments within the building of the College of Surgeons; these, however inconvenient they might be in some respects, furnished him with unusual facilities for pursuing his work by night as well as day in the museum, dissecting rooms, and library of that institution.

Owen's life of scientific activity may be divided into two periods, during each of which the nature of his work was determined to a considerable extent by the circumstances by which he was environed. Each of these periods embraces a term of very nearly thirty years. The first, from 1827 to 1856, was spent at the Royal College of Surgeons; the second, from 1856 to 1884, in the British Museum. It was in the first that he mainly made his great reputation as an anatomist, having utilised to the fullest possible extent the opportunities which were placed in his way by the care of the Hunterian Museum. For many years he worked in that institution under the happiest of auspices. From the routine and drudgery which always take up so large a portion of the time of a conscientious museum curator, he was relieved by the painstaking, methodical William Clift; the far more gifted son-in-law being thus able to throw himself to his heart's content into the higher work of the office. This at first mainly consisted in the preparation of that monumental *Descriptive and Illustrated Catalogue of the Physiological Series of Comparative Anatomy*, founded upon Hunter's preparations, largely added to by Owen himself, which was published in five quarto volumes between the years 1833 and 1840. This work, which has been taken as a model for many other subsequently published catalogues, contains a minute description of nearly four thousand preparations. The labour involved in preparing it was greatly increased by the circumstance that the origin of a large number of them had not been preserved, and even the species of the animals from which they were derived had to be discovered by tedious researches among old documents, or by comparison with fresh dissections. It was mainly for aid in this work that he engaged upon the long series of dissections of animals which died from time to time in the Gardens of the Zoological Society, the descriptions of which, as published in the Proceedings and Transactions of the Society, form a precious fund of information upon the comparative anatomy of the higher vertebrates. The series commences with an account of the anatomy of an orang-utan, which was communicated to the first scientific meeting of the Society, held on the evening of Tuesday, 9th November 1830, and was continued with descriptions of dissections of the beaver, suricate, acouchy, Thibet bear, gannet, crocodile, armadillo, seal, kangaroo, tapir, toucan, flamingo, hyrax, hornbill, cheetah, capybara, pelican, kinkajou, wombat, giraffe, dugong, apteryx, wart-hog, walrus, great ant-eater, and many others.

Among the many obscure subjects in anatomy and physiology on which Owen threw much light by his researches at this period were several connected with the generation, development, and structure of the Marsupialia and Monotrema, groups which always had great interest for him. It is a curious coincidence that his first paper, communicated to the Royal Society (in 1832), "On the Mammary Glands of the *Ornithorhynchus paradoxus*," was one of a series which only terminated in almost the last which he offered to the same Society (in 1887), being a description of a newly excluded young of the same animal, published in the *Proceedings*, vol. 42, p. 391.

On the completion of the *Catalogue of the Physiological Series* his curatorial duties led him to undertake the catalogues of the osteological collections of recent and extinct forms. This task necessitated minute studies of the modifications of the skeleton in all vertebrated animals, and researches into their dentition, the latter being finally embodied in his great work on *Odontography* (1840-45), in which he brought a vast amount of light out of what was previously chaotic in our knowledge of the subject, and cleared the way for all future work upon it. Although recent advances of knowledge have shown that there are difficulties in accepting the whole of Owen's system of homologies and notation of the teeth of Mammals, it was an immense improvement upon anything of the kind which existed before, and a considerable part of it seems likely to remain a permanent addition to our means of describing these organs. The close study of the bones and teeth of existing animals was of extreme importance to him in his long continued and laborious researches into fossil forms; and, following in the footsteps of Cuvier, he fully appreciated and deeply profited by the dependence of the study of the living in elucidating the dead, and *vice versâ.* Perhaps the best example of this is to be seen in his elaborate memoir on the Mylodon, published in 1842, entitled *Description of the Skeleton of an Extinct Gigantic Sloth (Mylodon robustus,* Owen), *with Observations on the Osteology, Natural Affinities, and Probable Habits of the Megatheroid Quadrupeds in General,* a masterpiece both of anatomical description and of reasoning and inference. A comparatively popular outcome of some of his work in this direction was the volume on *British Fossil Mammals and Birds,* published in 1844-46, as a companion to the works of Yarrell, Bell, and others on the recent fauna of our island. He also wrote, assisted by Dr. S. P. Woodward, the article " Palæontology " for the *Encyclopædia Britannica,* which, when afterwards published in a separate form, reached a second edition in 1861.

To this first period of his life belong the courses of Hunterian Lectures, given annually at the College of Surgeons, each year on a fresh subject, and each year the means of bringing before the world new and original discoveries which attracted, even fascinated, large audiences, and did much to foster an interest in the science among cultivated people of various classes and professions. They also added greatly to the scientific renown of the College in which they were given. To this period also belong the development and popularisation of those transcendental views of anatomy—the conception of creation according to types, and the construction of the Vertebrate archetype—views which had great attractions and even uses in their day, and which were accepted by many, at all events as working hypotheses around which facts could be marshalled, and out of which grew a methodical system, of anatomical terminology, much of which has survived to the present time. The recognition of homology and its distinction from analogy, which was so strongly insisted on by Owen, marked a distinct advance in philosophical anatomy. These generalisations, first announced in lectures at the College of Surgeons, were afterwards embodied in two works : *The Archetype and Homologies of the Vertebrate Skeleton* (1848), and *The Nature of Limbs* (1849).

The contributions which Owen made to our knowledge of the structure of Invertebrate animals nearly all belong to the earlier period of his career, one of the most important being his admirable and exhaustive memoir on the Pearly Nautilus, founded on the dissection of a specimen of this, at that time exceedingly rare, animal sent to him in spirit by his friend Dr. George Bennett, of Sydney. This was illustrated by carefully executed drawings by his own hand, and published in the year 1832, when he was only twenty-seven years of age. The Cephalopoda continued to engage his attention, and the merits of a memoir on fossil Belemnites from the Oxford Clay, published in the *Philosophical Transactions* in 1844, was the cause assigned for the award to him of the Royal Medal in 1846. He contributed the article "Cephalopoda," to the *Cyclopædia of Anatomy and Physiology* (1836), catalogued the extinct Cephalopoda in the Museum of the Royal College of Surgeons (1856), and wrote original papers on *Clavagella* (1834), *Trichina spiralis* (1835), *Linguatula* (1835), *Distoma* (1835), *Spondylus* (1838), *Euplectella* (1841), *Terebratula* (in the introduction to Davidson's classical *Monograph of the British Fossil Brachiopods*, 1853), and many other subjects, including the well-known essay on "Parthenogenesis, or the Successive Production of Procreating Individuals from a Single Ovum" (1849).

In 1843 his *Lectures on the Comparative Anatomy and Physiology of the Invertebrate Animals*, in the form of notes taken by his pupil, Mr. W. White Cooper, appeared as a separate work. Of this, a second expanded and revised edition was published in 1855. By this time, as the Royal Society's *Catalogue of Scientific Papers* shows, he had been the author of as many as 250 separate scientific memoirs.

In 1856, when Owen had reached the zenith of his fame, and was recognised throughout Europe as the first anatomist of the day, a change came over his career. Difficulties with the governing body of the College of Surgeons, arising from his impatience at being required to perform what he considered the lower administrative duties of his office, caused him readily to take advantage of an offer from the Trustees of the British Museum to undertake a newly-created post, that of Superintendent of the Natural History Departments of the Museum. It was thought that hitherto these departments, being under the direct control of a chief who had been invariably chosen from the literary side of the establishment, and whose title in fact was that of " Principal Librarian," had not obtained their due share of attention in the general and financial administration, and that if they were grouped together and placed under a strong administrator, who should be able to exercise influence in advocating their claims to consideration, and who should be responsible for their internal working, their relative position in the establishment would be improved. Owen was accordingly placed in this position, and bade farewell to the College of Surgeons, its museum, and its lectures. At the British Museum, however, he encountered the difficulties which are nearly always experienced by an outsider suddenly imported into the midst of an existing establishment without any very well-defined position. The Principal Librarian, Sir A. Panizzi, was a man of strong will and despotic character, and little disposed to share any of his authority with another. The heads of the departments, especially Dr. J. E. Gray, Keeper of Zoology, preferred to maintain the independence to which they were accustomed within their own sphere of action, and to have no intermediary between them and the Trustees, except the Principal Librarian, who, though perhaps with little sympathy, had also, from lack of special knowledge, but little power of interference in detail. Hence Owen found himself in a situation the duties of which were little more than nominal, probably for him the best that could have been, as it gave his indomitable industry full play in the directions for which his talents were best fitted, and with the magnificent material in the collections of the Museum at his command, he set to work with great vigour upon a renewed series of researches, the results of which for many years taxed the resources of most of the scientific societies of London to publish. It followed from the nature of the materials that came most readily to his hand, and the smaller facilities for dissection now available than those afforded by the College of Surgeons, that his original work was henceforth mainly confined to osteology, and chiefly to that of extinct

animals. The rich treasures of the palæontological department were explored, named, and described, as were also the
valuable additions which poured in from various parts of the
world, attracted in many cases by Owen's great reputation.
The long series of papers on the gigantic extinct Birds of New
Zealand, begun in the year 1839, at the College of Surgeons,
with the receipt of the fragment of a femur, upon which the
first evidence of their existence was based, was now continued
at intervals as fresh materials arrived.[1] The Marsupials of
Australia, the Edentates of South America, the Triassic Reptiles
from South Africa, the *Archæopteryx* from Solenhofen, the
Mesozoic Mammals from the Purbeck, the Aborigines of the
Andaman Islands, the Cave remains, human and otherwise, of
the South of France, the Cetacea of the Suffolk crag, the
gorilla and other Anthropoid apes, the dodo, great auk, and
Chiromys, and many other remarkable forms of animal life,
were all subjects of elaborate memoirs from his untiring pen.

---

[1] Much misconception as to the real nature of such researches exists in the
popular mind, and is perpetually reproduced in quasi-scientific literature. It is
commonly said that "Owen evolved a complete extinct creature from the fragment of a single bone." The "restoration" of *Dinornis* especially is quoted as
a triumph of inductive reasoning, whereas it was simply the empirical application of existing knowledge. Direct comparison of the fragment with the bones
already in the Museum having shown that it most closely resembled the femur
of the existing ostrich, but was of larger size, all that was really "predicted" or
"evolved" was that a bird having a thigh bone like that of an ostrich, but
of larger size, had existed in New Zealand. Another well-known but more
daring instance of inference as to the general characters and habits of an
extinct animal from a very slender foundation was less fortunate. A mutilated
skull from Australia, with a tooth of very remarkable form, was described in 1859,
as indicating "one of the fellest and most destructive of predatory beasts," and
named accordingly *Thylacoleo carnifex.* Upon the same empirical principle as
in the former case, the tooth had been compared with various teeth of known
animals, and appeared to resemble one which is characteristic of many recent
carnivorous animals, especially lions and tigers. The resemblance, however,
was a very superficial one, and the search for its nearest analogue was unfortunately not carried far enough, or it would have been found in a group of Marsupials
of which the existing members are harmless, vegetable feeders. When the
remainder of the dentition was discovered it was seen that the famous *Thylacoleo*
really belonged to this group, and had no claims to the attributes assigned to it.
Nevertheless, notwithstanding the abundant demonstration of its fallacy, Owen
continued, with characteristic tenacity, to maintain to the last his original
assumption, never acknowledging that, however valid his reasoning might have
been, it was entirely vitiated by its foundation upon false premises.

These were adorned in every case with a profusion of admirable illustrations, drawn as often as possible of the full size of nature. His contributions to the publications of the Palæontographical Society, mainly upon the extinct Reptiles of the British Isles, fill more than a thousand pages, and are illustrated by nearly three hundred plates.

He now also found leisure to perform the pious duty of vindicating the scientific reputation of his great predecessor, John Hunter, by arranging and revising for publication a large collection of precious manuscripts containing records of dissections of animals and observations and reflections upon numerous subjects connected with anatomy, physiology, and natural history in general. These were published in 1861, in two closely-printed octavo volumes, entitled *Essays and Observations in Natural History, Anatomy, Physiology, Psychology, and Geology, by John Hunter, being his Posthumous Papers on those subjects.* The original manuscripts had been destroyed by Sir Everard Home in 1823, but fortunately not before William Clift had taken copies of the greater part of them, and it was from these copies that the work was compiled. Its publication shows that Hunter, while occupied with a large and anxious practice—in itself labour enough for an ordinary man—while cultivating with a passionate energy the sciences of physiology and pathology, while collecting and arranging a museum such as has never been formed before or since by a single individual, had also carefully recorded a series of dissections of different species of animals which, as his editor justly says, " if published *seriatim*, would not only have vied with the labours of Daubenton, as recorded in the *Histoire Naturelle,* of Buffon, or with the *Comparative Dissections* of Vicq d'Azyr, which are inserted in the early volumes of the *Encyclopédie Méthodique* and the *Mémoires de l'Académie Royale de France,* but would have exceeded them both together."

In 1866 were published the first and second volumes, and in 1868 the third volume, of Owen's own great book on the Anatomy and Physiology of the Vertebrates.

This is the most encyclopædic work on the subject accomplished by any one individual since Cuvier's *Leçons d'Anatomie Comparée,* and contains an immense mass of information mainly based upon original observations and dissections. It is in fact a collection of nearly all his previous memoirs arranged in systematic order, generally in the very words in which they were originally written, and unfortunately sometimes without the revision which advances made in the subject by the labours of others would have rendered desirable. Very little of the classification adopted in this work, either the primary division of the Vertebrates into Hæmatocrya and Hæmatotherma, or the divisions into classes and sub-classes, has been accepted by other zoologists. The division of the Mammalia into four sub-classes of equivalent value, upheld by Owen, not only in this work, but in various other publications issued about the same time (Rede Lecture, etc.), founded upon cerebral characteristics, was especially open to criticism. Though the separation of the Monotremes and Marsupials from all the others as a distinct group (Lyencephala) is capable of vindication, the three other sub-classes, Lissencephala, Gyrencephala, and Archencephala, grade so imperceptibly into each other that their distinction as sub-classes cannot be maintained. The proposed definition of the distinguishing characters of the brain of Man (Archencephala) from that of other Mammals gave rise to a somewhat acute controversy, the echoes of which reached beyond the realms of purely scientific literature. On the other hand, the radical distinction between the two groups of Ungulates, the odd-toed and the even-toed, first indicated by Cuvier, when treating of the fossil forms, was thoroughly worked out by Owen through every portion of their organisation, and remains as a solid contribution to a rational system of classification.

The chapter called "General Conclusions" at the end of the third volume is devoted to a summary of his views on the principal controverted biological questions of the day, especially in relation to the teaching of Darwin, just then coming into great prominence. Although from the peculiarly involved style of Owen's writing, especialy upon these subjects, it is sometimes difficult to define his real opinions, it appears that before the publication of the *Origin of Species*, he had "been led to recognise species as exemplifying the continuous operation of natural law, or secondary cause, and that not only successively but progressively." Darwin's special doctrine of " natural selection," however, he never appreciated, and his strong opposition to it caused him, though quite erroneously, to be looked upon by those outside the world of science as a supporter of the old-fashioned and then more "orthodox" view of special creation. His most distinct utterance upon this subject is contained in the following paragraph : " So, being unable to accept the volitional hypothesis, or that of impulse from within, or the selective force exerted by outward circumstances, I deem an innate tendency to deviate from parental type, operating through periods of adequate duration, to be the most probable nature, or way of operation, of the secondary law, whereby species have been derived one from the other." (*Op. cit.* vol. iii. p. 807.)

His career as a lecturer did not entirely cease with his connection with the College of Surgeons, as, by permission of the authorities of the Museum of Practical Geology in Jermyn Street, he gave several courses on the fossil remains of animals, in the theatre of that institution, and he held in the years 1859, 1860, and 1861, in conjunction with his office at the British Museum, the Fullerian Professorship of Physiology in the Royal Institution. On the revival of the annual lecture on Sir Robert Rede's foundation in the University of Cambridge, in 1859, he was appointed to give the first, and took for his subject the Classification of the Mammalia. He also occasionally lectured at the Royal Institution on Friday evenings, his last appearance there being on 26th April 1861, when he delivered the discourse, " On the Scope and Appliances of a National Museum of Natural History," to be presently referred to. In April 1862 he gave four lectures on Birds at the London Institution.

While at the College of Surgeons he had been a member of a Government Commission for inquiring into the health of the Metropolis ; and subsequently (1849) of one on Smithfield and the other meat markets, in which he strongly advocated the entire suppression of intramural slaughter-houses, and the concomitant evil of the passage of droves of sheep and cattle through the streets of London. For the Great Exhibition of 1851 he was on the Preliminary Committee of Organisation, and he acted as chairman of the jury on raw materials, alimentary substances, etc., and published an elaborate report on their awards. He also delivered to the Society of Arts a lecture on "Raw Animal Products, and their Uses in Manufacture." Similar services were performed by him for the Exposition Universelle of Paris in 1858.

It has been already said that Owen took scarcely any part in the details of the administration of the British Museum, but one subject relating to that establishment did largely engage his attention from his first connection with it. That the accommodation afforded by the rooms devoted to natural history in the Museum at Bloomsbury was painfully inadequate for the purpose was evident to him as well as to every one else. Space must be obtained somewhere, even for the proper conservation and display of the existing collections, to say nothing of the vast additions that must be expected if the subject were to be represented in anything like the way in which it deserved to be in his eyes, and Owen in this respect had very large views. The scientific public, the officers of the Museum, and the Trustees, were much divided as to whether it would be better to endeavour to obtain this space in the neighbourhood of the existing Museum, or to remove a portion of the collection to a totally distinct locality. After some apparent hesitation, Owen threw himself strongly on the side of those who took the latter view, being the one which seemed to him to have the best chance of leading to a successful result, and he strongly urged upon the Government, and upon the public generally, in annual museum returns, lectures, and pamphlets, the desirability of the scheme. In his address as President of the Biological Section of the British Association at the York meeting in 1881, he has given a history of the part he took in promoting the building of the new museum at South Kensington, including his success in enlisting the sympathy of Mr. Gladstone, by whose powerful aid the difficulties and opposition with which the plan was met in Parliament were mainly overcome. His earlier views upon the subject are fully explained in a small work entitled *On the Extent and Aims of a National Museum of Natural History*, published in 1862, being an expansion of the lecture

he gave at the Royal Institution in the previous year.   Much
controversy arose about this time as to the best principle of
museum organisation, Owen adhering to the old view of a
public exhibition on a very extensive scale, while the greater
number of naturalists of the time preferred the system of
dividing the collections into a comparatively limited public
exhibition, the bulk of the specimens being kept in a manner
accessible only to the researches of advanced students.   The
Royal Commission on the Advancement of Science, of which
the late Duke of Devonshire was chairman, investigated the
subject fully, and reported (in 1874) in favour of the latter
view; but in the new building in the Cromwell Road there
was, unfortunately, little provision made for carrying it out
in a satisfactory manner.

As long ago as 1859, in one of his reports on the subject
to the Trustees, Owen recommended that the new museum
building, " besides giving the requisite accommodation to the
several classes of natural history objects, as they had been by
authority exhibited and arranged for public instruction and
gratification, should also include a hall or exhibition space
for a distinct department, adapted to convey an elementary
knowledge of the subjects of all the divisions of natural
history to the large proportion of public visitors not specially
conversant with any of those subjects."   The same idea, in a
later publication, is thus described : " One of the most popular
and instructive features in a public collection of natural history
would be an apartment devoted to the specimens selected to
show type-characters of the principal groups of organised and
crystallised forms.   This would constitute an epitome of
natural history, and should convey to the eye, in the easiest
way, an elementary knowledge of the sciences."   In every
modification which the plans of the new building underwent,
a hall for the purpose indicated in the above passages formed
a prominent feature, being in the later stages of the development of the building, called, for want of a better name, the
" Index Museum."   Though Owen gave the suggestion and
designed the general plan of the hall, the arrangement of its
contents was left to his successor to carry out.

In another part of his original scheme he was less success-

ful.   The lecture theatre, which he had throughout urged with
great pertinacity as a necessary accompaniment to a natural
history museum, was, as he says in the address referred to
above, "erased from my plan, and the elementary courses of
lectures remain for future fulfilment."

On several other important questions of museum arrangement Owen allowed his views, even when essentially philosophical as well as practical, to be overruled.   As long ago as
December 1841, he submitted to the Museum Committee of
the Royal College of Surgeons the question of incorporating
in one catalogue and system of arrangement the fossil bones of
extinct animals with the specimens of recent osteology, and
shortly afterwards laid before the Committee a report pointing
out the advantages of such a plan.   Strangely enough, though
receiving the formal approval of the Council, no steps were
taken to carry it out as long as he was at the College.   He
returned to the question in reference to the arrangement of
the new National Museum, and although no longer advocating
so complete an incorporation of the two series, apparently in
consideration of the interests of the division into "departments" which he found in existence there, he says, "The
department of Zoology in such a museum should be so located
as to afford the easiest transit from the specimens of existing
to those of extinct animals.   The geologist specially devoted
to the study of the evidence of extinct vegetation ought, in
like manner, to have means of comparing his fossils with the
collections of recent plants."[1]   Provision for such an arrangement is clearly indicated in all the early plans for the building, in which the space for the different subjects is allocated,
but not a trace of it remained in the final disposition of the
contents of the museum, as Owen left it in 1883.

Another essential feature of Owen's original plan, without
which, he says, "No collection of zoology can be regarded as
complete," is a gallery of physical ethnology, the size of which
he estimated (in 1862) at 150 ft. in length by 50 ft. in
width.   It was to contain casts of the entire body, coloured
after life, of characteristic parts as the head and face,

---

[1] *On the Extent and Aims of a National Museum of Natural History*, 2nd
edit. 1852, p. 7.

skeletons of every variety arranged side by side for facility of
comparison, the brain preserved in spirit showing its characteristic size and distinctive structures, etc.   "The series of
zoology," he says, "would lack its most important feature were
the illustrations of the physical characters of the human race
to be omitted."

An adequate exhibition of the Cetacea, both by means of
stuffed specimens and skeletons, also always formed a prominent
element in his demand for space.   "Birds, shells, minerals,"
he wrote, "are to be seen in any museum; but the largest,
strangest, rarest specimens of the highest class of animals can
only be studied in the galleries of a national one."   And
again: "If a national museum does not afford the naturalist
the means of comparing the Cetacea, we never shall know
anything about these most singular and anomalous animals."

When, however, the contents of the museum were finally
arranged, nominally under his direction, physical anthropology
was only represented by a few skeletons and skulls placed in
a corner of the great gallery devoted to the osteology of the
Mammalia, and the fine series of Cetacean skeletons could only
be accommodated in a most unsuitable place for exhibition in
a part of the basement not originally destined for any such
purpose.   The truth is that the division of the museum
establishment into four distinct departments, each with its
own head, left the "superintendent" practically powerless, and
Owen's genius did not lie in the direction of such a re-
organisation as might have been effected during the critical
period of the removal of the collections from Bloomsbury and
their installation in the new building.   Advancing age, also,
probably indisposed him to encounter the difficulties which
inevitably arise from interference with time-honoured traditions.
At length, at the close of the year 1883, being in his eightieth
year, he asked to be relieved from the responsibilities of an
office the duties of which he had practically ceased to perform.

The nine remaining years of his life were spent in peaceful
retirement at Sheen Lodge, an ideal residence for one who had
such a keen enjoyment of the charms of Nature in every form,
for, though so large a portion of his active life had been
passed among dry bones, anatomical specimens, microscopes,

and books, he retained a genuine love for out-door natural history, and the sight of the deer and other animals in the park, the birds and insects in the garden, the trees, flowers, and varying aspects of the sky, filled him with enthusiastic admiration. He also had his library around him, and it is needless to say that the habit of strenuous work never deserted him till failing memory and bodily infirmity made it no longer possible to continue that flow of contributions to scientific literature which had never ceased during a period of sixty-two years, his first and last papers being dated respectively 1826 and 1888. His wife and only child had died some time before, but the son (who had held an appointment in the Foreign Office) left a widow and seven children, who, coming to reside with him at Sheen, completely relieved his latter days of the solitude in which they would otherwise have been passed. During the autumn of 1892 his strength gradually failed, and he died on the 18th of December, literally of old age. In accordance with his own expressed desire, he was buried in the churchyard of Ham, near Richmond, in the same grave with his wife, a large and representative assemblage of men of science being present at the funeral ceremony.

It may be thought that the prodigious amount of work that Owen did in his special subjects would have left him no time for any other occupations or relaxations, but this was by no means the case. He was a great reader of poetry and romance, and, being gifted with a wonderful memory, could repeat by heart, even in his old age, page after page of Milton and other favourite authors. For music he had a positive passion; in the most busy period of his life he might constantly be seen at public concerts, listening with rapt attention, and in his earlier days was himself no mean vocalist, and acquired considerable proficiency in playing the violincello. He was also a neat and careful draughtsman; the large number of anatomical sketches he left behind him testify to his industry in this direction. His handwriting was unusually clear and finished, considering the vast quantity of manuscript that flowed from his pen, for he rarely resorted to dictation or any labour-saving process. Only those who have had to clear out rooms, official or private, which have been long occupied by him can have any idea of the quantity of memoranda and extracts which he made with his own hand, and most of the books he was in the habit of using were filled with notes and comments.

Owen's was a very remarkable personality, both physically and mentally. He was tall but ungainly in figure, with massive head, lofty forehead, curiously round, prominent, and expressive eyes, high cheek bones, large mouth and projecting chin, long, lank, dark hair, and during the greater part of his life, smooth-shaven face, and very florid complexion. Though in his general intercourse with others usually possessed of much of the ceremonial courtesy of the old school, and when in congenial society a delightful companion, owing to his unfailing flow of anecdote, considerable sense of humour, and strongly-developed faculty of imagination, he was not only an extremely adroit controversialist, but could say as hard things as any one of an adversary or rival. It was to Owen's many-sided and very complex character that the remarkably different estimates formed of him personally by his contemporaries was due. Unfortunately he was often engaged in controversy, a circumstance which led to a comparative isolation in his position among men who followed kindred pursuits, which was doubtless painful to himself as well as to others. It was this, combined with a certain inaptitude for ordinary business affairs, which was the cause of his never having been called to occupy several of the distinguished official positions in science to which his immense labours and brilliant talents would otherwise have fairly entitled him. Over the British Association he presided at the meeting at Leeds in 1858, and he had his full share of those honours and dignities to which a scientific man can aspire which involve no corresponding duties or responsibilities. He was made a C.B. in 1873, and a K.C.B. on his retirement from the Museum in 1884. He received the Prussian Order "Pour le Mérite" in 1851, the Cross of the French Legion of Honour in 1855, and was also decorated by the King of Italy with the Order of St. Maurice and St. Lazarus, and by the Emperor of Brazil with the Order of the Rose. He was chosen one of the eight foreign Associates of the Institute of France in 1859. The

Universities of Oxford, Cambridge, and Dublin conferred upon him their honorary degrees, and he was an honorary or corresponding member of nearly every important scientific society in the world. The Geological Society presented him with the Wollaston Medal in 1838, and the Royal College of Surgeons with its Honorary Gold Medal in 1883. He was the first to receive the gold medal established by the Linnean Society at the centenary meeting of that body in 1888. The Royal Society, of which he became a Fellow in December 1834, and on the Council of which he served for five separate periods, awarded him one of the Royal Medals in 1846, and the Copley Medal in 1851.

REMINISCENCES OF PROFESSOR HUXLEY [1]

THERE is no intention in this paper of giving either a biographical notice of Professor Huxley or an estimate of his position in science, philosophy, or literature. Both have been done over and over again in numerous journals and magazines that have appeared since his death. The main facts of his career, and his great contributions to human knowledge, must be perfectly familiar to the readers of this *Review*. I have, however, in response to an appeal from the Editor, put down a few personal reminiscences, gathered during a friendship of nearly forty years, which may throw some additional light upon the character and private life of one in whom all English-speaking people must take a deep interest. In doing this I fear I have been obliged to introduce myself to the notice of the reader more frequently than I should wish, but this seems inevitable in an article of this nature, and I trust will be forgiven for the sake of the main subject.

When Huxley returned to London from his four years' surveying cruise in H.M.S. *Rattlesnake*, under the command of Captain Owen Stanley, R.N., one of the first men of kindred pursuits who took him by the hand was George Busk, then surgeon to the Seaman's Hospital, the *Dreadnought*, lying in the Thames off Greenwich. About this time Busk removed from Greenwich to Harley Street, and although doing some practice as a surgeon, and even attaining to the position of President of the Royal College of Surgeons, his main occupation and chief pleasure were in purely scientific pursuits, and his

---

[1] *The North American Review*, September 1895.

great interest in and familiarity with microscopic manipulation, especially as applied to the structure of lowly organised animal forms—then rather in its infancy—was a strong bond of sympathy with Huxley. In 1852-54 they translated and edited jointly Kölliker's *Manual of Human Histology*, published by the Sydenham Society. This fact shows that Huxley had already made himself proficient in the German language, as he had also, while on board the *Rattlesnake*, taught himself Italian, with the main object of being able to read Dante in the original, so wide were his interests and sympathies.

It was through Busk that I first became acquainted with Huxley. This was shortly before his marriage, the incidents connected with which were of a somewhat romantic character. When the *Rattlesnake* was in Sydney Harbour the officers were invited to a ball, and young Huxley among the number. There for the first time he met his future wife, whose parents resided at Sydney. A few days after they were engaged, and the ship sailed for the Torres Straits to complete the survey of the north coast of Australia, all communication being cut off for months at a time, and then returned direct to England. At the end of seven years, on Huxley's appointment to the School of Mines, he was in a position to claim his bride, and welcome her to their first home in St. John's Wood. He often used to say that to engage the affections of a young girl under these circumstances, knowing that he would have to leave her for an indefinite time, and with only the remotest prospect of ever marrying, was an act most strongly to be reprobated, and he often held it out as a warning to his children never to do anything of the kind, and yet they all married young and all happily. Huxley's love at first sight and constancy during those seven long years of separation were richly rewarded, for it is impossible to imagine a pair more thoroughly suited. I cannot help relating a little incident which clings to my memory, though it happened full thirty years ago. A rather cynical and vulgar-minded acquaintance of mine said to me one day: "I saw Huxley in a box at the Drury Lane Theatre last night. Can you tell me who was the lady with him?" After a few words of description I said: "Oh, that was Mrs. Huxley."—"Indeed," he said; "I thought it could not be his wife, he was so very attentive to her all the evening." As intimate friends knew, they had at first many household troubles and cares to contend with, a large family of young children, much ill-health, and not very abundant means, but through it all Huxley's patience and sweetness were admirable. The fierce and redoubtable antagonist in the battlefield of scientific or theological controversy was all love and gentleness at home.

The fact that he had sailed under Captain Owen Stanley, who died when in command of the *Rattlesnake* in Australia, brought him into very friendly communication with the Captain's brother, Arthur Stanley, the late dean of Westminster, *the* Dean, as many of us always used to, and still do, call him, just as the first Duke of Wellington was always called *the* Duke. Notwithstanding the great differences of their interests and pursuits, they remained intimate until Stanley's death, and to be with them when they met was a rare occasion of hearing much delightful talk and many displays of playful wit. If I had the faculty of a Boswell, I should have much worth narrating of many charming little dinner parties at one or the other of our houses, when Huxley and the Dean were the principal talkers. I remember a characteristic *rencontre* between them which took place on one of the ballot nights at the Athenæum. A well-known popular preacher of the Scotch Presbyterian Church, who had made himself famous by predictions of the speedy coming of the end of the world, was up for election. I was standing by Huxley when the Dean, coming straight from the ballot boxes, turned towards us. "Well," said Huxley, "have you been voting for C?"—"Yes, indeed, I have," replied the Dean. "Oh, I thought the priests were always opposed to the prophets," said Huxley. "Ah?" replied the Dean, with that well-known twinkle in his eye, and the sweetest of smiles. "But, you see, I do not believe in his prophecies, and some people say I am not much of a priest."

Speaking of Dean Stanley, I am reminded of a very interesting meeting which took place at my house, in Lincoln's Inn Fields, on 26th November 1878, just after his return from his visit to the United States. He had a great wish to see Darwin, who was one of the few remarkable men of the age with whom he was not personally acquainted. They moved in totally different circles, Darwin having, owing to ill-health, long given up going into general society. He had, however, a great admiration for the Dean's liberality, courage, and character, and was glad of the opportunity of meeting him. So we arranged that they should both come to lunch. They were mutually pleased with each other, although they had not many subjects in common to talk about. Darwin was no theologian, and Stanley did not take the slightest interest in nor had he any knowledge of any branch of natural history, although his father was eminent as an ornithologist and President of the Linnean Society. I once took him over the Zoological Gardens. His remarks were, of course, original and amusing, but the sole interest he appeared to find in any of the animals was in tracing some human trait, either in appearance or character. The Dean enjoyed intensely the broader aspects and beauties of nature as shown in scenery, but the details of animal and plant life were entirely outside his sympathies.

Another introduction consequent upon Huxley's voyage in the *Rattlesnake* was to Dr. Vaughan, then Headmaster of Harrow. Mrs. Vaughan was Captain Owen Stanley's sister, and soon after Huxley's return he was asked to dine and pass the night at Harrow. This was a new experience. The young, rough sailor-surgeon was at first quite out of his element in the refined, scholastic, ecclesiastical society he found himself plunged into. Among those who were present was an Oxford don (the first of the class Huxley had ever met), whose great learning, suave manner and air of superiority during dinner, greatly alarmed and repelled him, as he afterwards confessed. Bed-time came, and both stood upon the staircase, lighted candle in hand. They looked straight into each other's eyes, and the don addressed a few words directly to Huxley for the first time. He was much interested, and an animated conversation ensued. Instead of bidding each other "good night" they adjourned to a neighbouring room, sat down and talked till two o'clock in the morning. This was the beginning of Huxley's life-long friendship with the late Master of Balliol, Dr. Jowett.

It may surprise some people to know, but that he has told it himself in an exceedingly interesting and delightfully written short autobiographical sketch prefixed to his works, that Huxley was not in early life anything of what is commonly called a naturalist. Most men who have distinguished themselves in the field of zoology or palæontology have loved the subject from their early boyhood, a love generally shown by the formation of collections of specimens. Huxley never did anything of the kind. His early tastes were for literature and for engineering. He attributed the awakening of his interest in anatomy to Professor Wharton Jones's lectures at Charing Cross Hospital, where he received his medical education. Wharton Jones was one of the pioneers of microscopic research in this country; a great enthusiast in his work, but a man of modest and exceedingly retiring disposition, and very little known outside a small circle of friends. He published several papers on histology in the *Philosophical Transactions*, and made a specialty of ophthalmic surgery. Perhaps of his various contributions to the advancement of his subject, not the least important was that of making a scientific anatomist of Huxley.

The next man who had a real influence upon Huxley's professional career was Sir John Richardson, a very keen zoologist, at that time principal medical officer at Haslar Hospital, near Portsmouth, where the naval assistant surgeons first proceeded on appointment. It was through him that Huxley was appointed to the surveying ship, *H.M.S. Rattlesnake.* He was not naturalist to the expedition, as has been sometimes said; indeed, he would at this time have been hardly qualified for such a post, for, although he had published a short paper on the microscopic structure of the human hair, he had as yet done no zoological work. Moreover, the ship did carry an accredited naturalist, John Macgillivray, who published a *Narrative of the Voyage of H.M.S. "Rattlesnake,"* *during* 1846-50, in two volumes (1852).

Huxley's official duties were only with the health of the crew, and as he had a surgeon above him, he had plenty of leisure at his command. How this leisure was employed in laying the foundation upon which his future distinction rested has often been told. He had his microscope with him, and he threw himself with the greatest ardour into the investigation of the structure of the lowly organised but beautiful forms of animal life which abounded in the seas through which the ship sailed, and which the surveying operations in which she was engaged gave ample opportunities for observing under the most favourable conditions. This was almost a new field of research. He became fascinated with it, and his success in its pursuit was the main cause of his adopting zoology as the principal subject to engage his energies during the rest of his life.

As said before, Huxley, unlike many other zoologists, was never a collector, and had not the slightest tincture of the spirit of a museum curator. He cared for a specimen according to the facilities it afforded for investigation. He cut it up, got all the knowledge he could out of it, and threw it away. I believe he never made a preparation of any kind, and he cared little for dissections sealed down in bottles.

When, in 1862, he was appointed to the Hunterian Professorship at the College of Surgeons, he took for the subject of several yearly courses of lectures the anatomy of the vertebrata, beginning with the primates, and as the subject was then rather new to him, and as it was a rule with him never to make a statement in a lecture that was not founded upon his own actual observation, he set to work to make a series of original dissections of all the forms he treated of. These were carried on in the workroom at the top of the college, and mostly in the evenings, after his daily occupation at Jermyn Street (the School of Mines, as it was then called) was over, an arrangement which my residence in the college buildings enabled me to make for him. These rooms contained a large store of material, entire or partially dissected animals preserved in spirit, which, unlike those mounted in the museum, were available for further investigation in any direction, and these, supplemented occasionally by fresh subjects from the Zoological Gardens, formed the foundation of the lectures, afterwards condensed into the volume on the Anatomy of Vertebrated Animals, published in 1871. On these evenings

it was always my privilege to be with him, and to assist in
the work in which he was engaged.   In dissecting, as in
everything else, he was a very rapid worker, going straight to
the point he wished to ascertain with a firm and steady hand,
never diverted into side issues, nor wasting any time in unnecessary polishing up for the sake of appearances, the very
opposite, in fact, to what is commonly known as " finikin."
His great facility for bold and dashing sketching came in
most usefully in this work, the notes he made being largely
helped out by illustrations.   He might have been a great
artist, some of his anatomical sketches reminding me much
of Sir Charles Bell's, but he never had time to cultivate his
faculties in this direction, and, I believe, never attempted any
finished work.   His power of drawing on the black-board
during the lectures was of great assistance to him and to his
audience, and his outdoor sketches, made during some of his
travels, as in Egypt, though slight were full of artistic feeling.
His genius was also conspicuously shown by the clever
drawings, often full of playful fancy, which covered the paper
that happened to be lying before him when sitting at a council
or committee meeting.   On such occasions his hand was rarely
idle.

   It is very singular that although, as admitted by all who
heard him, he was one of the clearest and most eloquent of
scientific lecturers of his time, he always disliked lecturing;
and the nervousness from which he suffered in his early days
was never entirely overcome, however little apparent it might
be to his audience.   After his first public lecture at the Royal
Institution he received an anonymous letter, telling him that
he had better not try anything of the kind again, as whatever
he was fit for, it was certainly not giving lectures !   Instead
of being discouraged, he characteristically set to work to mend
whatever faults he had of style and manner, with what success
is well known.   Nevertheless, he often told me of the awful
feeling of alarm which always came over him on entering the
door of the lecture-room of the Royal Institution, or even the
College of Surgeons, where the subject was most familiar and
the audience entirely sympathetic.   He had a feeling that he
must break down before the lecture was over, and it was only

by recalling to his memory the number of times he had
lectured without anything of the kind happening, and then
drawing conclusions as to the improbability of its occurring
now, that he was able to brace himself up to the effort of
beginning his discourse.   When once fairly away on his
subject all such apprehensions were at an end.   Such experiences are, of course, very common, but they were probably
aggravated greatly in Huxley's case by the ill health, that
miserable, hypochondriacal dyspepsia which, as he says himself, was his constant companion for the last half century of
his life.   Bearing in mind the serious inroad this made in the
amount of time available for active employment, it is marvellous
to think of the quantity he was able to accomplish.   When
the time comes for forming a just estimate of the value of his
scientific work, and if quality as well as quantity be fairly
taken into account, it will, without doubt, bear comparison
with, if it will not exceed, that of any of his contemporaries.

If, instead of taking up medicine and afterwards science
as a profession, he had gone to the bar, he must infallibly
have achieved the highest measure of success.   As an advocate
he could scarcely have been surpassed.   His clear, penetrating
insight into the essentials of an intricate question, the rapidity
with which he swept aside all that was irrelevant, and the
forcible way in which he could state the arguments for his
own side of a case, and his brilliant power of repartee, would
have been irresistible in a court of justice.   He was also free
from a quality which paralyses the effective action of many
men of great mental capacity—the faculty of seeing something at least of both sides of a case at the same time.
When he took up a cause he took it up in thorough earnest,
and it must be admitted that there was then very little
chance of his feeling any sympathy for the other side.   He
had some strong prejudices against doctrines, against institutions,
and against individuals, and as his nature was absolutely
honest and truthful, he never cared to conceal them.   On the
other hand, no man was more loyal to the causes he approved
of or the people he liked.   He could always be relied upon to
carry out to the uttermost of his power anything he had
undertaken to do.   To the younger workers in his own fields

of research nothing could exceed his generous assistance, sympathy, and encouragement. These qualities were, above all others, the main causes of the devoted attachment he won from every one who was brought much into personal contact with him.

In one of the recent biographical notices which have appeared of Huxley it is said that " no man of more reverent religious feeling ever trod this earth." This statement has much of truth in it. If the term " religious " be limited to acceptance of the formularies of one of the current creeds of the world, it cannot be applied to Huxley, but no one could be intimate with him without feeling that he possessed a deep reverence for " whatsoever things are true, whatsoever things are honest, whatsoever things are just, whatsoever things are pure, whatsoever things are lovely, whatsoever things are of good report," and an abhorrence of all that is the reverse of these, and that, although he found difficulty in expressing it in definite words, he had a pervading sense of adoration for the infinite, very much akin to the highest religion.

XXIV

EULOGIUM ON CHARLES DARWIN[1]

THE Council of the Linnean Society has honoured me with the request that I would say some words regarding the life and work of our illustrious member Charles Darwin, whose name, it may be said with truth, is more widely known throughout the civilised world than any other that has been enrolled upon the list of Fellows of the Society.

Darwin has, moreover, special claims for consideration from us on such an occasion as this, inasmuch as a large and very important portion of his work was first communicated to the world by means of papers read at our meetings and published in our Journal.

Here, on the 1st of July 1858, was read the celebrated essay "On the Variation of Organic Beings in a State of Nature, on the Natural means of Selection, on the Comparison of Domestic Races and True Species."

Here also were first made known, in a succession of memoirs, extending over many years, those remarkable investigations into the structure and life-history of plants, "any one of which, taken on its own merits" (I quote the words of one of our leading authorities in this department of science), "would alone have made the reputation of any ordinary botanist."

Darwin's life and Darwin's work are, however, so familiar to every one here, and have been so recently and so exhaustively treated of, in every aspect in which they can be viewed, that to attempt to say anything new upon them, or even to clothe what is well-known in any original form, would be for me a hopeless task.

[1] Centenary Meeting of the Linnean Society, 24th May 1888.

The brevity with which I will speak will therefore be not a measure of our appreciation of the subject or of the man, but of a conviction that few words are needed to express what we all know and all feel.

The recently published *Life and Letters* has brought before a wide circle of readers a most vivid presentiment of what Darwin really was.

A character so simple, so transparent, so unaffected, duly recognising its own strength, and at the same time fully conscious of its own imperfections, a life so singularly consistent, so steadily uniform throughout in its aims, and so undeviatingly honest to all its convictions: such a character and such a life, already well known to his intimate friends, is now before the whole world revealed, as one may say, to its very depths.

Nothing more of any importance, either of character or life, will ever be known. Any additional detail of incident or adventure that can ever be brought to light, any further publication of his voluminous correspondence, would only fill in little vacuities that may be left in the picture, but will never alter the outlines, or the colour, or the tone. The picture, as already drawn in that book, will remain, substantially, the same, for it is that of the man himself, and, as I have said, of a man singularly free from the complexities and contradictions which make up the composite character of many whose names have risen conspicuous above those of their fellow-men. To the admirable qualities of his domestic life, his modesty, his graciousness, his geniality, his generous appreciation of the work and opinions of others, justice has been fully rendered, even by the least sympathetic critics of his scientific work. One of the most recent of these is constrained to say, " To know Darwin was to feel attracted to him, to know much of him was to love him."[1]

It concerns us here to speak rather of the one great characteristic which, throughout the whole of his lengthened career, dominated all others, and made him what he was,— the consuming, irrepressible longing to unravel the mysteries of living nature, to penetrate the shroud which conceals the

[1] The Duke of Argyll.

causes and methods by which all the wonders and all the diversity, all the beauty—yes, and all the deformity too—which we see around us in the life of animals and plants have been brought about.

Against our ignorance on these subjects his life was one long battle, and in reading its history and seeing the gradual development of his plan of operations, one is continually reminded of a great strategist directing a vast army spread over a wide and varied field of operations ; now surveying the whole at a glance, now pressing on his various forces wherever an opening presents itself anywhere along the line, now carefully scrutinising the weak and the strong points of every position ; omitting no precaution where danger threatens, now bringing one branch of the forces to bear, followed up and supported, if need be, by others of a different kind, one after another, in close and telling array; masses of facts, experiments. observations, and arguments thrown in to stop a breach or strengthen any menaced or wavering post, and all arranged, grouped, marshalled, and handled with the skill and vigilance with which a successful general handles a living army in the conduct of a great and complicated campaign.

To all this, most of the work which we others do is but irregular, guerilla warfare, attacks on isolated points, mere outpost skirmishing, while his was the indefatigable, patient, unintermittent toil, conducted in such a manner and on such a scale that it could scarcely fail to secure victory in the end.

The main victory gained by his work was, as we all know, the destruction of the conception of species as being beyond certain narrow limits fixed and unchangeable, a conception which prevailed almost universally before his time. That this has been gained chiefly by means of Darwin's work and writing, there can be no doubt. Let us admit that others had prepared the way, that the work was carried on simultaneously by many others also, that if the present generally accepted view is true, it must have made its way if Darwin had not lived or spoken ; I say, grant all this to the fullest, and the fact remains that he was the main agent in the conversion of almost the whole scientific world from one to a totally opposite conception of one of the most important operations of nature.

Such a revolution as this, with all its momentous consequences to the study of zoology and botany, effected in so short a space of time, is, as has often been said, without a parallel in the history of science, and it is one the full significance of which those who have not lived through it, and been workers at biology in both the pre-Darwinian and post-Darwinian epochs, must find difficulty in realising.

There is, moreover, no doubt but that this rapid conversion was much facilitated by the fascinating nature of the theory of the operation of natural selection in intensifying and fixing variation, as originally propounded in these rooms independently and simultaneously by Darwin and by Wallace. This theory has been subjected to keen criticism, and difficulties have undoubtededly been shown in accepting it as a complete explanation of many of the phenomena of evolution. That other factors have been at work besides natural selection in bringing about the present condition of the organic world, probably every one now admits, as, I need not say, Darwin did himself. This, however, is not the occasion to enter into a critical examination of this large and complex subject. Indeed, the time seems scarcely yet come when we can do so with the necessary calmness and impartiality. Prejudices on the one hand and on the other, and the cloud of side-issues which were aroused when the theory was first promulgated, and which prevented many from understanding what was really implied by it, still hover around, and many of us deem it best to rest with suspended judgment not only upon this, but upon the various other hypotheses put forward to account for the origin of species, and to turn again with increased interest and zeal to investigate the facts upon which these hypotheses are based. No one can deny that, whatever opinion may ultimately prevail regarding Darwin and his works, the controversies that have gathered round them have proved a marvellous stimulus to research, and have given new life to investigations into a great variety of subjects,— subjects so diverse as palæontology, morphology, embryology, the geographical distribution, the habits, and the life-history of all living things,—into every branch, in fact, of biological science.

They have made us also realise in fuller measure than ever before the depth of the still unfathomed mysteries that confront us everywhere. The endeavour to penetrate these mysteries, to solve some of these problems which lie everywhere in our path in wandering through the field of nature, is surely a most legitimate employment for the faculties of man ; and he who has devoted to this endeavour a life of patient, eager, and, above all, honest toil, undaunted by constant physical weakness and suffering, and has steadily persevered to the end in his one great aim, alike through evil report and good report, deserves our gratitude and our reverence.

Though Darwin did not tear down the curtain which obscures our gaze into the past and lay bare to our vision the birth of life, and all its various manifestations upon earth, as has been too rashly said by some of his enthusiastic disciples, he lifted the veil here and there, and gave us glimpses which will light the path of those who follow in his steps, and, even more than this, he showed by his life and by his work, beyond any one of the age in which we live, the true methods by which alone the secrets of nature may be won.

THE END

Printed by R. & R. CLARK, LIMITED, *Edinburgh.*

BOOKS FOR
STUDENTS OF NATURAL HISTORY.

BY THE SAME AUTHOR.

AN INTRODUCTION TO THE OSTEOLOGY OF THE MAMMALIA: being the substance of the course of Lectures delivered at the Royal College of Surgeons of England in 1870. By Sir WILLIAM HENRY FLOWER, F.R.S., F.R.C.S. Illustrated. Third Edition. Revised with the assistance of HANS GADOW, Ph.D. Crown 8vo. 10s. 6d.

THE CAMBRIDGE NATURAL HISTORY. Edited by S. F. HARMER, M.A., Fellow of King's College, Cambridge, Superintendent of the University Museum of Zoology; and A. E. SHIPLEY, M.A., Fellow of Christ's College, Cambridge, University Lecturer on the Morphology of Invertebrates.

Vol. II.—WORMS, LEECHES, ETC.

Flatworms. By F. W. GAMBLE, M.Sc. Nemertines. By Miss L. SHELDON. Threadworms, etc. By A. E. SHIPLEY, M.A. Rotifers. By MARCUS HARTOG, M.A., D.Sc. Polychaet Worms. By W. BLAXLAND BENHAM, D.Sc. Earth-worms and Leeches. By F. E. BEDDARD, M.A., F.R.S. Gephyrea, etc. By A. E. SHIPLEY, M.A. Polyzoa. By S. F. HARMER, M.A.

NATURAL SCIENCE.—"Certain to prove a most welcome addition to English zoological literature. . . . The whole is admirably illustrated."

Vol. III.—SHELLS.

Molluscs and Brachiopods. By the Rev. A. H. COOKE, M.A., A. E. SHIPLEY, M.A., and F. R. C. REED, M.A. Illustrated. Medium 8vo. 17s. net.

FIELD.—"We know of no book available to the general reader which affords such a vast fund of information on the structure and habits of molluscs."

Vol. V.—INSECTS AND CENTIPEDES.

Peripatus. By ADAM SEDGWICK, M.A., F.R.S. Myriapods. By F. G. SINCLAIR, M.A. Insects. Part I. By DAVID SHARP, M.A. Cantab., M.B. Edin., F.R.S. Fully Illustrated. Medium 8vo. 17s. net.

NATURE NOTES.—"When the following volume is completed, Mr. Sharp's work will, we think, last for many years as the standard text-book on Entomology in England."

NATURAL HISTORY AND ANTIQUITIES OF SELBORNE. By GILBERT WHITE. With Notes by FRANK BUCKLAND, a Chapter on Antiquities by Lord SELBORNE, and new Letters, illustrated by P. H. DELAMOTTE. New and Cheaper Edition. Crown 8vo. 6s. In Two Volumes. Crown 8vo. 10s. 6d.

A HANDBOOK OF BRITISH LEPIDOPTERA. By EDWARD MEYRICK, B.A., F.Z.S., F.E.S., Assistant Master at Marlborough. Illustrated. Extra Crown 8vo. 10s. 6d. net.

AN INTRODUCTION TO THE STUDY OF SEAWEEDS. By GEORGE MURRAY, F.R.S.E., F.L.S., Keeper of Botany in the Natural History Department of the British Museum. Illustrated. Crown 8vo. 7s. 6d.

THE NATURAL HISTORY OF THE MARKETABLE MARINE FISHES OF THE BRITISH ISLANDS. Prepared by order of the Council of the Marine Biological Association especially for the use of those interested in the Sea-fishing industries. By J. T. CUNNINGHAM, M.A. Oxon. With a Preface by Prof. E. RAY LANKESTER, LL.D., F.R.S. Medium 8vo. 7s. 6d. net.

MACMILLAN AND CO., LTD., LONDON.

www.ingramcontent.com/pod-product-compliance
Lightning Source LLC
Chambersburg PA
CBHW032312280326
41932CB00009B/783